D1711152

Improving Primary Education in Developing Countries

Improving Primary Education in Developing Countries

Marlaine E. Lockheed and Adriaan M. Verspoor

with

Deborah Bloch, Pierre Englebert, Bruce Fuller, Elizabeth King,
John Middleton, Vicente Paqueo, Alastair Rodd,
Ralph Romain, and Michel Welmond

Published for the World Bank
Oxford University Press

Oxford University Press

OXFORD NEW YORK TORONTO DELHI
BOMBAY CALCUTTA MADRAS KARACHI
PETALING JAYA SINGAPORE HONG KONG
TOKYO NAIROBI DAR ES SALAAM
CAPE TOWN MELBOURNE AUCKLAND
and associated companies in
BERLIN IBADAN

Manufactured in the United States of America
First printing December 1991

Library of Congress Cataloging-in-Publication Data

Lockheed, Marlaine E.
Improving primary education in developing countries / Marlaine E. Lockheed, Adriaan M. Verspoor.
p. cm.
Includes bibliographical references and index.
ISBN 0-19-520872-2
1. Elementary education—Developing countries. 2. Education—Developing Countries—Statistics. I. Verspoor, Adriaan, 1942– . II. Title.
LC2608.L63 1991 91-30421
372.9172'4—dc20 CIP

In the new century, the dividends from knowledge will grow as dramatically as the penalties of ignorance increase. Much of the gulf between misery and opportunity, squalor and hope, can be bridged by education, by investing in the bright inquisitiveness of children. . . . Education, and human resource development more broadly, must be a central focus of the development effort in the 1990s.

Barber B. Conable
President, World Bank

Contents

Foreword *xiii*

Preface *xv*

Acknowledgments *xviii*

1 Primary Education and Development *1*

The Contribution of Education to Development *2*
The Importance of Cognitive Competencies *5*
The Shortcomings of Education in the Developing World *10*
Inadequate Human Resources as a Consequence of Poor Education *16*
Primary Education: The First Step *17*
Notes *19*

2 A Brief History of Primary Education in Developing Countries *21*

Expanding Opportunities for Education *22*
The Search for "Relevant" Content and Better Delivery of Education *29*
Dealing with Austerity and Adjustment *31*
Conclusions *37*
Notes *38*

3 Improving Learning Achievement *39*

A Model of Educational Effectiveness *40*
Effective Schools *41*
Improving the Curriculum *45*
Providing Learning Materials *47*
Time for Learning *57*
Effective Teaching *62*
Teachability: Children's Learning Capacity *72*
Summary *86*
Notes *88*

4 Improving the Preparation and Motivation of Teachers 90

General Academic Background *92*
Pedagogical Skill Development *97*
Motivation *101*
Summary *115*
Notes *116*

5 Strengthening Institutional Capabilities *117*

Constraints on Managerial and Institutional Effectiveness *118*
Strategies for Addressing Managerial and Institutional Weaknesses *127*
Summary *142*
Notes *144*

6 Improving Equitable Access *145*

Disparities in Schooling *145*
Strategies to Promote Equitable Access *154*
Summary *167*
Notes *170*

7 Strengthening the Resource Base for Education *171*

Problems and Issues *171*
Using Existing Primary Education Resources Efficiently *178*
Mobilizing Domestic Resources for Primary Education *185*
Ensuring Equitable Financing *200*
Summary *203*
Notes *204*

8 International Aid to Education 207

Patterns of Giving to Primary Education *207*
Why Donors Have Neglected Primary Education *214*
New Priorities for Donor Support *215*
Summary *217*
Notes *218*

9 Education Reform: Policies and Priorities for Educational Development in the 1990s *219*

Challenges in Educational Development *219*
Priorities for Reform *224*
Implementation Strategies: Steps toward Successful Reform *231*

Appendix 233

Bibliography 389

Index *417*

Boxes

1-1. How Primary Education Helps Farmers Solve Problems 6
2-1. Rural Education Centers in Burkina Faso 32
3-1. The Problem with Textbooks in Developing Countries 48
3-2. The Book Flood Experiment for Teaching English as a Second Language in Fiji 50
3-3. Cross-Age Peer Tutoring 66
3-4. Teacher Training through Distance Education 69
3-5. Interactive Radio Instruction 73
3-6. The Head Start Program in the United States 82
4-1. Recruitment Incentives: Zimbabwe's Integrated Teacher Education Course 93
4-2. The Use of Career Ladders to Promote and Retain Talented Teachers 114
5-1. Textbook Agencies in Indonesia and the Philippines 129
5-2. Improving School Supervision in Pakistan and Bangladesh 131
5-3. School Clusters in Sri Lanka, Thailand, and Papua New Guinea 133
5-4. Management Training in Malaysia 135
5-5. Establishing Accountability by Testing Achievement in Thailand 140
5-6. Educational Research Institutes in the Republic of Korea 143
6-1. Gender Discrimination in Costa Rican Textbooks 151
6-2. Improving Rural Education in Colombia through Escuela Nueva 160
6-3. Addressing the Needs of Child Workers in India and Kenya 162
6-4. Improving Schooling for Girls and Rural Children in Bangladesh 165
6-5. Improving Equity through Bilingual Education in Guatemala and Nigeria 168
6-6. Providing Education to the Handicapped in Zambia 169
7-1. Efficiency Gains in Northeastern Brazil 176
7-2. Taxes for Education 190
7-3. Arguments for and against Earmarking Funds 192
7-4. Financing Education in Zimbabwe: Partnerships between the Government and the Community 194
7-5. Using Poll and Property Taxes for Education 199
7-6. Financing Good Primary Education for All: The Korean Experience 202

Figures

1-1. The Primary School Cycle in 169 Countries, 1985 7
1-2. Students Reaching the Last Year of Primary School, by Country Income Level, 1970–85 11
1-3. Estimated Number of Scientists and Engineers in Research, by Country Income Level, 1985 18

2-1. Population Age 6–11 and Enrollment in Primary Schools in the Ninety-Nine Largest Developing Countries, 1965–85 *22*
2-2. Average Annual Population and Enrollment Growth Rates, by Country Income Level, 1975–85 *25*
2-3. Population Age 6–11 and Enrollment in Primary Schools in Low- and Middle-Income Countries, 1965–85 *26*
2-4. Public Expenditure for Education in Selected Intensely Adjusting Countries, 1980 and 1985 *36*
3-1. A Model of Effective Schooling *42*
3B-1. Gains in Children's Readiness for School as a Result of Head Start, 1969–70 *83*
4-1. Index of Recurrent Spending per Primary Teacher, Selected Years, 1970–85 *104*
4-2. Change in Real Average Salary of Teachers in Seven Sub-Saharan African Countries, 1980–85 *105*
4-3. Change in Real Teacher Salaries and Average Manufacturer Salaries, 1980–85 *106*
7-1. Average Annual Net Transfer from Government to Nonfinancial State-Owned Enterprises in Selected Countries, 1978–85 *186*
8-1. Average Annual Aid Disbursed to 143 Countries for Primary Education, 1981–86 *209*
8-2. Aid Dispersed for Primary Education by Country Income Level, 1981–86 *210*
9-1. Gross and Net Enrollment Ratios of Children Age 6–11 in Low-Income Economies, 1985 *226*
9-2. Gross and Net Enrollment Ratios of Children Age 6–11 in Lower-Middle-Income Economies, 1985 *228*
9-3. Gross and Net Enrollment Ratios of Children Age 6–11 in Upper-Middle-Income Economies, 1985 *230*

Tables

1-1. Median Repetition Rates of Primary Students for Selected Years, 1965–85 *12*
1-2. Test Performance of Students in Three Subjects, Various Years *13*
1-3. Average Years of Education Completed by Adults in Sixty-Five Countries, 1980 *17*
2-1. Projected and Actual Enrollment in Primary Schools, 1970 and 1980 *23*
2-2. Gross Enrollment Ratios and Females as a Percentage of Total Enrollment, Selected Years, 1965–85 *27*
2-3. Net Enrollment Ratios, Selected Years, 1965–85 *28*
2-4. Population, School Places, and Enrollment in Ninety-Nine Developing Countries, 1985 *29*
2-5. Median Public Expenditure on Education, Selected Years, 1965–85 *33*

2-6. Median Public Recurrent Expenditure per Primary Student, Selected Years, 1965–85 33
2-7. Median Public Recurrent Expenditure per Primary Student as a Percentage of GNP per Capita, Selected Years, 1965–85 34
3-1. Curriculum Time Devoted to Major Content Areas in Ninety Countries, 1980s 45
3B-1. Mathematics Skills Taught in Developing Countries Compared with Achievement in the United States, 1986 48
3B-2. Reading Skills Taught in Developing Countries Compared with the Curriculum in the United States 49
3-2. Availability of Textbooks in Selected Countries, Various Years 54
3-3. Percentage of Classrooms with Instructional Material in Brazil, 1985, and Somalia, 1984 56
3-4. Official Instructional Time in 110 Countries, 1980s 59
3-5. Official and Actual Instructional Time in Indonesia, 1978 60
3-6. Student-Teacher Ratio by Country Income Level, Selected Years, 1965–85 61
3B-3. Effectiveness, Cost, and Cost-Effectiveness of Three Educational Interventions in the United States, 1980s 66
3-7. Academic Achievement through Interactive Radio Instruction, Various Years 72
3-8. Percentage of School-Age Children with Nutritional Deficiencies and Parasites, Various Years 78
3-9. Median Preschool Enrollment Rate, Selected Years, 1965–80 81
3-10. Interventions to Improve Learning Achievement and Estimated Annual Marginal Cost per Student 87
4-1. New Teachers Required to Achieve Universal Access to Primary School by 2000 91
4-2. Length and Content of Primary Teacher Training Programs in Selected Countries 96
4-3. The Annual Cost of Teacher Training as a Multiple of the Annual Cost of General Secondary Education in Selected Countries 97
4-4. The Curriculum in Teacher Training Colleges in Haiti, Nepal, and Yemen 99
4-5. Factors in the Attractiveness of Teaching Most Often Cited by Teachers in Liberia and Zimbabwe 102
4-6. The Average Salary of Primary School Teachers in Relation to the Average Nonagricultural Wage and to Gross National Product per Capita in Selected Countries 103
5-1. The Locus of Decisionmaking Authority in Primary Education Systems in the Republic of Korea, Nigeria, and the Philippines 119
5-2. The Distribution of Primary Education Budgets, Selected Countries 121

6-1. Regional Disparities in Gross Enrollment in Primary Schools, Selected Countries, Various Years *147*
6-2. Student-Classroom Ratio in Four Cities in Malawi, 1986 *148*
6-3. Gross Enrollment by Gender in Primary Schools, Selected Countries, Various Years *149*
6-4. Official Instructional Time in Single- and Multiple-Shift Primary Schools in Selected Countries *157*
7-1. The Costs and Achievement Gains Associated with Selected Educational Inputs in Second Grade in Northeastern Brazil, 1985 *177*
7-2. Incremental Cost Effects of Increasing Expenditure per Student in Low- and Lower-Middle-Income Countries *177*
7-3. Cost per Student Place of Local and International-Grade Construction Materials in Six Sub-Saharan African Countries *179*
7-4. Cost Effects of Specific Inputs *181*
7-5. Median Difference between the Length of the Primary School Cycle and the Time Needed to Produce a Graduate in Developing Countries, Selected Years, 1970–85 *182*
7-6. The Cost of Producing a Primary School Graduate and the Recurrent Savings from Reducing Repetition and Dropout Rates *184*
7-7. Community Support for Self-Help Primary School Projects in Zambia, 1979–84 *197*
8-1. Aid Disbursed to 143 Countries for Primary Education, 1981–86 *208*
8-2. Average Annual Contribution of Major Bilateral and Multilateral Donors to Primary Education, 1981–86 *211*
8-3. World Bank Lending Commitments for Primary Education, Fiscal Years 1963–90 *212*
8-4. Type of Aid for Primary Education by Country Income Group and Region, 1981–86 *213*
8-5. Aid for Selected Educational Support Activities, 1981–86 *214*
9-1. Challenges for Countries at Different Stages of Educational Development *225*

Foreword

A nation's children are its greatest resource. In only a few decades the prosperity and quality of life of all nations will be determined by today's children and their ability to solve the problems that face them, their families, their communities, and their countries. Education unlocks this ability, and investment in children's learning is the most important contribution a nation can make to a better future.

The centrality of children's learning is widely recognized. It is no coincidence that 1990 was International Literacy Year, a year in which world leaders convened in Bangkok to discuss strategies for meeting the basic learning needs of children and a year in which the World Bank recommitted itself to expanding its annual investment in education over the next decade.

The World Bank has long acknowledged the vital relationship between education and economic development and the central importance of primary education for both. Since 1963, when the Bank began lending for education, it has aimed to assist developing countries expand and improve their education systems. The world economic crisis has impeded the development of national systems of primary education that enable children to reach acceptable levels of learning. To realize the potential contribution of education to development, nations must find ways to use resources more effectively and efficiently in their pursuit of learning for all.

This study places learning at the center of programs to improve the education of children. It synthesizes the results of four years of research and consultation on the effectiveness and efficiency of primary education in developing countries, a program conducted by the Education and Employment Division of the Population and Human Resources Department of the World Bank. It draws on comprehensive reviews of the research and evaluation literature, on commissioned studies, and on original research conducted in the division. It has benefited from consultations with policymakers in developing countries, representatives of donor agencies, and experts in the field, and it was the basis for *Primary Education* (World Bank 1990a).

ANN O. HAMILTON
Director, Population and Human Resources Department
The World Bank

Preface

This book is about children, learning, and primary schools. Specifically, it is about how to improve the ability of primary schools to teach children and help them learn effectively. By placing learning at the center of this volume, we hope to remind readers of several truths. First, learning takes place in schools and classrooms among teachers and children, not in ministries of education or finance. Second, the output of the education process—learning—is paramount; inputs—teacher training, instructional materials, and so forth—may be important but only if they help children learn. Third, the cost and financing of education must be examined from the perspective of their impact on learning.

This book is not about out-of-school learning, although much learning does take place in families and communities. It is not about nonformal education, even though alternatives to school do exist in some parts of the world. Nor is it about all purposes of primary education.

Primary education serves several purposes. One is to teach students basic cognitive skills. A second is to develop attitudes and skills that children need to function effectively in society. A third is to advance nation building. While recognizing the importance of these objectives, this book focuses specifically on problems that primary schools in developing countries face in teaching children literacy, numeracy, and problem-solving skills.

We frequently hear that "the quality of education in developing countries is eroding." The truth of this statement depends on the indicator of quality used. If, on the one hand, one considers inputs as the sole indicator, quality appears to have declined, since the inputs available per child have declined. If, on the other hand, one considers the output of education—children's learning—as the indicator of quality, we are hard pressed to say anything since so little research has been done comparing children's learning over time.

We believe that the quality of primary education in a country must be judged by the learning of all its children. Improving the quality of primary education therefore means ensuring that more of a nation's children complete the primary cycle having mastered what was taught.

This book is concerned with strategies to improve the learning of all children. It explores five key areas for improvement: the inputs necessary for children to learn, methods for improving teachers and teaching, management requirements for promoting learning, ways to extend effective education to traditionally disadvantaged groups, and the means to afford enhanced education.

In writing this book, we encountered enormous difficulties in demonstrating that educational quality, as indicated by children's learning, is unsatisfactory. Cross-national evidence of children's learning is limited to a few studies, and evidence of change in learning is virtually nonexistent. Some countries keep records of the proportion of children taking an examination who ultimately pass it, but changes in the pass rate may simply reflect the wider range of ability among those tested that results from expanding educational access. We therefore strongly encourage governments to engage in regular national assessments of children's learning and to better monitor their own educational progress.

The book itself was developed through a process that integrated the World Bank's operational experience, policy analysis, and research with a review of scholarly literature and extensive consultations with educational leaders in developing countries, World Bank staff, external experts, and donor agencies. Although these individuals and agencies made substantial contributions, the findings, interpretations, and conclusions of the study are the responsibility of the authors and should not be attributed to the World Bank, its board of directors, or any of its member countries.

Marlaine E. Lockheed
Adriaan M. Verspoor

Acknowledgments

This book is the product of a team effort and would not have been possible without the contributions of all members of the team. Together they developed the overall theme of the book, and as individuals they prepared drafts of chapters based on their experience and particular areas of expertise. The intent was to produce not a collection of readings by various authors, with the inevitable duplications in coverage and differences in approach and style, but a unified treatment of the subject from a single point of view. Therefore, most of these contributions were, in varying degrees, revised, reorganized, expanded, or reduced to fit the requirements of the book.

We gratefully acknowledge our extensive debt to the following individuals: Ralph Romain and Pierre Englebert for chapter 2, Bruce Fuller and Alastair Rodd for chapter 4, John Middleton and Deborah Bloch for chapter 5, Deborah Bloch for chapter 6, Vicente Paqueo and Elizabeth King for chapter 7, and Pierre Englebert and Vicente Paqueo for chapter 8. A special note of appreciation is due to Deborah Bloch, who extensively edited several chapters in addition to the two she helped draft. Michel Welmond had principal responsibility for preparing the statistical appendix, with data generously provided by Unesco (the United Nations Educational, Scientific, and Cultural Organization).

We are indebted to numerous individuals for their valuable comments and assistance with various aspects of this book. Wadi Haddad and George Psacharopoulos, both of the World Bank, supported initial programs of research and review. Other colleagues at the Bank—Alan Berg, Nancy Birdsall, Paul Blay, Joan Claffey, Sidney Chernick, Ralph Harbison, Stephen Heyneman, Donald Holsinger, Emmanuel Jimenez, Martin Karcher, Bruno Laporte, Judith McGuire, Costas Michalopoulos, Peter Moock, Eileen Nkwanga, João Oliveira, John Oxenham, Barbara Searle, Christopher Shaw, James Socknat, Cecilia Valdivieso, Alex ter Weele, Michael Wilson, Laurence Wolff, George Za'rour, and Adrian Ziderman—provided useful reviews of earlier drafts or comments at seminars. So too did Marv Alkin (UCLA—University of California, Los Angeles), Eva Baker (UCLA), Françoise Caillods (IIEP—International Institute for Educational Planning), Robert Calfee (Stanford University), Gabriel Carron (IIEP), William Cummings (Harvard University), Thomas Eisemon (McGill University), William Firestone (Rutgers University), Gabriele Gottelmann (IIEP), Habib Hajjar (IIEP), Jacques Hallak (IIEP), Eric A. Hanushek (University of Rochester), Dean T. Jamison (UCLA), Kenneth King (University of Edinburgh), Beryl Levinger (American Field Service), Lars Mählick (IIEP), Noel McGinn (Harvard University), John Nkinyangi (UCLA), Hilary Perraton (Commonwealth Secretariat), Ernesto Pollitt (University of California, Davis), Neville Postlethwaite (University of Hamburg), Susan Roper (California State University at San Luis Obispo), Bikas Sanyal (IIEP), Lewis Solmon (UCLA), Thomas Tilson (Education Development Center), and Merlin Wittrock (UCLA).

The book benefited in particular from the suggestions of a review panel convened by the World Bank in May 1989 to comment on the manuscript; this panel comprised Shahnaz Wazir Ali (State Ministry for Education, Pakistan), Anil Bordia (Ministry of Education and Human Resource Development, India), João Calmon (Senate, Brazil), A. Babs Fafunwa (National Primary Education Commission, Nigeria), Enríque Ipina Melgar (Ministry of Education and Culture, Bolivia), Yung-Dug Lee (Seoul National University), Weifang Min (Beijing University), Joyce Mpanga (State Ministry for Education, Uganda), Santanina Rasul (Senate, Philippines), Patricio Rojas Saavedra (formerly of the Ministry of the Interior and Education, Chile), Mina Siagura (Ministry of Education, Papua New Guinea), Ekavidhya Na Thalang (Ministry of Education, Thailand), Mark Blaug (London University), Martin Carnoy (Stanford University), Chester Finn (Vanderbilt University), Isahak Haron (University of Malaya), and Torsten Husén (University of Stockholm). Helpful comments were also received at meetings of the International Working Group on Education in November 1988, the European Education Donor Agencies in April 1989, the World Congress of Comparative and International Education Societies in July 1989, and the World Conference on Education for All in March 1990.

Our understanding of education efficiency was sharpened at a meeting convened by the World Bank in February 1987 to discuss the internal efficiency of education in developing countries. The review panel consisted of Joan Claffey (USAID—United States Agency for International Development), Joseph Farrell (Ontario Institute for Studies in Education), Eric Hanushek (University of Rochester), Lucila Jallade (Unesco), Henry Levin (Stanford University), Donton Mkandawire (Malawi Certificate Examination and Testing Board), Carlos Muñoz Izquierdo (Centro do Estudios Educativos), Richard Murnane (Harvard Graduate School of Education), John Ryan (Unesco), Awang-Had Salleh (Northern University of Malaysia), Gajendra Shrestha (Tribuvan University, Nepal), Anthony Somerset (University of London), Edita Tan (University of the Philippines), Gareth Williams (University of London), and Ayotunde Yoloye (University of Ibaden).

Beryl Levinger and Rae Galloway contributed key information on education and nutrition, and Carol Copple provided invaluable help in summarizing the research on preschools. USAID generously supported Bruce Fuller's contribution to chapter 4.

Deborah Bloch lent her editorial expertise to the entire manuscript. Carol Copple, Joseph DeStefano, Pierre Englebert, Rae Galloway, Alastair Rodd, and Michel Welmond provided valuable assistance in preparing the boxes, tables, and graphs. The help of Alastair Rodd and Linda Larach during the final stages of the project was inestimable.

Cynthia Cristobal and Teresa Hawkins provided quick and efficient word processing and administrative services that assisted greatly in the preparation of this manuscript. Elizabeth R. Forsyth and Kathryn Kline Dahl edited the manuscript for publication.

As a final note, we would like to acknowledge the special contributions of John A. Lockheed, Sr., and Antoinette Verspoor, who have given us unswerving support and encouragement over the course of many decades.

Marlaine E. Lockheed
Adriaan M. Verspoor

1

Primary Education and Development

EDUCATION IS A CORNERSTONE of economic and social development; primary education is its foundation. It improves the productive capacity of societies and their political, economic, and scientific institutions. It also helps reduce poverty by mitigating its effects on population, health, and nutrition and by increasing the value and efficiency of the labor offered by the poor. As economies worldwide are transformed by technological advances and new methods of production that depend on a well-trained and intellectually flexible labor force, education becomes even more significant.

Primary education has two main purposes: to produce a literate and numerate population that can deal with problems encountered at home and at work and to serve as a foundation on which further education is built. In many countries in the developing world, education systems are unable to meet their objectives. First, they do not teach children already in school the core skills contained in their national curriculum; second, they do not provide all school-age children, particularly girls, with the opportunity to attend school. As a result, these primary education systems are ineffective and jeopardize national efforts to build a base of human capital for development.

To address these shortcomings, the first priority for primary education should be to increase children's learning in school so that most students master the curriculum and complete the primary cycle. Second, access to school must be provided for all school-age children. Both goals are important. School attendance without learning is meaningless, and development opportunities are lost when a large fraction of the school-age population has no access to schooling.

Developing countries can progress toward these goals only when available resources are allocated to the most cost-effective inputs. Most middle-income countries could significantly improve educational effectiveness by reallocating funds within the existing primary education budget.[1] In most low-income countries, however, the scope for such reallocations is limited, and additional resources for primary education are essential. The continuing economic crisis jeopardizes the ability of many countries, particularly in the developing world, to maintain the present level and quality of their educational services. Countries and groups with low education levels increasingly miss out on the gains made possible by technological innovations and new production processes. As the gap between the educated and the uneducated widens, the uneducated are progressively less able to enhance and shape their own economic and social development. This process constrains international competitiveness and confines a large proportion of the world's population to poverty.

The future development of the world and of individual nations hinges more than ever on the capacity of individuals and countries to acquire, adapt, and advance knowledge. This capacity depends, in turn, on the extent to which the population has attained literacy, numeracy, communications, and problem-solving skills. To move forward, all developing countries must improve the education and training of their labor force. Advanced education and training must rest on the solid foundation of good primary education.

The Contribution of Education to Development

Completed primary education helps alleviate poverty and advance economic and social development. A diverse body of literature demonstrates that the adults in developing countries who have higher levels of educational attainment have more paid employment, higher individual earnings, greater agricultural productivity, lower fertility, better health and nutritional status, and more "modern" attitudes than adults who have lower educational attainment. They are also more likely to send their children to school. These characteristics are dimensions of development. Primary education has other benefits for individuals and society as well. Education forges national unity and social cohesion by teaching common mores, ideologies, and languages. It also improves income distribution, increases saving and encourages more rational consumption, enhances the status of women, and promotes adaptability to technological change.[2]

Economic Development

Research and experience demonstrate that an educated labor force is a necessary, albeit not sufficient, condition for economic development (Schultz 1961; Denison 1962; McMahon 1984). Across countries, the correlation between national investment in education and economic growth is striking.[3] The industrialized economies of the late nineteenth and early twentieth centuries were based on a relatively well educated and skilled labor force. Peaslee

(1965; 1969) examined the relationship between growth in primary school enrollment and gross national product (GNP) per capita over a 110-year period (1850–1960) for thirty-four of the richest countries. He found that none had achieved significant economic growth before attaining universal primary education.

Benavot (1985) studied a more recent period, 1930–80, and found that primary education had a significant positive effect on the economic growth of 110 developed and developing countries. For the period 1945–80, Lau, Jamison, and Louat (1991) found economic growth powerfully affected by primary education in twenty-two East Asian and Latin American countries and by secondary education in fifty-four East Asian, Latin American, African, and Middle Eastern countries. Virtually all the newly industrialized economies with dramatic growth in the past twenty-five years, such as Hong Kong, Israel, Japan, the Republic of Korea, and Singapore, achieved universal or almost universal primary enrollment by 1965. The most successful of those economies also had high enrollment in secondary school and a labor force that was almost universally literate just before rapid and sustained industrial growth began (World Bank 1987b).

EARNINGS. Education has a significant effect on earnings, and the rate of return to education is high.[4] Psacharopoulos (1985) found very high social rates of return and estimated that the returns to completed primary education are 27 percent and the returns to secondary education are 15–17 percent. Private returns to education are significantly higher than social returns, reaching 49 percent for primary and 26 percent for secondary education (McMahon 1984).[5] The returns to education in developing countries are higher than those in more advanced countries, although they appear to diminish over time (Psacharopoulos 1985; Ryoo 1988). Education remains profitable, however, and its social rate of return still exceeds that of other investments by a considerable margin. For example, even informal, nonfarm family enterprises that are engaged in so-called modern production realize positive returns to education (Ryoo 1988).

PRODUCTIVITY. Earnings provide an indirect measure of productivity, but physical productivity is the best measure of education's economic impact. Workers and farmers with more education are physically more productive than those with less. Of particular importance is the productivity of farmers, since much of the labor force in developing countries works in subsistence agriculture. The effect of education on agricultural production can be assessed by comparing the agricultural output of farmers with different levels of educational attainment. Lockheed, Jamison, and Lau (1980) summarized the findings of eighteen studies containing thirty-one data sets from thirteen developing countries. They concluded that four years of primary education increased the productivity of farmers 8.7 percent overall and 10 percent in countries undergoing modernization (largely in Asia). Education increased the ability of farmers to allocate resources efficiently and enabled them to improve their choice of inputs and to estimate more accurately the effect of those inputs on their overall productivity.

Social Development

REDUCED FERTILITY. In many countries, rapid population growth has made raising the standard of living difficult, and excessive fertility has adversely affected maternal health and child survival rates. A growing population of young people strains education budgets, and in many low-income countries an increasingly large number of young people, often the children of rural and low-income families, receive no education at all. The disparity between urban and rural areas and between the emerging urban middle class and the poor seems to be growing. Reducing fertility rates must therefore be an important part of any development program, and reduced fertility depends heavily on educating women (Cochrane 1979). Fertility levels are determined most immediately by age of marriage, length of breast-feeding, and use of contraceptives, which are influenced in turn by the socioeconomic circumstances of the individual. One of these circumstances is the education of the mother (Haverman and Wolfe 1984). Educating women ultimately reduces fertility; even though fertility in Africa and Asia apparently increases with a few years of education, it declines thereafter with more schooling (Cochrane 1986). One explanation for this pattern may be the type of skills learned at different levels. Elementary education improves hygiene and nutritional practices, which improve both child survival and fertility, while further education highlights the advantages of controlling family size (Holsinger and Kasarda 1975).

IMPROVED CHILD HEALTH AND NUTRITION. The education of women is also closely related to child health, as measured by nutritional status or by infant and child mortality. Children of educated mothers live healthier, longer lives. One year of maternal education translates into a 9 percent decrease in child mortality. Apparently, the more education a woman attains, the more likely she is to seek professional health care, which diminishes child mortality. In Africa, a difference of one percentage point in the national literacy rate is associated with a two-year gain in life expectancy (Cochrane 1986). One determinant of life expectancy, and a major influence on mental and physical capacity, is nutrition. The scant evidence available suggests a strong positive relationship between the mother's education and her child's nutrition. Education apparently changes the mother's preference for foods and increases her influence in decisions about how food is distributed in the household.

ATTITUDINAL MODERNITY. A social consequence of greater education is the adoption of more "modern" attitudes. Becoming modern involves adopting rational, empirical, egalitarian beliefs, which are a precondition for functioning effectively in the political and economic institutions required for development. Research in eleven developing countries found that individuals scored higher on attitudinal measures of individual modernity as their years of education increased (Inkeles and Smith 1974; Holsinger and Theisen 1977). The influence of school on the adoption of modern practices was greater than that of other factors such as home environment, urbanization, and factory experience. Although modern attitudes were initially believed to be transmitted through the formal structure of schools, the influence of curriculum seems to be more direct (Armer and Youtz 1971). Students attending schools with a modern curriculum had more modern attitudes than those attending schools with a traditional one. For example, students who were taught the arts and sciences held more modern attitudes than students who attended teacher education institutions, polytechnics, or schools with a limited, religious curriculum.

The Importance of Cognitive Competencies

Education affects development through its enduring impact on various dimensions of cognitive competence: literacy (reading and writing), numeracy, modernity, and problem solving (Scribner and Cole 1981). These cognitive skills affect an individual's productive behavior and ability to use the products of technological change, such as pesticides and medicines, correctly. When schools are good and educate many children well, the process of development occurs relatively quickly; when schools are bad and educate few children well, education's impact on development is relatively slow. To increase the pace of economic and social development in developing countries, schools must teach most school-age children the essential skills targeted by the primary curriculum.

The few empirical studies that have examined how the cognitive consequences of education affect the earnings of individuals have found important effects.[6] Not only are literate and numerate individuals more likely to enter the modern wage labor market, they are also more likely to earn higher wages than less literate and less numerate workers with equal years of schooling.

Numeracy and literacy are also valuable skills for workers outside the modern wage labor market, such as rural farmers. Farmers who can read, write, and understand numbers can allocate inputs efficiently and thus increase productivity (Jamison and Moock 1984). Numeracy helps farmers estimate the profitability of past activities and the risk of future ones. Reading and writing help farmers keep records and properly apply modern agricultural technologies, such as agricultural chemicals, inorganic fertilizers, and new seed varieties. Cotlear (1986) stresses the relationship between education and technological innovation by emphasizing the importance of noncognitive aspects of education, such as receptivity to new ideas, that put the educated farmer more easily in contact with new technologies.

Education also affects production by developing analytic modes of problem solving. Cotlear (1986) notes that education increases the ability of farmers to think abstractly, which enables them to recognize the causal relation between technology and output. An example of this comes from Eisemon's (1989) survey of farmers in Kenya, which examined the effect of primary education on the cognitive skills of farmers. Data were obtained from heads of household who had not been to school and from those who had completed up to seven years of primary education; both men and women were surveyed. Farmers who had been to school were able to construct causal models of events in the natural world and to demonstrate how these events could be controlled by humans. They were able to observe, diagnose, and correct common agricultural problems better than farmers with fewer years of education. They actively sought to solve problems, while unschooled farmers did not (see box 1-1).

Box 1-1. How Primary Education Helps Farmers Solve Problems

Farmers in Kenya were shown a diseased plant and asked about the cause of the disease and the measures that might control or prevent it. An unschooled Kenyan farmer mentioned three possible causes of damage to maize—weeds, birds, and hailstones—none of which was responsible for producing the symptoms in the specimen shown. The farmer gave these as generic causes of crop damage and did not attribute particular kinds of damage to specific causes.

By comparison, a Kenyan farmer with seven years of primary education made a complex causal model that correctly identified the cause of damage and a possible solution:

> This is what Amodonde, the stalk borer [bug], does. It attacks the stem and makes it wither at the buds, sometimes without you knowing it. You buy chemicals from the store and apply when the maize is small, 2 or 3 feet. You spray the buds after the first weeding, from the top, when it is about to rain so the chemicals don't dry up. You can also put sulphur ammonia. It is also good for top dressing.

This farmer's understanding of how technology could improve productivity was enhanced by a primary curriculum that taught science in conjunction with farming practice and that emphasized scientific theory over memorization.

Source: Eisemon 1989.

The Development of Cognitive Competencies as a Goal of Primary Education

The official primary school cycle in most countries spans six years (see figure 1-1), which is ample time for students to meet the curriculum objectives. For virtually all countries these objectives include developing cognitive competence, which encompasses both basic literacy and numeracy and the ability to apply basic skills to new problems. Few schools achieve this goal, and the reason for their failure is rooted in the origin of mass education systems.

Modern public education developed from two distinct educational traditions, one concerned with educating the elite and one with educating the masses. As the noted American educator Lauren Resnick explains (1987, p. 5), the objective of elite systems of education has been to "train an intellectual elite, drawn largely from privileged social strata, in capabilities of reasoning, rhetoric, mathematical and scientific thought and other skills that today carry the higher order [thinking] label." In contrast, virtually all mass education systems were established with the aim of producing minimal levels of competence in the general population. Resnick points out that schools for the masses

> concerned themselves with basic skills of reading and computation, with health and citizenship training, and the like. Routinized performance rather

Figure 1-1. The Primary School Cycle in 169 Countries, 1985

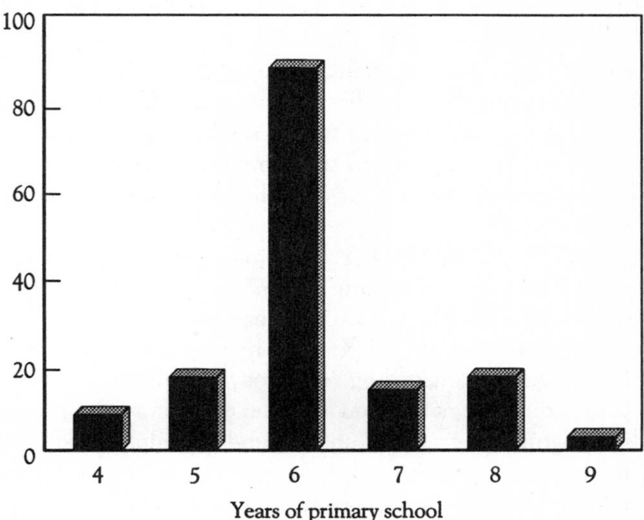

Source: Unesco 1987.

> than creative and independent thought was stressed. Mass education was, from its inception, concerned with inculcating routine abilities: simple computation, reading predictable texts, reciting religious or civic codes.

Thus teaching what are known as higher-order thinking skills in primary schools for the masses is a new idea:

> Although it is not new to include thinking, problem solving, and reasoning in *someone's* curriculum, it is new to include it in *everyone's* curriculum. It is new to take seriously the aspiration of making thinking and problem solving a regular part of a school program for all of the population . . . even the poor. (Resnick 1987, p. 7)

To varying degrees, all societies must update their education systems by charging them with developing the skills needed to solve the problems of the future.[7] Meeting this challenge will be difficult for all countries, but especially for developing ones. First, in many countries neither the intended nor the implemented curricula stress the acquisition of problem-solving skills or even skills such as comprehension, application, synthesis, and analysis; they emphasize instead rote memorization of facts. Second, although official curricular objectives, composition, and content are fairly uniform among countries, the conditions under which education takes place in many developing countries effectively keep students from learning higher-order thinking skills. And third, most children in developing countries do not remain in primary school long enough to develop basic or higher-order literacy and numeracy.

Virtually all countries incorporate the same subjects into the curriculum of their primary schools and give them the same emphasis (Benavot and Kamens 1989). These subjects include language arts (reading and writing), mathematics, science, social studies, and moral and aesthetic education. More than 50 percent of scheduled time is used to teach language skills and mathematics. This emphasis reflects the common objective of primary education: to impart both basic and higher-order literacy and numeracy skills.

BASIC LITERACY AND NUMERACY. A major objective of primary education is to teach the basic skills of reading and mathematics. Many educators still debate how reading should be taught. Most agree, however, that basic literacy requires several types of skills, which some believe are ordered in stages that range from prereading to decoding to fluency (Chall 1983). Prereading includes understanding the nature and function of print, reading common signs and labels, recognizing letters and some sounds (in alphabetic systems), and writing one's name. Decoding involves understanding the correspondence between letters and sounds (again, in alphabetic systems), identifying about 1,000 common words, and reading simple texts. Fluency involves integrating prereading and decoding knowledge and skills, recognizing about 3,000 familiar words and derivatives, relying on context and meaning as well as decoding to identify new words, and reading more fluently.

At all stages students must relate to what is being learned and connect it with their own experience. The more that students connect with the learning material, the more rapid their progress will be. The time required to attain literacy skills varies by the quality of instruction and the ability of the individual student, but some generalizations are possible. The speed at which learners progress can vary considerably between individuals in developed and individuals in developing countries. In developed countries, where learning resources are abundant and literacy skills are typically acquired in the student's first language, fluency takes from two to three years of conventional formal education. In developing countries, where resources and written materials are scarce and the acquisition of literacy skills is often complicated by a second language, fluency often takes longer.

A second cognitive competency is basic numeracy. Dossey and others (1988, p. 31) define basic numeracy as "a considerable understanding of two-digit numbers" and their addition and subtraction, some understanding of multiplication and division, knowledge of relations among coins, and the ability to read information from charts and graphs and use simple measurement instruments. Most children in developed countries acquire this rudimentary level of numeracy by the age of nine, or after three years of conventional formal education. A recent assessment of mathematics achievement in the United States found that by the fourth grade most children have acquired a command of basic arithmetic operations (98 percent of nine-year-olds knew simple arithmetic facts; Dossey and others 1988). A related study of mathematics achievement in six developed countries found that by the eighth grade, virtually all children could add and subtract two-digit numbers without regrouping and could solve simple number sentences involving these operations (Lapointe, Mead, and Phillips 1989). For example, all the students could solve the following:

> 29 = [ ] + 16.
> What number should go in the brackets to make the number sentence above TRUE?

Research suggests that children in developing countries can deal fairly easily with real-world mathematics problems that require basic numeracy. Carraher, Carraher, and Schliemann (1987, p. 95) note that "situations in which arithmetic problems are embedded may have a strong impact on how they are solved. This impact . . . seems to result from the meaning that problems have for children when they engage in problem solving. Situations that present quantities embedded in a meaningful transaction—such as calculating the amount of change after a purchase or the number of children in a school—seem to engage children in problem-solving procedures." For example, third-grade children in Brazil correctly solved 57 percent of word problems about a store, but only 38 percent of computational problems requiring the same operations and using the same numbers. Mathematics instruction has been improved by building on children's existing mathematics skills, but several years of formal education may be needed for students to achieve basic numeracy.

PROBLEM-SOLVING SKILLS. Basic literacy and numeracy are insufficient foundations for further learning, which requires the ability to apply basic skills to new and more complex problems (that is, "higher-order" literacy and numeracy). Higher-order literacy implies that reading is a tool for learning new information, ideas, attitudes, and values; the text goes beyond the language and knowledge that the reader has acquired through listening and direct experience. Many educators believe that children at all levels of basic literacy engage in higher-order thinking when they are required to understand and interpret what they read. As Resnick (1987, p. 7) notes, however, "It is a new challenge to develop educational programs that assume all individuals, not just an elite, can become competent thinkers." Achieving higher-order literacy requires, under the best conditions, formal education lasting five to six years, which is the international norm for the primary school cycle. In countries where many under- and untrained teachers work in impoverished schools, achieving higher-order literacy often takes longer.[8]

Higher-order numeracy is the ability to use basic operations to solve simple problems. One definition of mathematics proficiency is that learners at the "beginning problem-solving level" (a) have an initial understanding of the four basic operations, (b) can add and subtract whole numbers in one-step word problems and money situations, (c) can multiply a two-digit and a one-digit number, (d) can compare information from graphs and charts, and (e) are beginning to analyze simple logical relations. In developed countries, most children reach this level of numeracy by the eighth grade, or the age of thirteen; in the United States, for example, 73 percent of thirteen-year-olds can solve one-step word problems and analyze simple logical relations (Dossey and others 1988). In other developed countries, including Canada, Great Britain, and Spain, more than 85 percent of thirteen-year-olds have reached this level by the seventh or the eighth grade (Lapointe, Mead, and Phillips 1989). No directly comparable data on performance are available for developing countries, but students in poor countries will probably take at least as long to reach this level of numeracy as those in more affluent ones.

The Shortcomings of Education in the Developing World

While developed and newly industrialized countries have increased their investment in education and training and have concentrated on creating a population with broad cognitive competencies and problem-solving skills, developing countries have been unable to enroll and teach comparable proportions of their children.

Low Completion Rates

Despite relatively high gross enrollment rates worldwide, including developing countries, fewer than 60 percent of the children who enter school in the low-income countries and about 70 percent of those who enter school in the lower-middle-income countries reach the last year of primary school (see figure 1-2). Moreover, primary school completion rates declined over the past decade in the poorest countries. As a result, illiteracy remains widespread, and fewer than 30 percent of the adults in the labor force have completed primary school.

Low completion rates are the result of high rates of early dropout, which are partly due to poor academic achievement and high rates of repetition. On average, repetition rates in low- and lower-middle-income countries are two to five times higher than those in upper-middle- and higher-income coun-

Figure 1-2. Students Reaching the Last Year of Primary School, by Country Income Level, 1970–85

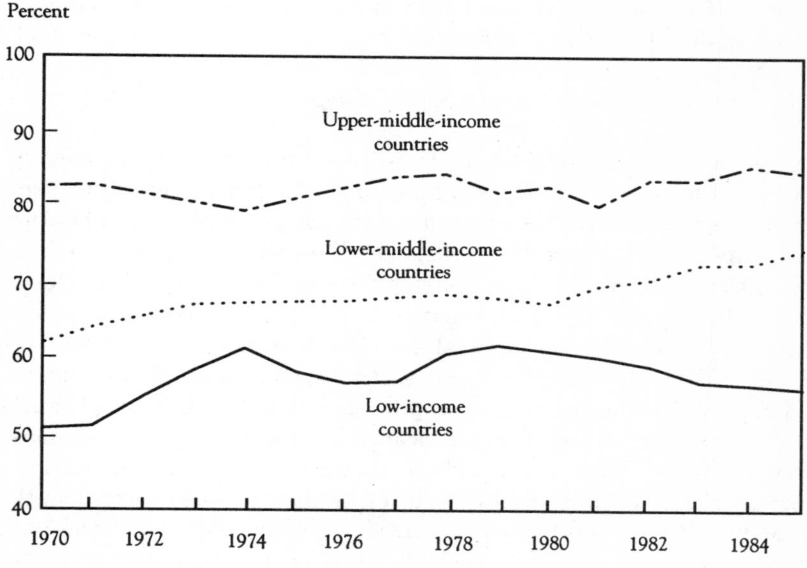

Source: Unesco 1989b.

Table 1-1. Median Repetition Rates of Primary Students for Selected Years, 1965–85
(percent)

Country income level	1965	1970	1975	1980	1985
Low	21.7	20.9	17.0	15.6	16.3
Lower middle	19.0	15.8	11.4	11.7	10.6
Upper middle	10.3	11.4	9.0	6.5	7.5
High[a]	0.0	1.6	2.5	2.0	1.5

a. Excluding oil exporters.
Source: See table A-11 in the appendix.

tries, as indicated in table 1-1. The countries with the lowest income have the highest repetition rates. For example, fourteen countries reported primary-level repetition rates of 20 percent or more in 1985; 60 percent of those were low-income countries.

The likelihood that a student will repeat a grade is not equal at all grade levels. In some countries, nearly all students spend two years in the first grade, which means that a disproportionate number of students are found in that grade. For example, in Bangladesh, Bhutan, Chad, Ethiopia, Haiti, Laos, and Nepal, more than 40 percent of all students enrolled in grades one to five are in the first grade. In many African countries, children intentionally repeat their last year of primary school in order to be more prepared for the secondary-level entrance examination.

The Undereducation of Primary School Completers

In low-income countries even the individuals who complete their education often fail to reach national or international standards of cognitive performance in mathematics, science, and reading comprehension. While national standards for subjects such as social studies (history, geography), morals, and aesthetics vary considerably from country to country, most countries seek—at least officially—to have the primary curriculum impart lasting literacy and numeracy skills, as well as a general scientific understanding of the world.

Since the early 1960s the International Association for the Evaluation of Educational Achievement (IEA), a private organization of research institutions, has conducted studies of educational achievement in approximately forty countries.[9] Those studies indicate that students in some low- and lower-middle-income countries learn much less of their national curriculum than expected. On tests of mathematics, reading, and science—key areas of the primary school curriculum—students in developing countries performed poorly, answering correctly only about 40 percent of the questions posed. (Table 1-2 presents the average test scores for national, random samples of students in the participating countries.) In all cases, students were asked questions that reflected their own national curriculum but were common to

Table 1-2. Test Performance of Students in Three Subjects, Various Years
(average percentage of correct answers)

Economy	*Reading comprehension (Grade 6, 1970)*	*Arithmetic (Grade 8, 1980–82)*	*General science (Grade 6, 1970)*	*General science (Grades 4–6, 1983)*
High-income economies				
Australia	—	—	—	54
Belgium				
Flemish-speaking	65	58	53	—
French-speaking	74	57	48	—
Canada				
English-speaking	—	56	—	57
French-speaking	—	—	—	60
England and Wales	71	48	56	49
Finland	74	46	57	64
France	—	58	—	—
Germany[a]	—	—	51	—
Italy	65	—	55	56
Japan	—	60	61	64
Luxembourg	—	45	—	—
Netherlands	69	59	48	—
New Zealand	—	46	—	—
Norway	—	—	—	53
Scotland	70	50	51	—
Sweden	72	41	60	60
United States	67	51	61	55
Average	70	52	55	57
Upper-middle-income economies				
Hong Kong	—	55	—	47
Hungary	70	57	53	60
Iran	39	—	32	—
Israel	—	50	—	50
Korea, Rep. of	—	—	—	64
Poland	—	—	—	50
Singapore	—	—	—	47
Average	55	54	43	53
Low- and lower-middle-income economies				
Chile	61	—	36	—
India	53	—	36	—
Malawi	34	—	42	—
Nigeria	—	41	—	33
Philippines	—	—	—	40
Swaziland	—	32	—	—
Thailand	—	43	47	—
Average	49	39	40	36
Median (all countries)	68.0	50.0	51.0	54.5

— Not available.

a. Data refer to the Federal Republic of Germany before unification.

Sources: Heyneman 1980; Livingstone 1985; IEA 1988.

the curricula of all the countries in the study. Students who performed poorly thus did not reach national (or international) standards of achievement.

The first IEA reading comprehension test was administered to sixth-grade students in eight industrialized and five developing countries; the median score for correct responses was 68 percent across all countries. The scores in three of the least-developed countries (India, Iran, and Malawi) were 53, 39, and 34 percent, respectively (Heyneman 1980a). Similarly, in tests administered between 1980 and 1982, eighth-grade students from twelve industrialized and six developing countries took an IEA mathematics test containing simple arithmetic problems as well as more complex problems using algebra (Robitaille and Garden 1989). The median score for correct responses across all countries was 50 percent for arithmetic and 43 percent for algebra. Students in the three developing countries solved 39 percent of the arithmetic problems and 31 percent of the algebra problems (Livingstone 1985). In 1983 primary students in nine developed countries, six upper-middle-income countries, and two lower-middle-income countries were tested on their knowledge of science. Across all seventeen countries, the median score for correct answers in general science was 54 percent. Correct answers for students in upper-middle-income countries ranged from 47 to 64 percent, which resembled the range of scores in developed countries. Students in Nigeria and the Philippines, the only low- or lower-middle-income countries to participate, performed less well—33 and 40 percent of their responses were correct, respectively (IEA 1988).

These country-level averages hide significant variations among schools within each country. For example, the average percentage of correct answers on the science test ranged from 17 to 88 percent for public primary schools in the Philippines, and sixty-seven schools (15 percent of those tested) exceeded the median score for all nations (Lockheed, Fonacier, and Bianchi 1989). Thus every developing country has some schools in which children complete primary education having mastered the skills targeted in the curriculum. Not all differences in student achievement between schools are caused by the schools themselves. Some apparently effective schools recruit unusually talented students; many are high-quality private schools with students whose families can afford to pay the tuition (Jimenez, Lockheed, and Paqueo 1991). While good public schools do exist and do teach children successfully under difficult conditions, there are simply too few such schools in most developing countries. The result is that many students do not acquire the numeracy and literacy skills they need to function effectively in their own society. Reading comprehension—a necessary skill—is particularly weak.

International achievement tests have been criticized for "comparing the incomparable," for testing with multiple-choice instruments that are unfamiliar to students in developing countries, and for implicitly expecting impoverished schools in developing countries to perform as well as richly endowed schools in developed ones. Nevertheless, internationally comparable indicators are invaluable for measuring progress against commonly accepted standards of achievement.

International comparisons are not the only nor perhaps the most important indication that many schools in developing countries do not teach students adequately. The performance of students on national examinations and other achievement tests administered within the country is also telling.

For example, according to an unpublished World Bank report, children in rural schools in a North African country read the following passage in Arabic: "*Hind celebrated her birthday. Her friends gave her nice gifts. She thanked them and served them sweets and drinks.*" Only 28 percent of the third graders and 56 percent of fourth graders correctly answered the question "*What did Hind celebrate?*"; and only 21 and 38 percent, respectively, correctly answered the question "*What did Hind serve her friends?*" By any standard, children who have completed three or four years of schooling should be able to read and comprehend this simple text.

In the 1971 IEA reading test, sixth-grade students read the following passage in their own language: "*The dog is black with a white spot on his back and one white leg. The color of the dog is mostly: (a) black, (b) brown, or (c) grey.*" In industrial societies, 90 percent of the students understood what they read and answered this question correctly, but in India 36 percent, and in Chile 26 percent, failed to answer correctly (Heyneman 1980b). After six years of schooling, children should not have this degree of difficulty in reading comprehension.

Performance is poorest on tasks that require students to apply knowledge to new problems (higher-order literacy and numeracy). For example, 1,300 sixth-grade students attending schools in two Francophone African countries took a mathematics examination that had been administered to all sixth graders in France the same year. While 70–80 percent of the students had achieved basic numeracy and were able to solve arithmetic problems involving one only of the four operations (addition, subtraction, multiplication, and division), only 26 percent of the students in one country and 44 percent of the students in the second country were able to solve a simple consumer problem that required them to combine operations of multiplication and division: "*A package of 4 costs 3.80 francs. A package of 8 will cost: (a) 30.40 francs, (b) 7.60 francs, (c) 6.60 francs.*" By comparison, 78 percent of sixth-grade students in France, studying the same official curriculum, were able to solve the problem (Orivel and Perrot 1988).

Similarly, in an Anglophone African country, seventh-grade students were asked this consumer problem: "*If one meter of cloth costs 3.15 kwacha, what is the cost of 604 meters of cloth?*" Only 27 percent of the students replied correctly. In another Anglophone African country, sixth-grade students were asked: "*Judy had 13 books. She gave 6 books to Joseph. How many books has Judy now?*" Only 13 percent of the respondents answered this question correctly, according to unpublished IEA data from the Second International Science Study.

On the IEA general science test, students were shown three pictures of lighted candles: one in a large closed box, one in a small closed box, and one in an open box. The question was: "*Three candles, which are exactly the same, are placed in different boxes as shown. Each candle is lit at the same time. In what order are the candle flames most likely to go out?*" Across all countries, the median score was 54 percent correct for students in grades 4–6; the question was answered correctly by 25 percent of fifth graders in the Philippines and by 20 percent of sixth graders in Nigeria (IEA 1988).

When education does *not* produce literacy sufficient to understand the directions printed on, for example, commercial fertilizers or pesticides, serious environmental and health dangers may result. One study asked thirty farmers in Kenya, most of whom had some secondary education, to read the instructions for using two popular fungicides and pesticides, Dithane M-45 and Murphy Dawa ya Mboga, and found that they tended to memorize the instructions rather than to draw information from them. The farmers correctly answered fewer than one-third of the ten questions asked. The author notes:

> The dangers of improper use of pesticides and fungicides were very poorly understood. Three questions were asked about the handling, storage, and health hazards of the products. The questions related to storing and handling Dithane M-45 could be answered by recalling information from the instructions, but these [previously mentioned] precautions and the dangers of using Dawa ya Mboga have to be inferred from the instructions that accompany the product. . . . [Moreover,] while well-schooled cultivators may be able to read instructions in English and Kiswahili for the use of agricultural chemicals, this does not necessarily imply an ability to utilize such modern agricultural technologies to increase production. (Eisemon 1988, p. 128)

Inadequate Human Resources as a Consequence of Poor Education

A country's ability to apply modern technology to agricultural and industrial production is determined largely by the quality of its workers. Despite the rapid expansion of their education systems over the past two decades, human capital in developing countries remains seriously undeveloped. Too small a proportion of the adult labor force has completed a primary education, and even those who have finished often lack the skills needed to meet the demands of rapid economic development. The fraction of adults in developing countries who are educated enough to produce, acquire, adapt, and apply modern technology to agricultural and industrial production is perilously low.

Table 1-3. Average Years of Education Completed by Adults in Sixty-Five Countries, 1980

Country income level	*Adults age 15–19*	*Adults age 25–29*	*Adults age 35–39*	*Adults age 45–49*
Low	2.4	2.9	2.3	1.7
Lower middle	5.5	4.8	3.7	2.7
Upper middle	8.5	9.0	7.8	6.9
High	9.4	10.1	9.0	8.0

Sources: Horn and Arriagada 1986; Komenan 1987.

An Uneducated Adult Labor Force

Because their school participation rates are low and their repetition and dropout rates are high, the adult labor force in developing countries has little formal education (see table 1-3; Horn and Arriagada 1986; Komenan 1987). The average adult fifteen years of age and older in low-income countries has completed fewer than three years of formal schooling, which is insufficient for acquiring and sustaining literacy and numeracy (Hartley and Swanson 1984). As a result, the vast majority of the adult labor force in developing countries cannot respond to or take advantage of technological changes in subsistence farming, industrial production, or communications.

Scarcity of Scientists and Engineers

The adult labor force of most developing countries consists predominantly of subsistence farmers. Few adults work in the formal labor sector or have high-level skills, particularly in areas such as science and engineering. Countries with a large number of scientists and engineers—especially if they are involved in research and development—have a much greater possibility of adapting and developing new technologies (Rosenberg 1982; Bianchi, Carnoy, and Castells 1988). Indeed, the countries that have shown the greatest advances in creating and adopting technology in recent years are those with the highest proportion of scientists and engineers in their populations. As indicated in figure 1-3, low- and middle-income countries produce relatively few scientists and engineers. These countries need to develop scientific personnel capable of understanding the latest technological advances and then adapting and applying them to the local production of goods and services.

Primary Education: The First Step

A poor system of primary education compromises the entire system of human capital development. It produces students who are poorly prepared for secondary- and tertiary-level education and adults who are illiterate. Most important, it does not produce enough truly educated parents, workers, and managers who can contribute to development.

This book presents policy options for improving the effectiveness and availability of schools in developing countries. Improving educational effectiveness means increasing the number of primary schools whose students master the core knowledge and skills of the curriculum. Emphasis should be on learning. Preventing failure and routinely promoting students should, and can, be standard practice. Most education systems in developed and in upper-middle-income developing countries graduate between 80 and 100 percent of the students who enter primary school (figure 1-2).

Although poor-quality education exists at all levels, improvement must begin at the primary level, where children develop their basic attitudes and approaches to learning. Improving the quality of education for students in primary schools is a prerequisite for developing the human resource base required to meet the changing technological demands of the twenty-first century. To initiate a deeply rooted and sustainable process of technological development, human capital formation must be broadly based and allow a progressively larger share of the general population to participate in the process of economic transformation.

Figure 1-3. Estimated Number of Scientists and Engineers in Research, by Country Income Level, 1985

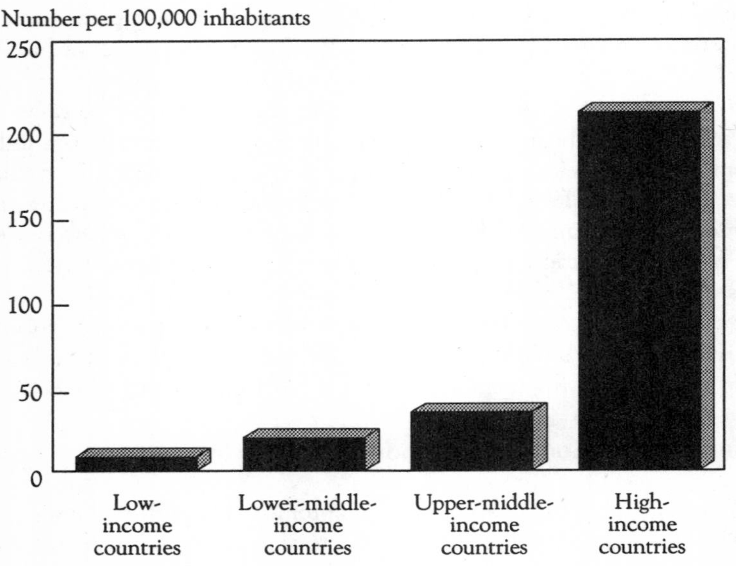

Source: Unesco 1988a.

Notes

1. In this book, countries are classified into four major income groups—low-income, lower-middle-income, upper-middle-income, and high-income economies—according to the 1988 *World Development Report* (World Bank 1988c). The thirty-nine countries with per capita incomes of less than $450 are classified as low-income economies. The thirty-four lower-middle-income economies have per capita incomes between $450 and $1,700. The twenty-six upper-middle-income economies have per capita incomes between $1,700 and $7,425. The seventeen high-income countries, which do not include the high-income oil exporters, have per capita incomes greater than $7,425. Countries that have populations of less than 1 million or that do not report GNP per capita are not included. The term "developing countries" refers to low- and middle-income countries.

2. Although most analysts agree that the process of education imparts not only knowledge and skills but also important attitudes, behaviors, and personal characteristics, consensus on this point is recent. Schultz (1961) first argued that education was the major source of acquired abilities, or human capital, that raised the productivity of workers. His analysis was quickly countered by scholars who noted that individuals with more ability were likely to obtain more education than those with less ability and that estimates of education's effect on productivity needed to control for initial ability (Psacharopoulos 1975). More recently, critics have argued that education does not contribute value-added knowledge or skills, but simply helps employers identify (or screen) individuals who possess either innate ability or personal characteristics—such as attitudes toward authority, punctuality, or high motivation—that are valuable to employers (see Woodhall 1987 and Winkler 1989 for reviews). However, recent research on the effect of schools on learning provides clear evidence that variations in the characteristics of schools are associated with variations in student outcomes (Creemers, Peters, and Reynolds 1989; Raudenbush and Willms 1991).

3. Most studies examine only the effect of years of formal schooling and not the effect of informal educational experiences nor of cognitive outcomes of formal schooling. Tilak (1989a) presents a recent review of this research.

4. The rate of return is the discount rate that equates the present value of the economic costs and benefits of an investment. It measures the profitability of alternative investments. The private rate of return is the discount rate that equates the private opportunity and direct cost of education with the individual's after-tax earnings. The social rate of return adds the cost of public subsidies and extends private gains to include taxes and any positive social externalities not captured by individuals. Earnings are generally used as a proxy for labor productivity, assuming that wages reflect the marginal productivity of labor, and few studies of the rate of return to primary education incorporate the opportunity costs of child labor, which are often sizable.

5. These figures are substantially lower when opportunity costs are included.

6. Most studies on how cognitive competencies mediate schooling deal with secondary school graduates. See, for example, Boissiere, Knight, and Sabot (1985).

7. The recognition of the relationship between technological change and the need to teach problem-solving skills is not new. In 1932 the Scottish educator J. Graham Kerr (1932, p. 84) wrote that "the reformed curriculum will have to be predominantly scientific. What is needed is the inculcation of the scientific method—by which the individual learns to observe and think for himself. Perhaps the most urgent reason for

developing the scientific habit in the masses of our citizens is to be found in these new applications of science."

8. Early on, Unesco noted that in developing countries at least four years of primary school were needed for the relatively select group of children then in school to become literate and numerate. That minimum had to be increased as schooling spread, the range of abilities being taught widened, and the average competence of teachers declined.

9. IEA's achievement tests, which are developed specifically for cross-national comparisons, examine the extent to which students have learned an agreed-upon curriculum; they also examine the degree to which the curriculum is taught.

2

A Brief History of Primary Education in Developing Countries

TRACING TRENDS IN EDUCATION across national, regional, and economic boundaries often obscures the unique context of each country. Nevertheless, the current condition of education in developing countries can be traced to three global trends that have been evident for the past twenty-five years: expanded availability, the search for relevance, and economic austerity.

First, the availability of primary education has expanded on a scale that is remarkable by any standard and reflects the strong determination of countries to provide their populations with universal access to schooling. As figure 2-1 indicates, the number of children enrolled in primary school in the ninety-nine largest developing countries increased nearly 200 million between 1965 and 1985, from 298 million to 482 million. During the same period, however, the school-age population (children six to eleven) also increased, from 372 million to 527 million. Because overage students filled a large proportion of school places, only 382 million children age six to eleven were enrolled in primary school in 1985, leaving another 145 million out of school.[1]

A second trend became evident in the late 1960s, when countries began to respond to the criticism that formal schooling was both irrelevant to the lives of the rural poor and overly expensive. The search for effective alternatives to the traditional content and delivery of primary education continued into the 1970s and 1980s, and many of the same concerns remain today.

Economic developments have been a third force shaping education. In the late 1970s and the 1980s most countries had to adjust to increasingly constrained financial circumstances, accompanied in most cases by growth—but in some cases by decline—in primary education. These constraints are likely

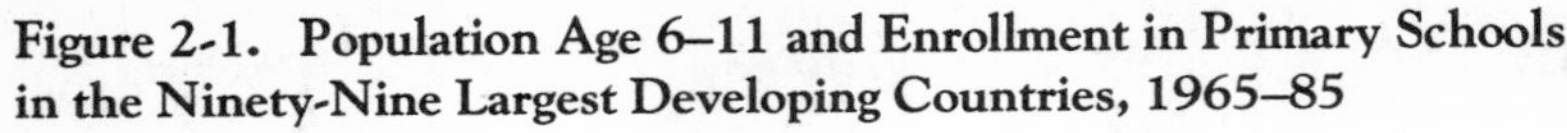

Figure 2-1. Population Age 6–11 and Enrollment in Primary Schools in the Ninety-Nine Largest Developing Countries, 1965–85

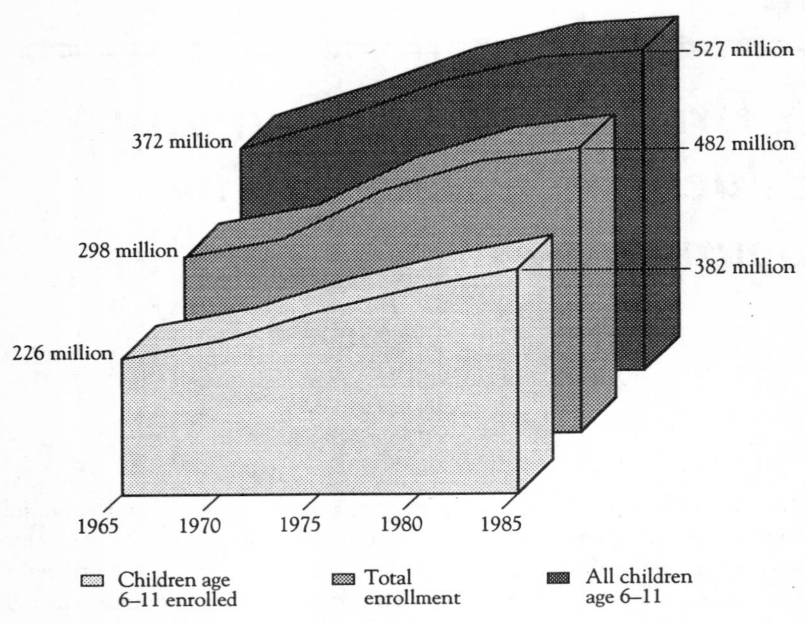

Source: See tables A-1, A-2, and A-3 in the appendix.

to continue for some time, posing a special challenge for nations seeking to expand and improve education over the next two decades.

Expanding Opportunities for Education

In 1960 the most striking aspect of primary education systems in most developing countries was their low capacity: the schools of many former colonies and traditionally independent countries served only a fortunate few. The United Nations (UN) had recognized this as a problem a decade earlier and had set the stage for expanding mass education in its *Universal Declaration of Human Rights* (United Nations General Assembly 1948), which stated: "Everyone has the right to education. Education shall be free at least in the elementary and fundamental stages. Elementary education shall be compulsory."

The benefits of education were clear to all. The developed countries were developed because they had, among other attributes, a large population of educated and trained workers. The pattern of social mobility in colonial societies also affirmed that the private returns to education were considerable. Virtually all newly independent countries in pursuit of economic development and social justice gave education, particularly primary education, top priority.[2]

Targets for Expansion

To translate the UN's principles into action, Unesco (the United Nations Educational, Scientific, and Cultural Organization) sponsored regional meetings to discuss the future of education. In 1956 a meeting was held in Lima on Free and Compulsory Education in Latin America and the Caribbean, a theme that was discussed again in Santiago in 1963. Asian countries met in Karachi in 1960 and in Tokyo two years later. They (excluding China) set themselves the goal of increasing gross enrollment ratios from about 70 percent in 1964 to about 90 percent in 1980. The Addis Ababa conference of 1961 set a goal for Africa: achieve universal enrollment in primary school by 1980.

Based on the projected growth of the school-age population and enrollment goals, the regional conferences projected primary enrollments for 1970 and 1980; these estimates are reported in table 2-1. The enrollment projections for the late 1960s (1965–70) were ambitious average annual growth rates: 7.6 percent for Africa, 9.6 percent for Asia, and 5.1 percent for Latin America and the Caribbean, all of which far exceeded the annual population growth rates of those regions.

Expansion Accomplished

TOTAL ENROLLMENT. The actual school enrollment achieved by the target dates 1970 and 1980 greatly surpassed the projections (table 2-1). By 1980, 59 million children were enrolled in school in Africa, compared with a projected 33 million, and 331 million children were in school in Asia, compared with a projected 220 million. Only in Latin America and the Caribbean was the actual enrollment in 1970 similar to that projected earlier (47 million

Table 2-1. Projected and Actual Enrollment in Primary Schools, 1970 and 1980
(millions)

	Projected enrollment		*Actual enrollment*	
Region	*1970*	*1980*	*1970*	*1980*
Africa	20.4	32.8	29.4	59.2
Asia	124.6	220.0	243.0	330.8
Latin America and the Caribbean	43.5	—	46.6	64.8

— Not available.

Sources: Unesco 1961, 1986, 1987; unpublished background material from the 1987 Regional Conference of Ministers of Education and Those Responsible for Economic Planning of Member States in Latin America and the Caribbean, in Bogotá.

and 44 million, respectively). No 1980 projections were made for Latin America and the Caribbean. Thus the 1960s and 1970s were decades of phenomenal expansion of primary school capacity; the annual growth rates of enrollment from 1965 to 1975 averaged 5.2 percent in low-income countries other than China and India, 4.1 percent in lower-middle-income countries, and 3.4 percent in upper-middle-income countries (see table A-2 in the appendix). From 1965 to 1986, the total number of schools doubled and that of teachers tripled in developing countries. Still, many countries did not achieve universal enrollment of school-age children in primary schools, largely because of unexpectedly high population growth rates and high rates of repetition that filled many school places with overage children. The situation was most acute in low-income countries, excluding China and India.

POPULATION AND ENROLLMENT GROWTH. Both enrollment in primary schools and the school-age population have grown substantially since 1965, with the annual growth rate of enrollment exceeding that of the population, on average, in all developing countries. Because the rates at which total enrollment grew during the late 1970s (4.2 percent in low-income countries without China and India and 5.4 percent in lower-middle-income countries) exceeded the rates at which the population grew (2.3 and 2.4 percent, respectively), the number of children not attending school in low- and lower-middle-income countries was expected to decline significantly during the 1980s.

This expectation, however, was not fulfilled. The average annual enrollment growth dropped steeply in both low-income and lower-middle-income countries in the 1980s (see figure 2-2), and the growth of total enrollment slowed to average annual rates of 2.7 and 2.2 percent, respectively. The sharpest declines, which often produced negative rates of enrollment growth, occurred in several low-income countries hard hit by war: Afghanistan (7.3 to –12.3 percent), Ethiopia (14.5 to 2.8 percent), Mozambique (6.0 to –0.9 percent), and Somalia (6.6 to –6.5 percent).

Even so, growth rates of total enrollment continued to exceed growth rates of the school-age population, which remained relatively stable from 1975 to 1985 in low-income countries other than China and India (about 2.4 percent), and fell from 2.2 percent to 1.8 percent in lower-middle-income countries and from 1.5 percent to 1.3 percent in upper-middle-income countries. From 1975–80 to 1980–85, enrollment growth rates declined in developing countries. Because enrollment growth remained about 0.5 percent higher than population growth, existing levels of coverage were sustained.

The growth rates of total enrollment declined for two reasons. First, as the countries that contributed most dramatically to the increase in overall enrollment in the 1970s arrived at or near universal primary education, their contribution to the overall increase in enrollment slackened. Mass education grows the fastest in countries where the educational system can accommodate 20–50 percent of the school-age population; it grows slowest in countries

Figure 2-2. Average Annual Population and Enrollment Growth Rates, by Country Income Level, 1975–85

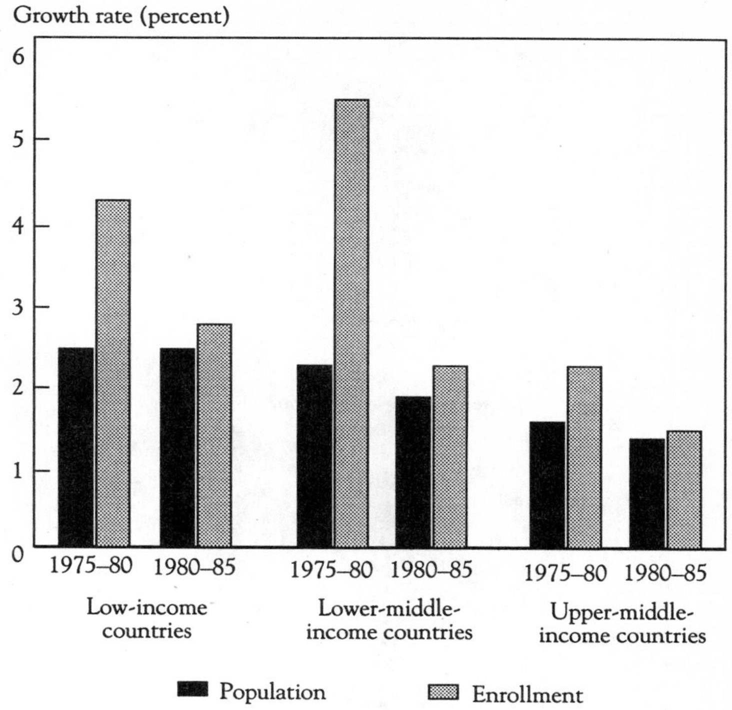

Source: See tables A-1 and A-2 in the appendix.

where more than 70 percent can be accommodated (Meyer, Ramírez, and Soysal 1989). Second, the financial constraints facing several countries in the late 1970s and early 1980s circumscribed both the public and the private capacity to expand and participate in education.

GROSS ENROLLMENT RATIOS. The gross enrollment ratio, which is the ratio of the total number of children enrolled—regardless of age—to the number of school-age children in a given country, indicates the capacity of the education system. Trends in the primary gross enrollment ratios for primary schools in developing countries are fairly consistent. Over time, gross enrollment ratios tend to peak at about 103 percent, after which they settle down to approximately 100 percent, presumably as the capacity of the secondary school system to absorb students improves. For all but the poorest countries, primary schools had the capacity to enroll virtually all school-age children by 1985 (see figure 2-3).

Figure 2-3. Population Age 6–11 and Enrollment in Primary Schools in Low- and Middle-Income Countries, 1965–85

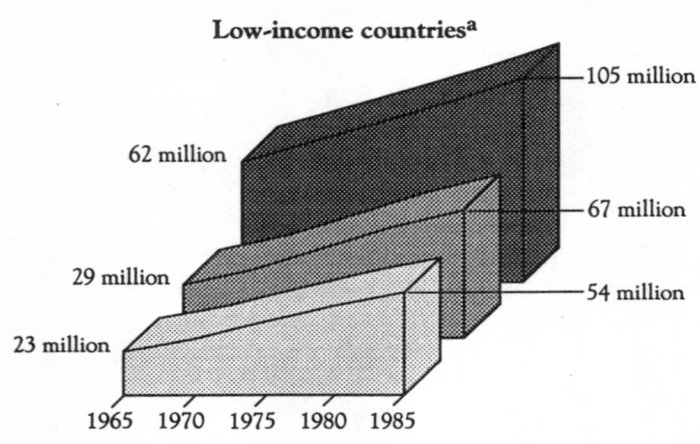

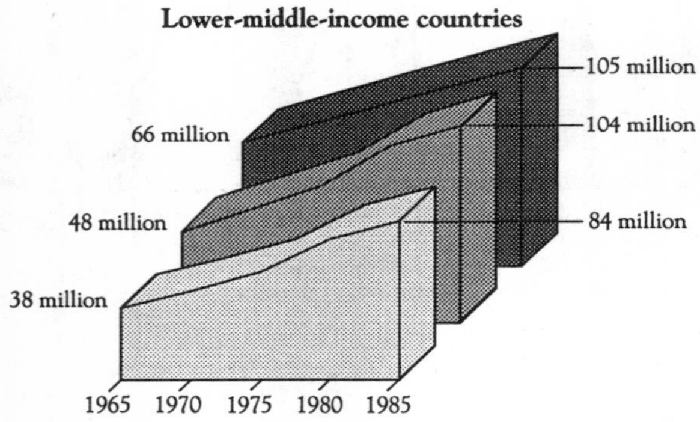

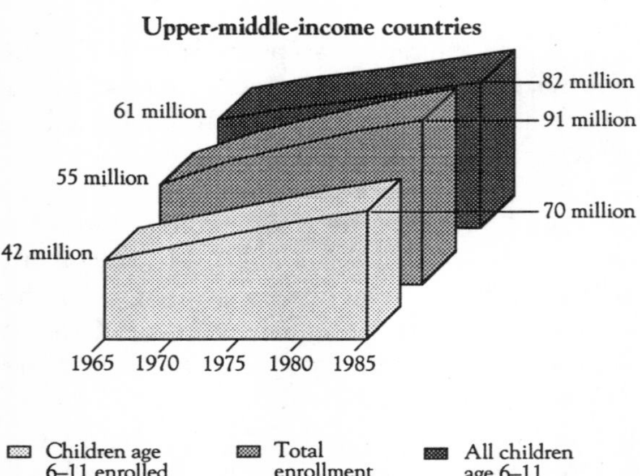

Children age 6–11 enrolled | Total enrollment | All children age 6–11

a. Excluding China and India.

Source: See tables A-1, A- 2, and A-3 in the appendix.

In 1985 the gross enrollment ratios of high-income and upper-middle-income countries ranged from 83 percent to well over 100 percent, with the weighted mean and median at or near 100 percent for both. The gross enrollment ratios of lower-middle-income countries ranged more widely, from 35 percent to more than 100 percent, but the weighted mean and median were about 100 percent, and only eight countries (23 percent) lacked the capacity to enroll at least 80 percent of the primary school–age group (Côte d'Ivoire, El Salvador, Guatemala, Liberia, Morocco, Nigeria, Papua New Guinea, and the People's Democratic Republic of Yemen).[3] For low-income countries, however, the picture was bleaker. Twenty-three of thirty-nine countries (59 percent) had gross enrollment ratios of less than 80 percent (but eight countries—China, Democratic Kampuchea, Lesotho, Madagascar, Myanmar, Sri Lanka, and Viet Nam—reported ratios of over 100 percent), and primary school places were available for only 67 percent of the school-age children. As a result, large numbers of primary school–age children remained out of school in 1985.

Moreover, the percentage of females enrolled has improved over time and been closely related to country income level (see table 2-2). The higher the level of development, the higher the proportion of girls enrolled in primary school. Although it is tempting to believe that improvements in gross enrollment ratios automatically benefit disadvantaged groups, this may not be the case for females in low-income countries. Between 1965 and 1985, the total gross enrollment ratios in these countries improved 23 percentage points, while the female share of enrollment increased only 5.

Table 2-2. Gross Enrollment Ratios and Females as a Percentage of Total Enrollment, Selected Years, 1965–85

Country income level	*1965*	*1970*	*1975*	*1980*	*1985*
Low (excluding China and India)					
Gross enrollment ratio	44.1	47.9	61.7	67.0	67.3
Females (percent)	38	41	42	42	43
China and India					
Gross enrollment ratio	94	85	106	101	110
Females (percent)	38	42	42	43	43
Lower middle					
Gross enrollment ratio	73.8	79.7	84.7	99.7	100.9
Females (percent)	44	45	45	46	47
Upper middle					
Gross enrollment ratio	95.4	105.5	98.3	102.4	103.3
Females (percent)	47	48	47	48	48
High					
Gross enrollment ratio	104.0	103.5	101.2	101.1	101.2
Females (percent)	48	49	49	49	49

Note: Ratios are weighted means.

Source: See tables A-2 and A-4 in the appendix.

Gross enrollment ratios indicate the capacity of the primary school system—how many school places exist—but not the coverage of the system—how many children of official school age are being accommodated. This information is provided by net enrollment ratios: the number of children within a given age range (typically six to eleven years old) who are in primary school, as a percentage of the total population between those ages.

NET ENROLLMENT RATIOS. Net enrollment ratios are often lower than gross enrollment ratios because many of the available school places are filled by students who are outside the appropriate age group for a particular country. Trends in net enrollment are difficult to assess, however, because of gaps in the time-series data and fluctuations in how the school-age population is defined; Unesco data on net enrollment ratios from 1965 to 1985 are available for less than 40 percent of all countries and for only four low-income countries. The net enrollment ratio may be approximated, however, from the number of six- to eleven-year-olds enrolled in school (regardless of the official school age in a particular country) divided by the number of six- to eleven-year-olds in the population. Table 2-3 shows that these ratios have risen steadily in low- and middle-income countries, whereas in high-income countries they have declined since 1965.

Out-of-School Children

According to estimates based on Unesco and World Bank data, some 145 million children age six to eleven in developing countries did not attend school in 1985 (table 2-4).[4] More than 90 percent of those children lived in low- and lower-middle-income countries and about 60 percent of them were girls; nearly 60 percent lived in four of the more populous countries (Bangladesh, India, Nigeria, and Pakistan). Given the average population growth rates from 1980 to 1985, the population six to eleven years old will reach 680 million by the year 2000, which means that 198 million more pri-

Table 2-3. Net Enrollment Ratios, Selected Years, 1965–85

Country income level	*1965*	*1970*	*1975*	*1980*	*1985*
Low (excluding China and India)	37.6	42.2	48.2	52.6	54.1
China and India	79.6	76.2	76.5	76.2	87.4
Lower middle	63.8	69.1	73.2	85.4	90.3
Upper middle	76.3	80.2	85.0	88.5	90.3
High	94.7	93.4	92.3	91.2	90.6

Note: Ratios are weighted means.
Source: See table A-3 in the appendix.

Table 2-4. Population, School Places, and Enrollment in Ninety-Nine Developing Countries, 1985

Country income level	Population age 6–11 (millions)	Primary school places (millions)	Children age 6–11 enrolled (millions)	Children age 6–11 out of school (millions)	Children age 6–11 out of school (percent)	"Excess" places[a] (millions)	Out-of-school children for whom places are available (percent)
Low (excluding China and India)	104.7	66.6	54.0	50.7	48.5	12.6	24.8
China and India[b]	235.4	220.2	174.3	61.1	26.0	45.9	75.1
Lower middle	104.9	104.4	83.6	21.3	20.3	20.8	97.7
Upper middle	81.8	91.1	69.5	12.3	15.0	21.6	175.9
Total	526.8	482.3	381.5	145.3	27.6	100.8	69.4

a. The number of primary school places minus the number of children age six to eleven enrolled.

b. These figures do not take into account the fact that six-year-olds in China are not expected to be in school.

Source: See tables A-1, A-2, and A-3 in the appendix.

mary school places will have to be created for universal primary enrollment to be attained.

The need for extra school places is a particular challenge for low-income countries, where meeting that need might not be feasible without the type of improvements and assistance discussed in subsequent chapters. The major obstacle is the sheer size of the required expansion. A total of 185 million school places were created in the developing world in the twenty years between 1965 and 1985, under positive economic conditions (see table A-2 in the appendix). Now the same number of places will have to be created under constrained economic growth in less than half the time.

There is some reason for hope, however. In 1985, middle-income countries had enough school places to enroll the existing out-of-school children; in upper-middle-income countries, nearly two times as many places were available as were needed. Even China and India had enough school places to enroll three-quarters of the children not in school. In other low-income countries, however, only 25 percent of the places needed to enroll all children were available. Altogether in 1985, sufficient school places existed to enroll 100 million of the 145 million children not in school.

The Search for "Relevant" Content and Better Delivery of Education

A 1978 internal review of education projects supported by the World Bank in seventeen developing countries found that in the late 1960s, planners in many of those countries were intensifying their efforts to expand primary education and reach all children. At the same time, they were concerned that the share of the national budget allocated to education in many countries was no longer growing and was, in some cases, declining. They had other concerns as well: (a) the strong, steady exodus of persons from rural areas, (b) the rise in unemployment among graduates of primary and even secondary schools who had traditionally been guaranteed jobs, (c) the difficulty of enrolling, or retaining in school, the traditionally disadvantaged groups in society (particularly women, rural dwellers, and the poor), (d) the marginal progress in making the education system more relevant to national needs (for example, producing more relevant textbooks or implementing curricular reforms that had been worked out on paper), and (e) the perception of many finance and planning ministries that the country's large investment in education was not being efficiently used and managed. Consequently, as the review found, plans prepared in the late 1960s and early 1970s emphasized qualitative objectives, social equity, rural needs, internal and external efficiency, and better planning and administration.

Curricular Relevance

In the late 1960s and early 1970s, policymakers and others were seeking ways to make the content of primary education more meaningful and the methods of delivering education more cost-effective within the context of nation building and economic development. Countries such as Kenya, Pakistan, the Philippines, Sierra Leone, and Trinidad and Tobago undertook sweeping and highly publicized reviews of their educational objectives and policies.

Throughout the developing world, countries were preoccupied with the role of the primary school in preparing children to participate actively and productively in nation building, economic development, and community life, particularly in rural communities. In 1967 President Julius Nyerere of Tanzania delivered his famous speech on "Education for Self-Reliance," in which he advocated establishing a system whose cost and impact on farm production and attitudes toward farming were appropriate to the country's needs and therefore more effective. In 1973 Tunisia undertook Initiation aux Travaux Manuels, a primary school program that sought to teach manual skills and eradicate the unfavorable attitudes that primary school graduates held toward manual work. Several countries adopted large-scale, comprehensive curriculum development programs that involved new curricula, new teaching materials, and the training and retraining of teachers and school supervisors. Some countries, such as Malaysia, established large curriculum development centers. Projects financed by the United Nations Development Programme in Bunumbu (Sierra Leone) and Kakata (Liberia) adopted an ad hoc arrangement based on a primary teacher training institution. The ruralization of programs was widely pursued. By 1980 several countries had begun using local or national languages to teach at the primary level, in many cases after conducting extensive research on language and literacy in the 1960s and 1970s.

Alternative Delivery Strategies

In an effort to overcome some of the logistical and financial difficulties of providing good-quality education for all, several countries embarked upon bold experiments and initiatives. For example, the government of Côte d'Ivoire launched a large-scale educational television program in 1971 to extend primary education throughout the country. The program was canceled in 1982, however, because of implementation problems and criticisms of cost and quality. Malaysia launched a similar program (1973), as did El Salvador. Radio was used in Nicaragua to teach mathematics and in Ethiopia, Jamaica, and Kenya to reach students and train teachers. Six countries—Bangladesh, Indonesia, Jamaica, Liberia, Malaysia, and the Philippines—experimented with using programmed materials and peer group instruction to provide primary education at a reasonable cost. The effectiveness of these programs was uneven, and few became institutionalized.

A major problem for poorer countries was that their school systems could not accommodate the entire school-age population. In the late 1960s and early 1970s, many researchers and policymakers felt that nonformal education could fill the gap (Coombs, Prosser, and Ahmed 1973).[5] Some countries even experimented with alternatives to formal primary schooling: for instance, youth clubs in Benin, rural education centers in Burkina Faso, and Koranic schools in Mauritania. These approaches proved to be unacceptable to the local communities (see box 2-1).

By the 1980s, serious and unresolved problems of ineffectiveness remained. Several of the initiatives and experiments had failed to point the way to change, because of cost or inadequate financing, insufficient attention to design or implementation, deficient capacity for implementation, novelty of the approach, or rejection by the communities concerned. The quest for answers continued, however, despite increased financial constraints.

Dealing with Austerity and Adjustment

The almost uniformly steady economic growth that had been achieved from 1960 to 1972 in most of the developing world provided a substantial source of revenue for financing national development programs.[6] Given the underdeveloped state of social services and infrastructure in most developing countries in the 1960s, the demands on their recurrent budgets could be managed reasonably well. Indeed, even though resources were scarce, most countries were able to devote a portion of public revenue to capital investments. This

Box 2-1. Rural Education Centers in Burkina Faso

An interesting example of alternatives to primary education that are rejected by parents comes from Burkina Faso, which embarked on a program of relevant education in the late 1950s. Concerned with their inability to attain universal primary education in the foreseeable future, colonial administrators proposed an abbreviated (three-year), low-cost course in French literacy, numeracy, and agricultural training for unschooled adolescents. This course was held in rural education centers, which the government planned to pay for by limiting the construction of conventional primary schools and having students grow cash crops and food for student meals. It was believed that rural education centers would accommodate all rural youth by the 1980s, but by 1972 the 737 centers were accommodating only 24,000 pupils, one-fifth of the total school population and one-sixth of the original target.

By 1973 the rural education program had become so unpopular—villagers were disappointed that no certificates were awarded—and attracting pupils had become so difficult that the government ended the experiment. The idea of rural education centers as a substitute for primary schooling was abandoned, and the centers became Young Farmers Training Centers run by the Ministry of Agriculture. In the 1980s, external evaluations found that rural, nonformal education programs were costly, ineffective, and unpopular. By the mid-1980s the government turned its attention back to making the formal system of primary education viable for transmitting basic education on a large scale.

The dual system of education ultimately failed for several reasons: (a) conventional primary schools continued to be built because of ardent public support for formal schooling; funds for the rural education centers were limited accordingly; (b) most of the available teaching time was used to introduce pupils to the language of instruction, French; (c) the level of attainment in literacy remained low; (d) the land available for farming was scant and of poor quality—often the worst in the village; (e) basic agricultural equipment was lacking; and (f) the conditions necessary for successful implementation were absent.

Sources: Sinclair 1980; Haddad and Demsky forthcoming.

situation did not last long. Most countries found themselves in a difficult and deteriorating financial situation beginning in the mid-1970s onward. The reasons for this state included (a) an oil crisis in the early 1970s, (b) worsening terms of trade in the 1970s and 1980s, (c) rising demands on public recurrent budgets as public services expanded and the demand for them increased, and (d) various natural disasters affecting agricultural productivity in particular. The impact on low- and lower-middle-income countries was particularly acute and further separated the ability of lower-income countries to educate all their citizens from that of higher-income ones.

Table 2-5. Median Public Expenditure on Education, Selected Years, 1965–85
(percentage of GNP)

Country income level	*1965*	*1970*	*1975*	*1980*	*1985*
Low	2.7	3.2	2.8	3.1	3.2
Lower middle	3.0	3.4	3.6	4.5	3.9
Upper middle	3.2	3.4	3.5	3.7	4.3
High	4.3	5.1	6.3	5.8	5.8

Source: See table A-15 in the appendix.

Austerity

School-age children represent a significantly higher proportion of the total population in low-income countries (13.6 percent) than in high-income countries (7.9 percent).[7] Since 1965, however, public expenditures on education, expressed as a percentage of GNP, have been lower in low-income than in middle- or high-income countries (see table 2-5).[8] Moreover, they have remained relatively constant—about 3 percent—in low-income countries, while increasing in middle- and high-income countries to 4–6 percent. Low-income countries have been unable to close the expenditure gap that separates them from higher-income countries. Low-income countries have been unable to close the expenditure gap that separates them from higher-income countries.

In real terms, the median public recurrent expenditures per student in primary school have been declining (albeit unevenly) in low-income countries while increasing steadily in middle- and high-income ones (see table 2-6); from 1965 to 1985, the median expenditures per pupil in low-income countries declined from $41 to $31, while in high-income countries they increased

Table 2-6. Median Public Recurrent Expenditure per Primary Student, Selected Years, 1965–85
(1985 dollars)

Country income level	*1965*	*1970*	*1975*	*1980*	*1985*
Low	40.7	38.7	40.7	29.4	30.9
Lower middle	72.5	71.3	81.6	75.5	101.7
Upper middle	194.3	197.9	258.9	255.7	296.6
High	824.5	841.7	1,117.6	1,382.8	1,551.4
Ratio of low-income country spending to high-income country spending	1:20	1:22	1:27	1:47	1:50

Source: See table A-20 in the appendix.

from $825 to $1,551. For low- and middle-income countries, the average expenditures per pupil were somewhat lower than this median figure, while for high-income countries they were nearly twice as high as median expenditures. The steady decline in median expenditure per pupil in low-income countries relative to that in high-income countries (from a ratio of 1:20 in 1965 to 1:27 in 1975 and 1:50 in 1985) is cause for serious concern, as it suggests that low-income countries may fall even further behind in educational development.

Moreover, public funding of per-pupil expenditures is greater in upper- than in lower-income countries (see table 2-7), and the effort in low-income countries has become lighter over time. In 1965 recurrent expenditures per pupil amounted to 20 percent of GNP per capita in low-income countries; by 1985 this figure had fallen below 12 percent. In lower-middle-income countries, there was little change over time, while in both upper-middle- and high-income countries, the effort increased significantly, from 10.6 percent to 12.5 percent and from 11.7 percent to 17.6 percent, respectively.[9]

Adjustment

The adjustment policies and programs that many countries introduced in the 1980s were intended to respond to various shocks, such as adverse weather or declining trade, and to rectify inappropriate policies that had hampered economic performance in the past. Adjustment policies are of two types: (a) stabilization policies, which generally rely on demand management and seek sustainable reductions in the current account of the balance of payments and the fiscal deficit, as well as reductions in the rate of price inflation, and (b) structural adjustment reforms of policies and institutions, which seek to improve resource allocation, increase economic efficiency, expand growth potential, and increase the country's resilience to shocks (World Bank 1988a). In the 1980s these policies included import restrictions, currency devaluations, increases in agricultural prices, curbed food subsidies, and, generally, reductions in public expenditures.

Table 2-7. Median Public Recurrent Expenditure per Primary Student as a Percentage of GNP per Capita, Selected Years, 1965–85

Country income level	*1965*	*1970*	*1975*	*1980*	*1985*
Low	20.0	14.0	13.0	11.7	11.9
Lower middle	10.3	11.0	9.2	9.5	10.9
Upper middle	10.6	8.9	8.1	9.1	12.5
High	11.7	11.5	14.4	15.9	17.6

Source: See table A-21 in the appendix.

Adjustment policies can reduce the resources allocated to education in general, to primary education in particular, to teacher salaries, or to nonsalary recurrent expenditures; they can also increase the cost of imported items. Available data on the effect of adjustment policies on primary education do not, however, provide conclusive evidence that any of these occurred. There are three reasons. First, adjustment policies can be implemented both with and without support for social services; when adjustment occurs without social service support, the consequences for education are likely to be more severe than when it occurs with such support. Second, adjustment policies are generally implemented in countries experiencing acute economic difficulties; the decline in support for education that follows adjustment, although observable, may be less severe than the decline that would have occurred if the country had not undertaken adjustment policies. Thus the policies may actually produce greater economic stability. Third, the available data are limited in scope, quality, time, and comparability.

REDUCED SPENDING FOR EDUCATION. The most serious effect that adjustment can have on primary education is to reduce the central government's allocation to education in general and to primary education in particular. One review found that education's share of total government expenditures declined between 1980 and 1986 in twelve out of thirteen intensely adjusting countries but in only three out of twelve nonadjusting countries with similar levels of economic development.[10] On average, education's share of total government spending declined from 15 to 12 percent in intensely adjusting countries, but increased from 10 to 12 percent in nonadjusting ones. In nine of the twelve intensely adjusting countries, per capita spending on education declined in constant terms (World Bank 1989).

Another study of countries undertaking structural adjustment from 1979 to 1983 found that a majority (68 percent) reduced government expenditures for education. In 22 percent of the cases, the percentage of spending reduced was less than the aggregate reduction; however, in another 46 percent, education was one of the most vulnerable sectors. Defense proved to be the most protected (Pinstrup-Anderson, Jamarillo, and Stewart 1987). In fifteen low- and lower-middle-income countries that underwent structural adjustment programs in the 1980s, public expenditure for education as a percentage of GNP declined from 4.22 percent in 1980 to 3.45 percent in 1985, almost twice the reduction experienced by low- and lower-middle-income countries in general (see figure 2-4). Other reviews have noted that the greatest impact of adjustment has been felt in Latin America and Sub-Saharan Africa (Cornia, Jolly, and Stewart 1987; Gallagher 1989; Ngomba and Oxenham 1989; Pfeffermann 1987).

Some analysts argue, however, that no reductions occurred as a result of adjustment (Amadeo and Camargo 1989; Demery and Addison 1987) or that

Figure 2-4. Public Expenditure for Education in Selected Intensely Adjusting Countries, 1980 and 1985

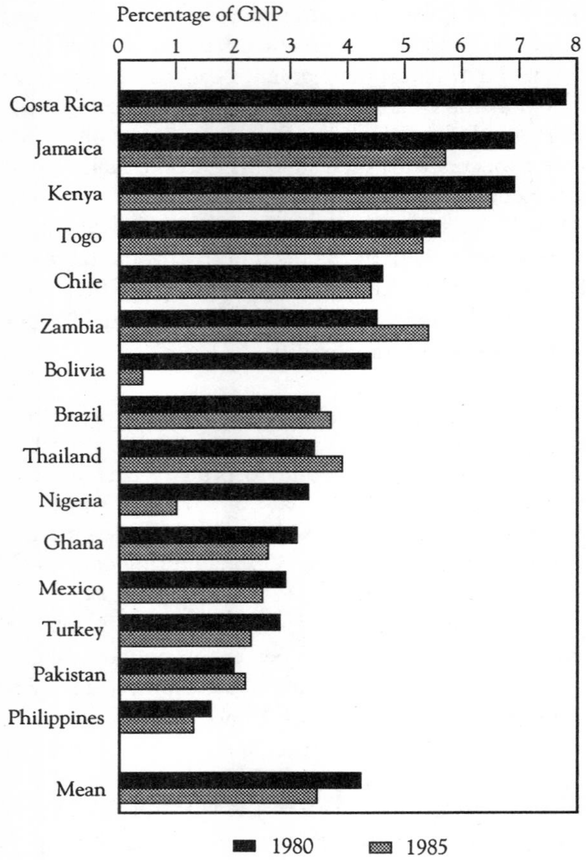

Source: Kakwani, Makonnen, and van der Gaag 1990; see also table A-15 in the appendix.

the observed reductions in overall allocations to education began before adjustment and therefore should not be considered a consequence of adjustment (World Bank 1986).

REDUCED SPENDING FOR EDUCATIONAL INPUTS. Even where education budgets did not decline, the expenditures per student and per teacher did. A 1985 study found that the typical response of five African countries to the economic crisis was not to redeploy resources systematically, but rather to curb spending for the items whose absence or reduction were unlikely to cause public embarrassment and to avoid, if possible, reducing salaries and student allowances and retrenching staff even nominally (Auerhan and others 1985). Hence, they reduced spending for administrative and pedagogical equipment and for supplies and travel first, which reduced the effectiveness of teachers and administrators. In Zambia, teacher salaries comprised 81 percent of the spending for primary education in 1974, 88 percent in 1980, and 91 percent in 1983—an increase achieved at the expense of educational supplies and administrative support; meanwhile, spending per student declined in primary schools 37 percent in real terms from 1979 to 1983 because higher enrollments were not accompanied by proportionate increases in the number of teachers. Even in Kenya, where education's share of the recurrent budget increased from 25 percent in 1978 to 35 percent in 1988, real recurrent spending per student declined and virtually eliminated expenditures on textbooks and teaching materials.

In many countries, teacher salaries declined in real terms (Colclough and Lewin 1990). In some cases, declines in teacher salaries increased teacher attrition. There was a reduction of some 7,000 teachers in Zaire in 1984, while in Ghana more than 4,000 qualified teachers left the system between 1977 and 1981 (Cornia, Jolly, and Stewart 1987), and in Somalia many teachers resigned as salaries, which did not increase between 1971 and 1988, became inadequate. One result of these cutbacks in intensely adjusting countries was a decline in the net enrollment ratio; between 1980 and 1985, net enrollment fell in fifteen out of twenty-five intensely adjusting countries but only nine out of thirty-three nonadjusting countries (World Bank 1989; Noss 1990).

Conclusions

The developing world has made tremendous strides in expanding primary education in the past three decades, and many countries have achieved universal primary enrollment. Most developing countries are, however, still a long way from achieving universal primary *completion.* With their populations growing faster than primary school enrollments, many countries will have to make a vigorous effort to reduce illiteracy over the next ten or fifteen years. Several education systems are in danger not only of failing to make quantitative and qualitative progress, but of losing ground in primary education.

The current financial crisis is expected to persist in many countries for the remainder of this century. This places a premium on the efficiency with which primary education is managed and produces its graduates. The knowledge gleaned from each country's efforts to improve its education system should be developed and disseminated. Particularly in the low-income countries, primary education may require substantial policy reform and assistance.

Notes

1. Different ways of defining the primary school–age population make it difficult to estimate the number of children not in school. Estimates of children six to eleven years old who do not attend school in developing countries range from 114 million to 145 million; the maximum estimate, which includes China, where six-year-olds are not expected to be in school, is used here.

2. By 1985, 131 out of 180 countries had made primary schooling compulsory for the entire nation; 97 percent of those without compulsory education were developing countries (Chang, Lussier, and Ventresca 1989).

3. In 1985, when these data were gathered, the Republic of Yemen was divided into the Arab Republic of Yemen and the People's Democratic Republic of Yemen.

4. This figure includes China, where children do not start primary school until age seven.

5. Coombs, Prosser, and Ahmed (1973, p. 11) define nonformal education as "any organized educational activity outside the established formal system that is intended to serve identifiable learning clienteles and learning objectives."

6. See, for instance, *Social Indicators of Development 1987* (World Bank 1987a). No country had a lower GNP per capita in 1973 than in 1965. *World Development Report 1988* (World Bank 1988c, pp. 224–25) shows that between 1965 and 1980 no country had negative growth in its gross domestic product (except war-torn Lebanon), while between 1980 and 1986, twenty-one countries did.

7. Calculations are based on data from World Bank (1987b) and table A-1 in the appendix.

8. Median figures are used to accommodate irregular patterns of missing data; complete time-series data are available only for ten low-income, nine lower-middle-income, ten upper-middle-income, and nine high-income countries. The nonmonetary share of GNP is higher in low- and lower-middle-income countries than in more developed ones. As a result, public expenditures on education (which are largely monetary) represent a higher proportion of the monetary share of GNP than of the total GNP.

9. Again, the nonmonetary component of GNP in developing countries may account for some of these differences.

10. The intensely adjusting countries received, prior to 1985, at least three structural adjustment loans. These are loans designed to support not particular projects but policy and institutional changes that seek to modify the economic structure in order to maintain both the growth and the viability of balance of payments in the medium term.

3

Improving Learning Achievement

SCHOOLING IN DEVELOPING COUNTRIES takes place under conditions that are very different from those in industrial countries. Primary students in industrialized countries are likely to attend classes in a modern, well-equipped building and to study a curriculum whose scope and sequence are well designed. On average they receive 900 hours of learning time and $52 of noncapital material inputs each year and have a teacher with sixteen years of formal education. Moreover, they share a teacher with fewer than twenty other children, most of whom are healthy and well fed. In many low-income countries, by comparison, students are likely to attend a shelterless school or one that is poorly constructed and equipped. Their curriculum is likely to be poorly designed. On average they receive only 500 hours of learning time and $1.70 of noncapital material inputs each year and have a teacher with only ten years of formal education. The learning environment typically has few resources, and classes consist of more than fifty children, many of whom are chronically undernourished, parasite ridden, and hungry. The job of educating children in developing countries is thus significantly more difficult than it is in developed countries.

Although alleviating the poverty of schools in developing countries is a desirable goal, not all improvements are affordable. Developing countries must therefore concentrate their resources on the improvements that are known to enhance student learning. In industrialized countries such as the United States, schools pay for inputs that are *not* systematically related to achievement (Hanushek 1986). The investment decisions of developing countries have not yet been researched fully. Nevertheless, education systems in developing countries have difficulty improving their efficiency for three reasons: (a) inadequate knowledge about the effectiveness of inputs, (b) inadequate knowledge about the cost of inputs, and (c) difficulty obtaining appropriate information (Lockheed and Hanushek 1988). The problem of inadequate knowledge is not limited to developing countries, however; in general, studies that assess the cost-effectiveness of alternative educational policies are extremely rare (Levin 1987). Nevertheless, the most promising avenues for improving learning, which are reviewed later in this chapter, take into account both cost and effectiveness.

To improve learning, resources must be distributed wisely and managed well. First, priority must be given to the inputs that make a difference in learning. Some inputs do not, even though they are highly attractive and popular. For example, an elaborate school facility is no more effective in producing learning than a modest facility. Second, the cost of a particular input or intervention must be considered. Some interventions effectively improve learning but are not cost-effective. Two examples are small classes and computer-aided instruction. Both can yield benefits, but the expenditure required to do so is not practical in developing countries. That is, students *do* learn better in small classes (of fewer than twenty students), and they *do* learn better when taught by a computer for a minimum of ten minutes per day. However, the level of resources required to achieve classes of twenty students or ten minutes of daily computer instruction is simply too high for developing countries. Ineffective inputs and impractical, expensive inputs are blind alleys for investment. Third, implementation strategies vary by the type of intervention selected. Some inputs, such as interactive radio instruction, are both effective and cost-effective but not widely implemented. These inputs—and the reasons for their limited use—must be considered carefully.

Interventions that are both cost-effective and feasible are the most desirable, since they generate the most learning for the resources spent. Unfortunately, little research clearly delineates the most cost-effective interventions for developing countries. Moreover, an intervention that is cost-effective in one context may not be in another, which makes generalizing difficult (Lockheed and Hanushek 1988). Although research on the feasibility of implementation is even more sparse, developing countries should consider policy interventions in five principal areas: (a) improving the curriculum, (b) increasing learning materials, (c) increasing instructional time, (d) improving teaching, and (e) increasing the learning capacity of students. In developing countries, the evidence is growing that these interventions raise student achievement. This chapter presents this evidence, indicates the extent to which particular inputs are available in developing countries, and makes policy recommendations. Blind alleys for investment are distinguished from promising avenues that merit serious attention.

A Model of Educational Effectiveness

Choosing effective inputs is the first step toward improving learning, but managing them well at the school level is also necessary. The objective of each primary school is to increase the number and quality of its graduates by increasing the probability that students will stay in school and be promoted to the next grade level on time. Increasing student learning is central to the process of producing more and better-educated primary school graduates, since student achievement is an important determinant of on-time promotion. A panel study of primary students in northeastern Brazil, a poor region, found that learning was a significant determinant of promotion (Harbison and Hanushek forthcoming). Galda and González (1980), in a study of how teachers promote students in Nicaragua, also found that prior learning achievement, among other critical factors, was positively associated with subsequent grade promotion. Internal World Bank data from the Philippines strongly suggest that the number of students completing the primary grade cycle increases as the school environment improves, as measured by the average cognitive achievement of pupils, especially the poor.

Improving educational inputs also appears to reduce grade repetition and dropout rates. In Sub-Saharan African countries, the correlation between the annual expenditure per pupil on educational materials and the time required for students to complete the primary cycle is significantly negative, when GNP per capita and the student-teacher ratio are held constant.[1] In addition, these measures increase the external efficiency of primary education.[2] Although the effect of learning on returns to education is difficult to estimate, a few studies strongly indicate that it is large (Behrman and Birdsall 1983; Boissiere, Knight, and Sabot 1985; Knight and Sabot 1990).

Figure 3-1 gives an overview of the process that links school inputs, learning, and primary school completion. Progression and completion are determined, first, by the positive effect of learning and, second, by the negative effect of variables such as the demand for child labor (which is totally exogenous to the education system) or restrictive promotion policies (which can be revised). Learning is determined by four factors of school input and process (curriculum, instructional materials, learning time, and teaching) and by the child's teachability. Family investments in health and nutrition and preschool experience play a role in determining the child's teachability, while the provision and effective use of school inputs are the responsibility of education management at all levels. The concept of effective schools encompasses the system of school and administrative relationships. Increased educational effectiveness produces greater educational efficiency; these relationships are discussed at length in chapter 7.

Effective Schools

The impact of enhanced inputs ultimately depends on how well schools use the available resources. Schools vary considerably in this respect. In spite of the barren conditions of many schools in the developing world, a few are able to teach students the knowledge and skills called for in the curriculum. These effective schools are able to transform their given inputs into student learn-

Figure 3-1. A Model of Effective Schooling

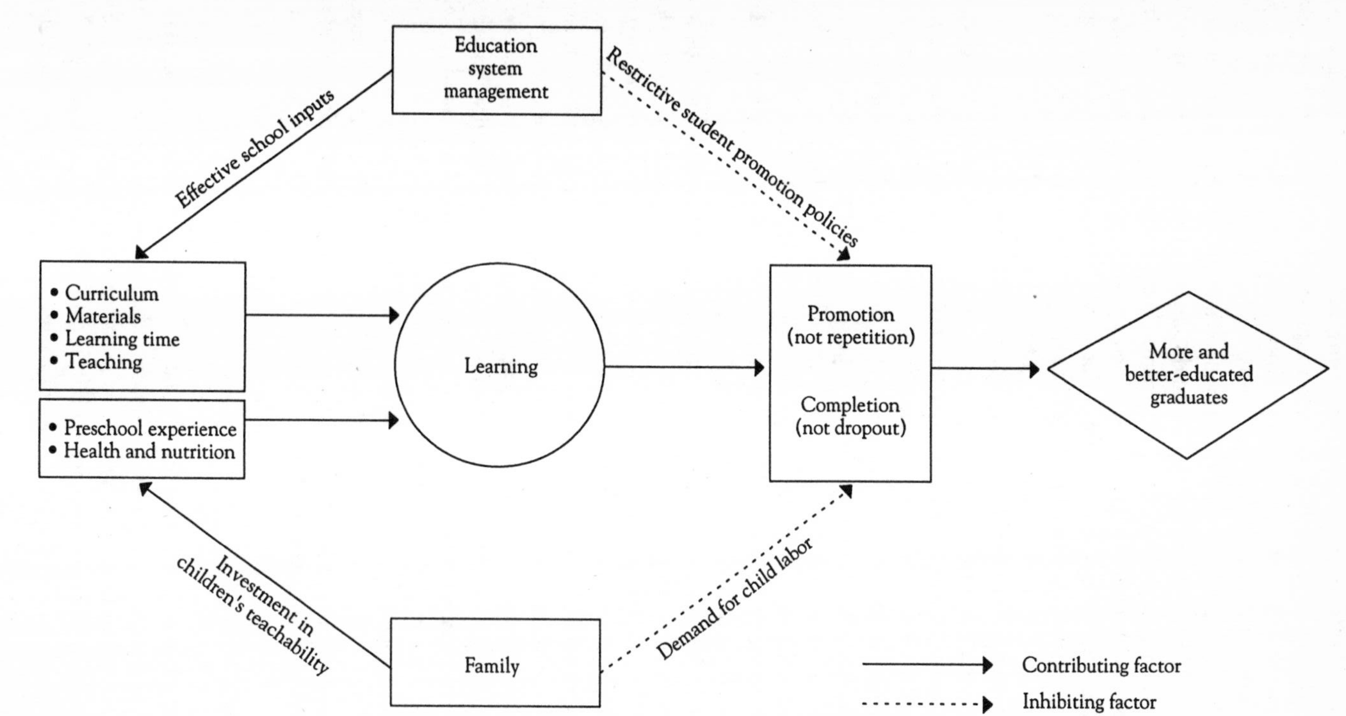

ing. In developed countries, where material inputs are readily available, research on the characteristics of effective schools has concentrated on a set of process factors that distinguish these schools: an orderly school environment, clear goals and high expectations, a sense of community, and strong instructional leadership (Purkey and Smith 1983). In addition, British studies suggest that involving students in school decisionmaking enhances attendance and encourages positive behavior (Reynolds 1983; Rutter 1983; Rutter and others 1979). For developing countries, the acquisition, distribution, and use of material inputs must be added to this list.

An Orderly School Environment

A student's ability to learn is heavily influenced by the school's environment. Learning occurs more easily if order prevails—if students and teachers attend classes regularly and according to an established timetable, if facilities are clean and in good repair, and if teaching materials are routinely provided. An orderly school environment is closely linked with effective classroom management: teachers gain and hold the attention of their students, maintain discipline, begin and end lessons on time, let students know what is expected of them, and monitor and evaluate student performance. Order within classrooms and within the school as a whole signals the seriousness with which a school approaches its task.

Despite the lack of empirical evidence on school environments in developing countries, ethnographic studies and anecdotal evidence suggest that in developing countries the basic elements of an orderly school environment are frequently missing: students and teachers are regularly absent, the stock of teaching materials is limited, and the physical surroundings are detrimental to learning. The conditions in many schools are so chaotic that it seems miraculous that learning occurs at all, and much of what does occur appears to be haphazard rather than the result of a deliberate focus on the content and process of instruction.

Academic Emphasis

Teachers and parents agree that in effective schools, academic achievement, basic skill acquisition, and appropriately structured learning activities are important. Effective schools establish clearly defined goals for academic achievement, and they concentrate their available resources and their operations on attaining them. For elementary school students to acquire basic and complex skills, schools must focus their curriculum on teaching those skills, provide sufficient time for teaching them, and coordinate instruction across grade levels. Effective schools continuously monitor student progress to determine whether their instructional goals are being met.

Academically successful schools also set high expectations for work and achievement. The concept of the school as a place of learning is communicated clearly to the students, and a commitment to learning is expected in every classroom. Expectations are manifested in the performance standards set by the school. Low standards reflect low expectations; high standards reflect high expectations.

Presently, academic instruction is not given the necessary priority in many schools in developing countries. Schools often operate in isolation from parents and the community, and teachers and students have many competing obligations. Students may have to work on the family farm. Teachers must leave class for family emergencies or administrative matters. Thus the academic focus presented by the official curriculum rarely corresponds to the reality of school life.

Instructional Leadership

Effective schools have strong principals who devote considerable time to coordinating and managing instruction, are highly visible in the school, and stay close to the instructional process. In many instances, effective principals adopt the management-by-wandering-about style of executives of successful large corporations (Peters and Waterman 1988). At the same time, instructional leadership is in many ways a shared responsibility. Effective leadership engenders a common sense of commitment and collegiality among the staff. Several studies suggest that teachers rate principals with a participatory style of school management as more effective than those who are more autocratic (Baker 1988a; Den Hartog Georgiades and Jones 1989).

In developing countries principals apparently function as the lower link in an organizational chain that extends from the school through district supervisors to the central ministerial staff. Principals are selected mainly for their seniority rather than for their personal traits or performance. They also operate under significant constraints, such as the chronic shortage of materials, operating funds, and staff development resources, that make instructional improvement extremely difficult. In addition, principals are overburdened with administrative tasks.

Creating effective schools requires change, and changing or improving organizations involves changing or improving the behavior of the smallest unit in the organization. In the education sector, the smallest units are the school and the classroom. National decisionmakers can establish standards for curricula and examinations, official timetables, teacher certification criteria, and attendance and promotion policies, but they cannot control what happens in each school or classroom. Yet teaching and learning occur at that level. The learning gains associated with the inputs discussed next can only be achieved if the inputs are, in fact, available and used in the schools.

Improving the Curriculum

Students learn the content of the curriculum they are taught; the more they are taught, the more they learn (Fraser and others 1987).

The Intended Curriculum

In primary schools, the composition of the curriculum is remarkably similar worldwide. A recent study of curricular emphasis in 130 countries, including 94 low- and middle-income countries, found that the curricula for primary schools not only contained the same subjects but gave the same relative importance to the various areas of curriculum content (Benavot and Kamens 1989). This consistency in emphasis was found across countries, regardless of the level of economic or educational development (see table 3-1); furthermore, it has held steady since the early 1960s. In primary schools more than 50 percent of the available time is spent teaching language skills (in the local, national, and official languages) and mathematics. Science, social studies, and the arts are given equal weight, about half that of mathematics and one-fourth that of language; other areas receive less attention.

Vocational subjects such as agriculture, domestic science, and business are more commonly taught in primary school in developing countries than in industrial countries, but the time allocated to such subjects is still minimal. In the 1980s about 6 percent of the total time was devoted to these topics, down from about 9 percent in the 1960s.

One important difference does exist between the curricular content of primary education in less developed and more developed countries. Although all countries spend about 36 percent of instructional time on languages, many

Table 3-1. Curriculum Time Devoted to Major Content Areas in Ninety Countries, 1980s
(percent)

Curriculum content	*Low-income countries*	*Lower-middle-income countries*	*Upper-middle-income countries*	*High-income countries*
Language	37	34	36	34
Mathematics	18	17	18	19
Science	7	9	8	6
Social studies	8	10	9	9
Moral education	5	6	4	5
Music and art	9	8	11	13
Physical education	7	6	7	9
Hygiene	1	2	2	1
Vocational subjects	6	7	3	1
Other	3	3	2	3

Source: Benavot and Kamens 1989.

low- and lower-middle-income countries divide this time between a national language and a nonindigenous, official language.[3] The result is that less time is available for instruction in either language. More developed countries concentrate instruction on a single national language that is also indigenous (for example, French in France). Language policy and bilingual education are politically sensitive and complex issues. Multiple-language instruction clearly decreases the attention paid to a single language, which could mean that children in multilingual developing countries acquire literacy slower than children in monolingual industrial countries (see chapter 5 for a discussion of issues related to bilingual education).

The Implemented Curriculum

The official, intended curriculum establishes broad guidelines for instruction, and teachers implement it using textbooks. In textbooks, the scope of the subject matter is defined and the sequence for instruction laid out. Textbooks are the major—if not the only—definition of the curriculum in most developing countries.

Unfortunately, the curricula presented in textbooks, particularly the scope and sequence of the material, are often poorly designed. Instructional design is important because inappropriately targeted curricula (which are too difficult or too easy) frustrate students and increase failure. In their study of the scope and sequence of textbooks for grades ones, three, and five in fifteen countries, Cope, Denning, and Ribeiro (1989) found that the material in the mathematics and reading textbooks for the first grade was appropriate, but the difficulty level increased greatly by third grade. In the fifth grade the mathematics texts were too difficult, while the reading texts were too easy.

Textbooks also suffer from factual inaccuracies, inappropriate illustrations, and problems with readability. In Pakistan, for example, textbooks contain a high proportion of factual and grammatical errors, significant deviations from the specifications set by the Curriculum Bureau of the Textbook Board, and language that differs sharply in difficulty from one grade level to another and from subject to subject among books at the same level (Cope, Denning, and Ribeiro 1989). Teachers in Pakistan reported that the textbooks were too hard for their students, who felt burdened and uneasy with the material. For example, because the children were not fluent in the Arabic style of Nasakh script used in the textbooks, their teachers had to read the textbooks and summarize the lessons for them (Pakistan 1983, p. 46).

The content of textbooks in many countries also fails to reinforce the development of higher-order thinking skills—that is, problem-solving skills and critical thinking. Cope, Denning, and Ribeiro (1989) found that the majority of the material presented in fifth-grade books was more appropriate for students in the second or third grade; only about one-third of the material challenged students to "learn the new" (see box 3-1). Although teachers can adjust their teaching methods to elicit thoughtful consideration of the text, textbooks that require more than memorization of problem solutions (as in mathematics texts) or that use written language to solve "real world" questions and problems also encourage the development of thinking skills. This type of curriculum reform and evaluation has rarely been conducted in developing countries, but Ogundare (1988) found that Nigerian students who were exposed to a "problems approach" curriculum not only learned more facts but also understood the material better and applied the knowledge to new problems better.

Policy Responses

Developing countries should examine policies to improve both the intended and the implemented curriculum used in their primary school classrooms.

BLIND ALLEY: ADJUSTING THE INTENDED CURRICULUM. Most curriculum reforms attempt to modify the intended curriculum by concentrating on the courses taught and the number of hours officially allocated to them. In extreme cases, courses on agriculture have been added to the class schedule in an effort to make school more appealing to children of farmers. Such changes in the intended curriculum are small, ineffective, and resisted by parents and teachers.

PROMISING AVENUE: IMPROVING THE IMPLEMENTED CURRICULUM. Recent efforts to reform curriculum have tackled the difficult issue of preparing a coherent, appropriately paced and sequenced instructional program for the subjects taught. Readability of material, accuracy of content, and instructional design are keys to this approach, which places the needs of the learner foremost. Instruction begins with simple concepts and builds to more difficult concepts; it clarifies the meaning of the material to the learner. Sophisticated instructional design is not easy, and translating curriculum specifications into good textbooks requires considerable expertise. Textbooks must have the appropriate content and reading level; be consistent in approach, method, and exposition; be properly sequenced; motivate the students; and, finally, be readily taught by less-qualified teachers yet allow good teachers to expand upon them (Neumann 1980). Throughout the world few individuals possess the expertise required for writing good textbooks, and most textbooks are therefore written by committees of experts. Improving the content of textbooks holds great promise for improving the learning of children in developing countries.

Providing Learning Materials

Instructional materials are critical ingredients in learning, and the intended curriculum cannot be easily implemented without them. Instructional materials provide information, organize the scope and sequence of the information presented, and provide opportunities for students to use what they have learned.

Effective Aids to Learning

The learning materials that enhance student achievement are textbooks, teacher guides, computers, and other learning aids.

TEXTBOOKS. Because textbooks deliver the curriculum, they are the single most important instructional material. "Nothing has ever replaced the printed word as the key element in the educational process and, as a result, textbooks are central to schooling at all levels" (Altbach 1983, p. 315). When textbooks are available, instructional time is not wasted while teachers and students copy text on and off the blackboard.

Box 3-1. The Problem with Textbooks in Developing Countries

Children who enter school in developing countries often spend more than one year in the first grade and progress slowly through the early grades. Why is this the case? One explanation is that the instruction contained in textbooks is not appropriately matched to the grade-level of the students. A study commissioned by the World Bank examined the level of difficulty in mathematics and reading textbooks used in the first, third, and fifth grades in fifteen developing countries. While expectations of student achievement were realistic in first grade, they were inappropriately high in fifth-grade mathematics and inappropriately low in fifth-grade reading. Furthermore, the steps from one concept to the next were very large and the intermediate stages few.

Table 3B-1. Mathematics Skills Taught in Developing Countries Compared with Achievement in the United States, 1986

	Percentage of textbook pages devoted to skill in developing countries			*Percentage of U.S. students mastering skill*	
Skill	*Grade 1*	*Grade 3*	*Grade 5*	*Grade 4*	*Grade 8*
Prenumbers	43	1	0	—	—
Simple arithmetic	41	6	2	98	100
Beginning skills and understanding	14	50	9	74	99
Basic operations, beginning problem solving	2	22	23	21	73
Moderately complex procedures, reasoning	0	21	67	1	16

— Not available.

Sources: Cope, Denning, and Ribeiro 1989; Dossey and others 1988.

Over the past decade, researchers have found that the availability of textbooks and other instructional materials has a consistently positive effect on student achievement in developing countries (Heyneman, Farrell, and Sepulveda-Stuardo 1981; Heyneman and Loxley 1983). Nicaraguan students in classes randomly assigned to receive textbooks scored significantly higher—by about one-third of a standard deviation—on a test of mathematics achievement than students in classes with no textbooks (Jamison and others 1981). In the Philippines, first- and second-grade children received textbooks under one of two conditions: a student-textbook ratio of 2:1 and a ratio of 1:1. A comparison group was drawn from students in school the previous year, when the ratio of students to textbook was 10:1. Textbooks had a substantial effect on learning: students who received textbooks in both conditions scored about one-third of a standard deviation higher than the comparison group on tests of science, mathematics, and Pilipino (Heyneman, Jamison, and Montenegro 1984). In Brazil the effect on student learning of

MATHEMATICS TEXTS. While most of the material in first-grade mathematics textbooks in developing countries is easy, material in grades three and five is quite difficult (see table 3B-1). More than 40 percent of third-grade textbooks cover skills mastered by only 20 percent of the fourth-grade children in the United States. More than 65 percent of fifth-grade textbooks cover material mastered by only 16 percent of the eighth-grade students in the United States.

READING TEXTS. Most of the material in first-grade reading textbooks in developing countries is typically taught before the first grade in the United States. While third-grade textbooks are generally on target for third graders, the difficulty level does not change for grade five. Over 60 percent of the reading material for grade five is more appropriate for grades two and three (table 3B-2).

Table 3B-2. Reading Skills Taught in Developing Countries Compared with the Curriculum in the United States, 1980s

	Percentage of textbook pages devoted to skill in developing countries			*Grade in which skill is taught in the United States*
Skill	*Grade 1*	*Grade 3*	*Grade 5*	
Prereading	59	1	2	Kindergarten
Decoding	33	12	0	1–2
Fluency	8	69	60	2–3
Learning new knowledge from reading	0	18	38	4–8

Sources: Chall 1983; Cope, Denning, and Ribeiro 1989.

adding basic instructional elements to poor rural schools was studied from 1981 to 1985; one of these elements was textbooks. Second- and fourth-grade students in schools receiving textbooks scored significantly higher on tests of mathematics and Portuguese than did students in schools without textbooks (Armitage and others 1986). Simply providing reading materials and time for reading was effective in Fiji (see box 3-2).

TEACHER GUIDES. Teacher guides that are well integrated with the textbook or other instructional materials can have a positive impact on student achievement. Particularly effective are guides that include information on what to teach and on how to teach it, diagnostic tests that help teachers monitor student learning and modify the daily lessons accordingly, suggestions on how to manage the classroom, and activities for classroom use.

COMPUTERS. In educational settings, the computer has replaced the television as the medium of modernity.[4] This is true in both developing and industrial countries (Hawkridge, Jaworski, and McMahon 1990). In industrial

Box 3-2. The Book Flood Experiment for Teaching English as a Second Language in Fiji

In 1980, fourth- and fifth-grade classes in eight rural schools in Fiji were inundated with a wide range of high-interest, illustrated storybooks in English. Each teacher received 250 books; half of these teachers were trained to use the shared book method, whereby they spend several days discussing the book, the pictures, and the title and then reading the text aloud while the children listen. The other half of the "book flood" teachers merely encouraged the children to read silently for thirty minutes each day. A control group of similar children were taught English using a conventional approach without any reading books.

Students were tested before the book flood began, after the program had been working for eight months, and again one year later. The book flood groups showed much larger improvements than the control group. In the first eight months, the shared book group and the silent reading group had each improved their reading level by fifteen months, while the control groups had advanced only six and a half months. After nearly two years with the enriched reading program, the book flood groups had increased their growth even more, and the effects had spread to writing, formal grammar, and other subjects.

The effects also persisted on national examinations. The pass rate on the two English examination papers was 76 percent for students from the shared book schools, 73 percent for students from the silent reading schools, and only 37 percent—a rate typical of small rural schools—for students from the control schools.

Source: Elley and Mangubhai 1983.

countries, computers are used to provide direct instruction in subjects such as reading, mathematics, language, and science, and research shows that appropriately designed computer-assisted instruction is effective. For example, a survey of thirty-two research studies examining computer-assisted instruction and the achievement of elementary school students in the United States found strong, significant, effects for students exposed to ten minutes of tutorial instruction on a computer daily (Levin, Glass, and Meister 1984).

Well-designed microcomputer programs can significantly reduce the time needed to teach basic skills such as reading and writing. For example, in approximately 100 hours, first-grade students in microcomputer classes in the Philippines learned to read and write English at a level achieved by only a fraction of children in conventional classes in the same schools. Of the computer students, 63 percent could read complete sentences and paragraphs, compared with 37 percent of the control group, and 36 percent of the computer group could write a story in which the ideas were well developed and well expressed, while less than 2 percent of the control group could do so. Computer students also spelled better than control group students (Semeo Regional Center 1987). There is little doubt that properly implemented computer-based learning programs are effective; however, research in North America concludes that computers are less cost-effective than other equally effective interventions (Levin, Glass, and Meister 1984).

Another tool that creates opportunities to practice is the electronic learning aid, which has been tried experimentally in developing countries. The most well-known electronic learning aid uses a keyboard, synthetic speech, and an eight-character alphanumeric display to offer interactive drill and practice in language and arithmetic skills. The effectiveness of this learning aid was evaluated in five primary schools in Lesotho, where its effect on learning was significant and positive, particularly among the less able students (Anzalone and McLaughlin 1984).

OTHER LEARNING AIDS. Other instructional materials can also facilitate the teaching and learning process. Some materials, such as filmstrips, posters, and audio tapes, help the teacher to communicate knowledge; some, such as pencils and paper, enable the student to practice what has been taught; and some—chalk and chalkboards, for example—do both (Baker 1988a). Although most educators agree that such learning aids are invaluable, the impact of specific aids has rarely been studied. As a result, few instructional materials have been shown to improve student learning or produce uniformly positive results. For example, in Brazil a package of writing materials (including chalk, notebooks, pencils, erasers, and crayons) boosted fourth-grade mathematics achievement—undoubtedly because the children then had an opportunity to practice mathematics. These materials had no effect on reading, however (Harbison and Hanushek forthcoming).

Availability of Instructional Materials

The availability and use of instructional materials in developing countries have not been widely documented, but the consensus is that primary students either lack textbooks and other instructional materials entirely or share them extensively with other students.

TEXTBOOKS. In the past, the lack of textbooks was the most serious educational deficit in developing countries. Students either had no textbooks or had no regular access to them (Paxman, Denning, and Read 1989). For example, the national student to French textbook ratio for French texts in the Central African Republic has been 10–20:1, according to World Bank data. In the Philippines, prior to a massive investment in textbooks, the student-textbook ratio was 10:1 (Heyneman, Jamison, and Montenegro 1984). In Brazil, before the implementation of a World Bank–assisted project to increase material inputs in rural schools, only 23 percent of all schools had a first-grade textbook (Armitage and others 1986); in the Dominican Republic, fewer than 20 percent of eighth-grade students in public schools had a mathematics textbook (Luna and González 1986); and in Botswana, fewer than 20 percent of primary school students had access to a science or social studies textbook (Botswana, Ministry of Finance and Development Planning 1984). One survey in Malawi found that fewer than 30 percent of primary students had their own textbook (Mundangepfupfu 1988).

Reducing the student-textbook ratio has been a central goal of investments in educational quality in many developing countries for the past decade. Successful projects have been undertaken in the Philippines, which achieved a student-textbook ratio of 2:1, and in northeastern Brazil, which tripled the number of primary schools receiving first-grade textbooks (Armitage and others 1986; Heyneman, Jamison, and Montenegro 1984). The availability of textbooks in developing countries has often improved as a result of donor activity (see table 3-2), but significant rural-urban differences in textbook availability remain. Little is known about how often teachers use the available textbooks, but it appears that the provision of textbooks does not guarantee their use. A 1983 study in the Philippines (Lockheed, Fonacier, and Bianchi 1989) reported that 32 percent of the fifth-grade science teachers used textbooks frequently, while a study of 127 primary classes in Botswana reported that the teachers used textbooks only 12 percent of the time (Fuller and Snyder 1991).

The availability of textbooks depends, to some extent, on their physical quality. Many developing countries opt for low-quality production to reduce the cost per unit. Low production standards and low costs, combined with state subsidies, can bring textbook prices within the reach of many poor parents. However, the useful life of such textbooks is short, often measured in months, and many parents must buy books several times during the year or have their children go without books. A recent study of textbook quality in twenty developing countries found that the physical quality of books was acceptable in fewer than half the countries studied—The Gambia, Ghana, Mexico, Mozambique, Papua New Guinea, and Sierra Leone (Paxman, Denning, and Read 1989).

TEACHER GUIDES. Teachers lack not only textbooks but also the teacher guides that supplement textbooks. Many teachers were educated in schools lacking good textbooks and may need assistance integrating textbooks into their instructional program. Teacher guides also help teachers boost student learning to higher cognitive levels by suggesting good exercises and questions. Unfortunately, teacher guides are seldom available in developing countries. An internal World Bank study found that schools in Guinea-Bissau had no teacher manuals for any grades or subjects other than grade one. A survey in Malawi reported that fewer than 15 percent of teachers had received a teacher guide for a subject other than English (Mundangepfupfu 1988). In rural Brazil only 44 percent of teachers had received teacher guides in 1983 (Armitage and others 1986).

COMPUTERS. Little up-to-date information exists on the availability of computers in primary schools in developing countries. Some countries, however—Argentina, Belize, Brazil, Chile, Colombia, Grenada, India, Korea, Malaysia, Mexico, Peru, the Philippines, Senegal, Sri Lanka, Swaziland, Thailand, and Trinidad and Tobago—have attempted to use computers in at least some schools (Anzalone 1991). Costa Rica has a major initiative for using computers in primary and secondary education. Other countries use computers in a variety of applications at the primary level, including to teach computer literacy (India, Mexico, and Tunisia), to assist conventional instruction (Grenada, Malaysia, and the Philippines), and to foster critical thinking, creativity, and problem solving (Argentina, Brazil, Kenya, and Senegal)(Hawkridge, Jaworski, and McMahon 1990).

OTHER LEARNING AIDS. In general, the availability of simple instructional materials has increased considerably in developing countries. For instance, most classrooms in these countries have a blackboard, and most parents provide their children with basic school materials, such as pencils and paper. However, other instructional materials, including dictionaries, globes, and maps, are frequently absent from primary classrooms (see table 3-3).

Policy Responses

Providing instructional materials has been a feature of international donor activity for the past decade. Of the 232 World Bank education projects approved between 1970 and 1983, 48 (21 percent) supported the preparation,

Table 3-2. Availability of Textbooks in Selected Countries, Various Years

Country	Year	Percentage of students without books	Distribution mechanism	Comment
Low-income countries				
Burkina Faso	1989	33	Purchased without subsidy.	The French book is too expensive for low-income families.
Central African Republic	1987	—	Provided by the state.	The national average is nine students per book, and no books are available for pupils in rural areas.
China	1986–87	0	Purchased with subsidy.	Every student receives a new textbook every semester in every subject.
Comoros[a]	1986	67	Purchased without subsidy.	One in three students has a French language book.
Equatorial Guinea[a]	1987	100	—	Books are available only in urban schools because distribution is poor.
Guatemala[a]	1988	75–100	—	One in four students had books in 1974, but ten years passed with no further production. Books are virtually absent in the classroom.
Haiti[a]	1985	75	Purchased without subsidy.	One in four children in rural areas has access to a textbook.
Madagascar	1989	—	Purchased with subsidy.	Most schools lack textbooks, and the available ones are old and in poor condition.
Mozambique	1988	—	Purchased with subsidy.	Supply is good in towns and bad in remote areas. Prohibitive fees impede student access to books.
Niger	1984	—	Provided by the state or purchased.	The state supplies one French text for every four students. For other subjects, only teachers receive free texts.

Pakistan[a]	1988–89	50	Purchased with subsidy.	Students in rural areas have limited access to textbooks.
Rwanda[a]	1988	87	Purchased with subsidy.	One pupil in eight has a set of books in Kinyarwanda, mathematics, and French.
Sierra Leone[a]	1986	25–65	Provided by the state.	Virtually no students in 100 provincial schools have textbooks.
Togo[a]	—	—	Purchased with subsidy.	Books are expensive, and twenty students frequently share one book.
Uganda[a]	1988	40	Provided by the state.	3,400 schools still did not have books after a World Bank project supplied texts to 5,400 schools.
Zaire	1988	—	Purchased with subsidy.	Many schools have no textbooks.
Middle-income countries				
Lesotho[a]	1985	—	Purchased with subsidy.	The ratio is about three students to every book.
Nigeria	1987	98	Provided by the state or purchased with subsidy.	The provision of books in Ibadan and Lagos is barely acceptable; the ratio in northern rural districts is two books to 100 students.
Paraguay[a]	1984	67	Provided by the state.	One in three students has a book; one in fifty has all four required books.
Peru[a]	1985	67	—	Two-thirds of the students in rural areas have no books.
Tunisia	1988	0	Purchased with subsidy.	In Tunis all students have copies of books the teachers asked them to buy. In rural areas the situation is unkown.
Turkey	1989	—	Purchased with subsidy.	The availability is good in large cities but not in rural areas.

— Not available.

a. Countries in which the International Bank for Reconstruction and Development supported textbook projects. Data are preproject.

Sources: Internal World Bank documents; Paxman, Denning, and Read 1989.

Table 3-3. Percentage of Classrooms with Instructional Material in Brazil, 1985, and Somalia, 1984

Instructional material	*Brazil, 1985*	*Somalia, 1984*
Exercise books and student notebooks	100	89
Pencils	89	93
Textbooks for all students	91	41
Blackboards	—	93
Chalk	95	95
Maps	—	23
Science equipment	—	9
Globes	—	11
Dictionaries	—	3
Libraries	—	4
Visual aids	—	28

— Not available.

Sources: Armitage and others 1986; IEES 1984c.

provision, or distribution of educational materials and textbooks. Most of these projects sought to decrease the student-textbook ratio (Romain 1985). In some instances, more sophisticated and expensive material inputs were provided. Although these materials may be effective when properly used, for the present they must be considered blind alleys for investment because of their high cost, including the cost of implementation.

BLIND ALLEY: COMPUTERS IN THE CLASSROOM. The new technologies that have elicited the most interest for primary educational applications are computers and microcomputers, which are effective instructional tools when used in settings with multiple machines and well-trained teachers. The cost of multiple computers makes widespread classroom use at the primary level unfeasible for most education systems in developing countries. Even in high-income countries, where microcomputers are widely available in primary schools, the cost of introducing computers has been higher than expected, principally because of the unanticipated need to train teachers. Evidence of how much computer education costs in developing countries is rare. One example comes from Belize, where the cost per student of a single computer education course at the secondary level amounted to 59–149 percent of the total public expenditure per pupil in secondary education (USAID n.d.).

Costs are high because hardware, software, infrastructure, and teacher training are expensive (Anzalone 1991). Hardware often must be imported, and although the cost of computer power per dollar has fallen, the absolute cost of computers has not declined rapidly. A second obstacle is that few developing countries can undertake expensive software development projects and must, therefore, adapt software developed for use in another country. Differences in language, curriculum, hardware, and operating systems could increase the cost of such adaptations. When portable computers are not readily available, buildings must be modified for computer use as well. Problems posed by electrical supply, heat, dust, and access to replacement parts can increase costs. Finally, the cost of retraining teachers to use computers must also be factored in. For example, a study in Belize of the cost of offering a computer literacy course to preservice teachers—a bare necessity for teachers using computers with their students—showed that the cost of the course per student represented 17–42 percent of the total public expenditure per pupil for higher education, depending on the relative number of computers, students, and teachers (USAID n.d.).

PROMISING AVENUE: GOOD TEXTBOOKS AND TEACHER GUIDES. Increasing student learning requires at least the provision of good textbooks and teacher guides. This begins with specifying the content—often done in the country to meet local needs—and writing the textbook—a difficult process that requires considerable expertise. The next step, deciding whether to publish books domestically (either by the government or by commercial publishers) or to purchase them from abroad, is determined by the existing level of publishing and printing capacity in the country, special requirements such as language, and the relative cost of the books, depending on the student-textbook ratio. In either case, the number of primary textbooks purchased or produced should reduce the student-textbook ratio to at least 2:1 for each subject taught.

Providing textbooks also entails distributing materials. Distribution of primary textbooks is often difficult in developing countries where schools are widely scattered and hard to reach by road. Warehousing and transportation are the most salient features of the distribution system. Textbooks require warehousing that is secure, weatherproof, well ventilated, equipped with shelves, clean, and organized. Transportation must be reliable and capable of meeting demand.

Countries should focus on providing their schools with good (pedagogically sound, culturally relevant, and physically durable) textbooks and on encouraging teachers to use them. Where the material is written or produced is less important. Technical assistance can help accomplish pedagogically significant objectives. When countries lack the appropriate technology for typesetting, printing, and binding textbooks internally, they should consider publishing them externally.

Time for Learning

Research from a variety of countries has shown that the amount of time available for teaching and learning academic subjects, and how well that time is used by students and teachers, is consistently related to how much children learn while they are in school.[5]

Effectiveness of Time

In general, when teachers devote more time to instruction, students learn more. In India, Iran, and Thailand students learned more science when the amount of time spent on instruction and reading increased (Heyneman and Loxley 1983). In Nigeria an increase in the amount of instructional time alone increased the amount of mathematics learned (Lockheed and Komenan 1989). The productivity of time, however, varies inversely with children's initial level of learning; additional instructional time is more important for low-performing students than for better students. In addition, sufficient instructional time is particularly important during the early grades; the time elasticities of achievement for the second grade are estimated to be 0.25 for mathematics and half that for reading (Brown and Saks 1987).

In-school learning time is also especially valuable for students from impoverished families, who usually work and spend relatively few of their out-of-school hours on learning. A 1981 study of factors related to ninth-grade mathematics achievement in Indonesia found that the time spent on classroom assignments was the most significant predictor of achievement for poor rural students, the second most significant predictor for poor, urban students, but only the eighth most important predictor for middle-class, urban students (Suryadi, Green, and Windham 1981).

Availability of Time

The annual number of hours available for children to study a given subject in school is determined by three factors: the hours in the official school year, the proportion of these hours assigned to the subject, and the amount of time lost because of school closings, teacher absences, student absences, and miscellaneous interruptions. Additional time for study can be provided by after-school study periods and homework assignments. The proportion of time children use effectively (time on task) is determined by their motivation and the teacher's management of available time.

OFFICIAL TIME FOR LEARNING. Worldwide, the official academic year for primary grades one through six averages 880 instructional hours. As indicated in table 3-4, this varies depending on the economic status of the country: low- and lower-middle-income countries have a shorter official school year than upper-middle- and high-income countries. However, in some developing countries the official academic year is substantially shorter than average (for example, 610 hours in Ghana), while in others it is longer (1,070 hours in Morocco).

Since curricular emphasis at the primary level is relatively consistent across countries (approximately 35 percent of the available time is spent on language—reading and writing—and 18 percent on mathematics), the offi-

Table 3-4. Official Instructional Time in 110 Countries, 1980s
(average annual hours)

Country income level	*Total*	*Language instruction*	*Mathematics instruction*
Low	870	322	157
Lower middle	862	293	147
Upper middle	896	323	161
High	914	311	174

Source: Benavot and Kamens 1989, updated tables.

cial number of hours available for instruction in these subjects is similar for industrial and developing countries. That is, official curricula in low-income countries allocate on average 322 hours to language and 157 hours to mathematics, which is about the same as the official time allocated to these subjects in industrialized countries (311 hours and 174 hours, respectively).

ACTUAL TIME FOR LEARNING. The amount of time lost because of unscheduled school closings, teacher and student absences, and disruptions is much greater in developing than in industrial countries. In Haiti the 1984 school year had 162 days, which was shorter than the international standard of 180 days; it was shortened even further by unofficial closings and delayed openings. The school day often began late, teachers were frequently absent on market days (Tuesdays and Fridays), and forty-eight public holidays were celebrated instead of the twenty-eight already built into the school year. The result was a functional school year of seventy days—43 percent of the country's official year and 40 percent of the international standard. In Malawi the school year is 192 days, but one-third of these days fall during the rainy season. Instruction is virtually impossible during a rainstorm (teachers cannot be heard above the noise of rain falling on the tin roofs of schools) and roads are impassable, keeping both students and teachers at home.

Teacher absences for administrative procedures are also common in developing countries. Many teachers must travel considerable distances to be paid, while others are assigned to schools far from their home; both situations contribute to teacher absences. Strikes in some countries also keep schools from functioning; months can be lost, as happened in Brazil in 1988. Teacher absences also occur when maternity leave policies do not provide substitute teachers. For example, Eisemon and Schwille (1991) observed that in a sixth-grade class in Burundi, students missed about two months of instruction when their teacher took maternity leave; a month passed before a substitute was placed in the classroom, and when the substitute became ill soon after, the students were left without a replacement teacher.

One observational study examined instructional time in Indonesia: not only was the local school week shorter than the national school week on the official calendar, but even fewer hours were actually taught (see table 3-5).

Table 3-5. Official and Actual Instructional Time in Indonesia, 1978
(hours per week)

Grade	*Official instructional time*		*Actual instructional time*
	Nationally	*Locally*	
1–2	26	13	9
3	33	22	—
4–6	36	24	18

— Not available.
Source: Shaeffer 1979.

Instructional time is also reduced when the school day itself is shortened (for example, to allow double shifts). A reduction in the length of the school day can affect instructional time differently for different subjects. At one extreme, it can shorten instructional time in all subjects equally, which reduces the hours of instruction in key subjects below acceptable norms. At the other extreme, retaining previous levels of instructional time in key subjects, such as reading or mathematics, could mean that the time allocated to subjects such as art, needlework, or vocational skills would have to be reduced or eliminated. When double shifts meant that the total instructional time would have to be reduced in Burundi, the number of hours devoted to practical agriculture and home economics was reduced from four hours to one hour a week; 92 percent of instructional time in the sixth grade was then devoted to the four subjects covered in the secondary school entrance examination (Schwille, Eisemon, and Prouty 1989).

Policy Responses

Developing countries should seek ways to improve and increase the time teachers devote to teaching and students devote to learning, particularly in primary schools.

BLIND ALLEY: LOWERING CLASS SIZE. Decreasing class size, although costly and possibly of little value, has its advocates. Certainly primary-level classes are very large in most developing countries, particularly in urban schools. Although no good international data are available on class size at the primary level, data are available on student-teacher ratios. In single-shift systems (where all students attend school at the same time rather than in shifts) the official class size is roughly equivalent to the student-teacher ratio, particularly in primary schools, where teachers have less time for in-school preparation. In 1985 the mean student-teacher ratio in the lowest-income countries was 39, and eight countries had student-teacher ratios exceeding 50 (see table 3-6). Actual class sizes may be far larger than the official size reported because of teacher absenteeism and specialization. Primary school classes often hold more than seventy students.[6] Typically, urban schools and the first and second grades have the largest

Table 3-6. Student-Teacher Ratio by Country Income Level, Selected Years, 1965–85

Country income level	*Ratio (weighted mean)* 1965	1970	1975	1980	1985	*Percentage change, 1965–85*
Low (excluding China and India)	42	41	42	42	39	–7.0
China and India	33	32	32	31	30	–9.0
Lower middle	37	33	32	32	29	–22.0
Upper middle	32	30	27	27	24	–25.0
High (excluding oil exporters)	28	26	23	21	20	–29.0

Source: See table A-9 in the appendix.

classes. In Senegal, for example, 27 percent of urban classrooms had between 70 and 120 students; Malawi had an average of 64 primary students per classroom in 1986, and in four urban regions class size exceeded 140.

From 1965 to 1985 the student-teacher ratio declined throughout the world: 7 percent in low-income countries and more than 20 percent in middle- and high-income countries. Developing countries frequently introduced policies intended to reduce the student-teacher ratio. They set targets to keep classes under thirty-five or forty students, even when this meant putting children on double or triple shifts. Such policies may be counterproductive when they cause a significant loss of instructional time.

On the whole, students in larger classes spend less time on task. Students in a class of five spend an estimated 90 percent of their time on task; that figure drops to 61 percent for students in a class of 20, and only 12 percent in a class of 100 (Mulligan 1984).[7] Increased time on task is responsible for the improved achievement that has been documented in classes of twenty or fewer students. However, achievement is not significantly boosted by reducing class size from forty or forty-five to thirty-five or forty students (Glass and Smith 1979). Moreover, since achievement can be high in classes of fifty or more (in Korea, for instance), effective teaching strategies apparently do exist for large classes. Still, large classes are often unruly, and teachers tend to emphasize rote learning rather than problem-solving skills. For these reasons, it may be worthwhile to reduce classes to no more than forty to fifty students.

However, if lowering class size further has little effect on learning until classes reach about twenty students, such policy is generally not a wise investment (Jamison 1982). Making primary school classes this small while maintaining the existing level of instructional time could require many countries to triple both the teaching force (which is rarely feasible because not enough teachers are available) and the number of classrooms. These investments would triple the national education budget for both capital and recurrent expenditures. Reducing classes to fewer than forty is, therefore, ineffective and prohibitively expensive, and a blind alley in most developing countries.

PROMISING AVENUE: SETTING AND MAINTAINING STANDARDS FOR INSTRUCTIONAL TIME. Time for learning can be effectively increased by setting and maintaining standards for instructional time. Although setting standards can be accomplished by national or state policy, maintaining standards requires local monitoring.

Setting standards is the responsibility of central policymakers. Two alternatives are available. On the one hand, policymakers can lengthen the school year if it is shorter than international standards (880 hours). In many countries this is an attractive alternative because it permits teachers to cover the conventional primary curriculum fully. On the other hand, in countries where double shifts require that the school year be shorter than 880 hours, policymakers may have to give priority to maintaining the minimum instructional time in core subjects, especially reading and mathematics, rather than to covering a wide range of subjects. Low-income countries that have an 880-hour school year and allocate time in the traditional proportions spend 322 hours on language, 157 hours on mathematics, and about 70 hours each on science, social studies, the arts, and physical education. If the instructional year were reduced by half to 440 hours, all instructional time would need to be allocated to language and mathematics in order to maintain international standards. Otherwise, achievement in those key subjects would be likely to decline.

Maintaining standards for instructional time requires local attention to ensure that (a) schools are open during official hours, (b) teachers are present and teaching during the official instructional periods, (c) temporary distractions, such as administrative or visitor interruptions, are avoided, and (d) appropriate arrangements are made for continuing instruction in routine inclement weather. Monitoring study periods and assigning homework are other local policy options that can extend the time available for students to study.

Teacher and student attendance can be increased by providing incentives for attendance (for teachers as well as students) and establishing local monitoring to confirm that school is actually in session and teaching is ongoing. Two important incentives for teachers are to provide adequate local housing and to pay salaries locally so that teachers need not travel to receive compensation. Providing teaching materials appears to be another incentive for teachers (Kemmerer 1990). When teachers are likely to be absent for long periods, substitute teachers must be assigned to their classes. Education budgets must include funds to pay for this. Determining local causes of teacher and student absences may be required before further action is taken.

Effective Teaching

Teachers are central to the delivery as well as the quality of education. The academic and professional training of teachers has a direct and positive bearing on the quality of their performance and consequently on the achievement of students (Avalos and Haddad 1981; Husén, Saha, and Noonan 1978; Schiefelbein and Simmons 1981). Effective teaching is determined by the individual teacher's knowledge of the subject matter and mastery of pedagogical skills.

Knowledge of the Subject Matter

Evidence from developed countries shows that the teacher's knowledge of the subject matter has a strong, positive effect on student achievement. Only a few studies from developing countries have examined this question, but their results are consistent (Fuller 1987). In Uganda, for example, the English language proficiency of teachers was found to have a positive effect on student achievement in both language and mathematics (Heyneman and Jamison 1980). In Iran the achievement of teachers taking a secondary school graduation examination was correlated with the achievement of their second-grade students (Ryan 1972). Level of formal education is often considered an indicator of a teacher's knowledge of the subject matter. Although the evidence is not consistent in all studies (Fuller 1987), teachers with less than a secondary education appear to be less effective than those who graduated from secondary school. In Pakistan, for example, students whose teachers had a secondary education did better on tests of mathematics than students whose teachers had only a primary education (McGinn, Warwick, and Reimers 1989).

Pedagogical Practices

Teaching is complex, and teachers must command a wide range of instructional strategies for teaching specific subjects and managing the classroom. At a bare minimum, effective teaching involves (a) presenting material in a rational and orderly fashion, pacing the class to the students' level and taking into account individual differences, (b) allowing students to practice and apply what they have learned, particularly in relation to their own experience, (c) letting students know what is expected of them, and (d) monitoring and evaluating performance so that students learn from their mistakes. Effective teaching strategies may differ by subject and by age. Appropriate teaching guides can enhance teaching effectiveness.

Little research on the effectiveness of classroom teaching has been conducted across nations, but the results are consistent with those from studies in developed countries (Anderson 1988; Anderson, Ryan, and Shapiro 1989). In general, teaching practices that enhance student learning (a) require students to participate actively, (b) allow students to practice what is being taught and apply it to their own experience, (c) monitor and evaluate student performance, and (d) give students appropriate feedback on their performance.

ACTIVE STUDENT PARTICIPATION. A variety of techniques can be used to generate the active participation of students. In preschool and the early grades of primary school, teachers can present carefully sequenced curriculum in small hierarchical steps by offering information, having the children repeat the information and answer questions about the material, and giving prompt feedback on their responses (Stallings and Stipek 1986). For teaching basic skills such as multiplication, a drill-and-practice format may be effective; in other cases, teachers must ask questions that stimulate the students' thinking. Most important, teachers must use the available classroom time as intensively as possible. Several studies examining the long-term effects of this type of instruction have concluded that positive academic effects are retained throughout secondary school, with students achieving higher levels of academic performance, dropping out and repeating a grade less often, and graduating more often than students who are taught more conventionally. Children, both younger and older, need to practice the comprehension and application of material; they also need to hear stories and interpret what is read to them.

In addition to basic skills, higher-order skills should be developed intensively (Cole 1990). For reading, this involves comprehension and application skills; for mathematics, it involves applying skills to new problems. To facilitate comprehension and knowledge acquisition, active student involvement is again essential and can be encouraged by asking students to think about the relationship between the text they are reading and their experience and knowledge (Wittrock 1986). Having children actively relate to the text sharply enhances their understanding of what they have read. For example, Au (1977) asked low-achieving, native Hawaiian primary school students to state in their own words several events in their lives that related to the stories teachers read to them. After one year of this type of instruction, children scored 40–60 percentile points higher in reading than control groups of students.

PRACTICE. Teachers can give students the opportunity to rehearse what is taught by assigning individual work sheets (at home, in class, during a study hall) and homework and by organizing group work and discussion. A study of 900 mathematics classes in Pakistan found that the number of exercises assigned was one of the best predictors of mathematics achievement (McGinn, Warwick, and Reimers 1989). Work sheets and homework are effective but often require additional material or out-of-class time that is not available.[8]

A more promising technique for developing countries is group work, which is known to be cost-effective. Small, cooperative group learning involves four to six students working together on a task (such as a science project). The task can be divided so that each member works on a different part or everyone works jointly on all parts of the task. Cooperative learning programs in developed countries showed significant gains in achievement for all group members (in both basic and problem-solving skills), but particularly for students who began as low achievers (Cohen 1986; Sharan and Shachar 1988; Slavin, Leavey, and Madden 1984). Recent research in the Philippines also found that group work led to higher levels of student achievement in science, with students who did group work outscoring students who did not by 57 percent (Lockheed, Fonacier, and Bianchi 1989).

Cross-age peer tutoring, an instructional alternative that relies on older students to tutor younger students, also offers students more opportunities to practice. It has been highly effective in industrial countries and used selectively in developing countries—for example, in the Escuela Nueva schools in Colombia. In experimental studies, both tutors and their pupils gain substantially in overall learning; cross-age peer tutoring is also one of the most cost-effective educational practices (Levin and Meister 1986). In all cases, active participation allows students to relate to the material and to make conceptual sense of it. Box 3-3 describes a North American cross-age peer tutoring program.

MONITORING AND EVALUATION. Teachers discover what students already know and what they still need to learn by monitoring student work through essays, quizzes and tests, homework, classroom questions, and standardized tests. Student errors on tests and in class are early warning signs of learning problems, which teachers can then correct before they worsen. Studies demonstrating the effectiveness of monitoring and evaluating student performance are beginning to emerge from developing countries, and the results are consistently positive (Arriagada 1981; Lockheed, Fonacier, and Bianchi 1989; Lockheed and Komenan 1989). For example, the frequency of teacher evaluations (progress reports) was positively related to achievement in Colombia, the time teachers spent correcting tests and exercises was related to achievement in Argentina and Colombia, the time they spent discussing exercises was related to science achievement in Paraguay, and the time they spent monitoring and evaluating student performance was related to mathematics achievement in Swaziland. In the Philippines, students whose teachers frequently tested them outperformed by one-quarter of a standard deviation students whose teachers tested them infrequently. Monitoring performance at the classroom level is particularly important and should be distinguished from national examinations that select students for admission to secondary school.

FEEDBACK. Research from developed countries indicates that the effectiveness of classroom tests and quizzes is closely linked to the immediacy of the feedback that students receive; students who receive immediate feedback on their quizzes outperform students who receive delayed feedback (Kulik and Kulik 1988). Immediate feedback also enhances their motivation.

Availability of Knowledgeable and Skilled Teachers in the Classroom

To function as an independent source of information, primary teachers need a general academic education that is superior to the level of the students they are teaching. This implies that they should have completed at least lower sec-

Box 3-3. Cross-Age Peer Tutoring

Cross-age peer tutoring has a long, informal history in education. In one-room schools, older students routinely helped teach younger students. The benefits of such an arrangement have been noted at least since the nineteenth century in Europe and North America. Tutoring often helps individuals who are not well served by instructional methods that address the entire class. It is also valuable for very large classes, such as those in the early grades in developing countries, where the benefits of individualized instruction can offset the drawbacks of oversized classes.

One well-implemented, cross-age peer tutoring program in Boise, Idaho, has been extensively evaluated for cost and effectiveness. In the Boise program, children in the upper elementary grades tutor students in the second and third grades. Research studies have found comparable achievement gains for both tutors and pupils. Compared with control groups, students in peer tutoring classes moved from the fiftieth to the eighty-fourth percentile in mathematics and from the fiftieth to the sixty-eighth percentile in reading (see table 3B-3).

Peer tutoring was both more effective and more cost-effective than either lowering class size from thirty-five to twenty students or providing ten minutes of computer-assisted instruction daily. The finding on cost-effectiveness should provide additional support for the use of peer tutoring.

Table 3B-3. Effectiveness, Cost, and Cost-Effectiveness of Three Educational Interventions in the United States, 1980s

	Increase in achievement (percentile points)		Cost	*Cost-effectivenesses*[a]	
Intervention	*Mathematics*	*Reading*	*(dollars)*	*Mathematics*	*Reading*
Peer tutoring	34	18	212	0.16	0.08
Computer-assisted instruction	5	9	119	0.04	0.08
Lowering class size from thirty-five to twenty students	9	4	210	0.04	0.02

a. Cost-effectiveness is expressed as the percentile increase per $100. Thus an investment of $100 in peer tutoring will yield about four times as much mathematics learning as a similar investment in establishing computer-assisted instruction or reducing class size.

Source: Levin, Glass, and Meister 1984.

ondary, if not upper secondary, school. When teachers lack this level of education, their teaching will need to be supplemented by tools such as detailed curriculum guides or interactive radio instruction (see pages 71–72 and box 3-5). In most developing countries, primary school teachers are required to have at least a junior secondary certificate, with specialized training in pedagogy (Dove 1986).

Many countries lack qualified personnel and therefore must recruit unqualified persons to teach primary school. Standards are lowered most often when the capacity of primary schools is expanding rapidly. The result is that persons with incomplete secondary education often staff primary schools. For example, in Nepal one-third of all teachers lack any secondary education, and in ten of the thirty-three Sub-Saharan African countries for which data are available, the majority of primary teachers lack a complete secondary education. In Nigeria only five years of primary education were required for an individual to enter teacher education in 1981; this was the lowest minimum educational requirement of all African countries at the time (Zymelman and DeStefano 1989). Where teacher education requirements are low, many primary school teachers are weak in the subjects they are teaching, particularly in upper primary grades.

At present, many of the teaching practices in developing countries are not conducive to student learning: (a) instruction for the whole class that emphasizes lectures by the teacher, has students copy from the blackboard, and offers them few opportunities to ask questions or participate in learning, (b) student memorization of texts with few opportunities to work actively with the material, and (c) little ongoing monitoring and assessment of student learning through homework, classroom quizzes, or tests. Although 40 percent of the teacher's time is devoted to lectures and 19 percent to student participation in U.S. classrooms, active student participation is rarely encountered in classrooms in developing countries (Pfau 1980). In Nepal 78 percent of fifth-grade science instruction was devoted to lectures and less than 7 percent to student participation (Pfau 1980). In Thailand 54 percent of fifth-grade mathematics instruction consisted of teacher lectures, explanations, or demonstrations; another 30 percent of instructional time was spent on written work, and only 4 percent was used for oral work of any kind, including discussion (Nitsaisook, cited in Avalos 1986). In Botswana students listened to lectures for 54 percent of the observed instructional time and spent another 43 percent of the time on oral recitation, which should not be confused with discussion (Fuller and Snyder 1991). In Jamaica, Jennings-Wray (1984) noted that teachers spent 59 percent of the classroom time talking; teachers dominated the lessons and posed few open-ended questions.

Group work, which encourages discussion, is rarely encountered. In Nepal no group work was observed; in the Philippines only 10 percent of fifth-grade science teachers used small groups for instruction (Lockheed, Fonacier, and Bianchi 1989). Group work is, however, infrequent in the United States as well, occurring only 10 percent of the time in fifth-grade classrooms (Lockheed and Harris 1982). Nor is classroom assessment prevalent in many developing countries. In the Philippines, for example, only 29 percent of fifth-grade science teachers frequently tested their students (Lockheed, Fonacier, and Bianchi 1989); in Nigeria, only 10 percent of primary school teachers used continuous assessment techniques (Ali and

Akubue 1988); and in Botswana students took tests only 1 percent of the time (Fuller and Snyder 1991).

Policy Responses

This chapter discusses three policy responses appropriate for improving classroom instruction: in-service training, interactive radio instruction, and programmed learning materials. Improving the teaching of prospective teachers is discussed further in chapter 4.

Different strategies may be appropriate for increasing the teacher's knowledge of the subject matter and for improving pedagogical skills. For example, increasing the knowledge of teachers may be accomplished most effectively through courses in the subject matter, whether offered far from or close to school, whereas improving the quality of teaching (pedagogical knowledge) may be achieved most effectively through school-based in-service training. In practice, knowledge of the subject matter is often taught concurrently with pedagogical skills. Many countries have also tried to improve the quality of instruction in primary classes directly, by supplementing teachers with interactive radio instruction and by providing them with highly structured curriculum materials.

PROMISING AVENUE: IN-SERVICE TRAINING. Three common forms of teacher in-service education are short-term residency programs, continuous training and visit programs, and distance education programs (correspondence courses, educational television, or radio programming). Although most in-service training emphasizes pedagogical practices, some programs are designed to provide the equivalent of a secondary school education.

The most effective forms of training are ongoing, rather than one-shot courses with no follow-up. In Bangladesh, efforts to strengthen primary education benefited greatly from the central use of recurrent, school-based, in-service teacher training. Each teacher and assistant teacher received approximately two months of intensive training on general topics. Then they received three days of training on common teaching problems every other month. Training concentrated on increasing learning achievement, covering such matters as practical methods of teaching each major subject, adapting the curriculum to the social and physical environment of the pupil, understanding how children develop and learn, evaluating teaching and learning, managing classrooms (especially multigrade classrooms), and building effective parent-teacher and community relations. Backup materials were also provided (Verspoor 1989).

In-service teacher training on locally effective practices for managing the classroom and using time can provide teachers who are unaware of effective instructional strategies with a new repertoire of skills. In India, Nigeria, and Thailand active training methods such as microteaching and simulation, role playing, and discussing case studies improved the participatory and discovery teaching of the teachers trained. Teachers in Thailand who were exposed to a six-day in-service training program emphasizing classroom management strategies significantly changed their teaching practices (Nitsaisook and Anderson 1989). In-service teacher education, particularly ongoing programs that monitor and evaluate teachers regularly, shows much promise for improving teaching in developing countries and has received support from the international donor community. Between fiscal years 1963 and 1984, 22 percent of World Bank–assisted education projects involved in-service training for primary school teachers.

Distance education is another effective method of upgrading the teaching skills of primary school teachers in developing countries. Distance education for teacher training has usually been implemented through correspondence courses supported by either short residential courses, such as the Zimbabwe Integrated Education Course (Taylor 1983) or tutorial activities under the auspices of strategically located learning centers, such as the Logos II program in Brazil (see box 3-4). Sometimes the printed materials are complemented

Box 3-4. Teacher Training through Distance Education

Developing countries attempting to increase the number of primary school places are often faced with a serious shortage of teachers. Creating school places (by constructing school buildings and providing equipment) can be easier than producing a sufficient number of trained teachers. Furthermore, expanding access to schooling has meant, in many instances, creating primary education programs in remote and isolated regions of the country, where staffing schools is particularly difficult. As a consequence of both situations, poorly trained and inexperienced teachers may be hired and the quality of instruction may suffer. In such cases, distance education can be an effective way to upgrade teachers' skills and thus improve the quality of education.

The Logos II project in Brazil is a good example of a distance education teacher training program. Logos II is a self-paced learning program that combines subject matter and pedagogical education. All students are unqualified primary school teachers who are currently employed. The curriculum consists of modules or short courses on specific topics, presented in separate pamphlets. The topics range from mastering the primary school curriculum to mastering specific pedagogical techniques. Students study the pamphlets at home and then return to learning centers to be tested on the modules. Other activities at the learning centers include microteaching sessions, special tutoring for difficult topics, socializing opportunities, and study groups. The entire Logos II program takes from thirty to fifty weeks to complete, after which students take a certification examination.

Source: Oliveira and Orivel 1981.

by radio or television programs, as is the case with the Radio Teacher Training Project in Nepal (Butterworth, Karmacharya, and Martin 1983). Some distance education courses also include occasional face-to-face contact with instructors.

In Latin America, two-thirds of all postsecondary distance education institutions provide teacher training (Nettleton 1991). In Africa, teacher education is the most successful use of distance education. Experience in Nepal indicates that to succeed, distance programs for teachers must be undertaken with a high degree of information about the needs of the recipients, including their language facility (Anzalone and Shyam 1989). Furthermore, the participants must be highly motivated and must receive continuous support (written commentary on performance, for example) from the program.

Students do tend to be highly motivated because many enroll in distance education programs in order to pass certification examinations and thus qualify for an improvement in salary or employment status. Working teachers also have an opportunity to apply what they have learned without delay. Furthermore, when the programs are self-paced, the students can adapt the content and rhythm to their own schedule and specific needs. Underqualified teachers need flexible and relevant programs for upgrading their skills, and distance education provides this.

The primary advantage of distance education for in-service teacher training lies in its cost-effectiveness. This is not to say that distance education is necessarily cheaper than more traditional forms of teacher training. In fact, distance education programs, such as those in southern Africa, are less expensive only when their enrollments are large. However, for the specific needs of unqualified teachers, especially in rural and remote areas, distance education is probably the most cost-effective method available. First, because the students do not need to stop teaching to participate, they do not forgo their salaries. Nor is hiring substitute teachers necessary. The opportunity cost is thus significantly lower for distance education than for more formal teacher training schemes. In addition, providing face-to-face training for the teachers who can be reached through distance education is often prohibitively expensive. The majority of these teachers work in remote regions where communication and transportation links with other areas are weak. More traditional forms of in-service and preservice teacher training are difficult to establish, staff, and administer in these regions and would thus require greater investments and incentives than similar institutions in urban areas. Consequently, for many countries distance education may be the most cost-effective method of reducing the number of underqualified teachers.

In Tanzania teacher training at a distance was not only more effective than an equivalent conventional residential program, but it cost four times less per graduate. Similar cost-effectiveness was found for the Logos II program in Brazil; moreover, it was six times more cost-effective than traditional residential teacher training methods (Oliveira and Orivel 1981). Distance education for teachers in Sri Lanka was also more cost-effective than either preservice teacher education or residential in-service training (Cummings and others 1990).

PROMISING AVENUE: INTERACTIVE RADIO INSTRUCTION. When teachers lack sufficient knowledge to instruct students correctly in a particular subject, such as mathematics or a foreign language, supplemental instructional media such as interactive radio can be useful. Instructional media typically pay more attention to the correct order and pacing of instruction than does a teacher. With interactive radio instruction (IRI), which is broadcast directly into the classroom and substitutes for the teacher, the instructional materials and delivery strategy are highly coordinated, and the teaching is highly effective. Interactive radio instruction depends on specific instructional principles: a segmented structure (about one dozen topics per lesson), active and frequent student responses, immediate reinforcement, and periodic practice. These principles are applied to the radio lessons and to the printed support materials. Most of the instruction is contained in the radio lessons, which are organized and presented for pedagogical effectiveness.

The achievement results of interactive radio have been positive in mathematics, language, reading, and writing. Field tests have demonstrated that in the formal school environment, children taught through IRI outscore traditionally schooled children on achievement tests. In informal environments, interactive radio students can keep up with their counterparts in conventional classrooms. Evaluations of five programs show that IRI students perform significantly higher than children in conventional classrooms (see table 3-7).

Interactive radio is also one of the most cost-effective educational interventions. It requires few printed materials, thereby reducing production costs and the need to maintain cumbersome systems for distributing textbooks and other bulky materials. Only minimal teacher training, usually no more than one day, is necessary. Once radio lessons have been developed, the annual cost per student is very low because the same lesson can be transmitted to thousands of new students at minimal cost.

Despite clear evidence that IRI is effective and teachers are enthusiastic about it, interactive radio has not been implemented widely. In at least two countries, however—the Dominican Republic and Thailand—interactive radio has been institutionalized on a national basis. In addition, successful pilot applications have taken place in a number of countries and in a variety of subjects (see box 3-5). These experiences demonstrate that interactive radio is feasible for developing countries.

Table 3-7. Academic Achievement through Interactive Radio Instruction, Various Years

Program	*Radio class (mean percentage of correct answers)*	*Conventional class (mean percentage of correct answers)*	*Effect size (d-statistic)*[a]
Bolivia Radio Math (grade 2)	66	47	0.91
Honduras Radio Mental Arithmetic[b] (grade 1)	52	44	0.80
Kenya Radio English (grades 1–3)	46	36	0.53
Nicaragua Radio Math (grade 1)	65	41	1.31
Thailand Radio Math (grade 2, rural)	58	44	0.58

a. The *d*-statistic is calculated by taking the mean score of the radio class, subtracting the mean score of the control group (the conventional class), and dividing by the standard deviation of the control group (Glass, McGaw, and Smith 1981).

b. Program includes new textbooks.

Sources: Friend 1989; Fryer 1989; Lockheed and Hanushek 1988.

PROMISING AVENUE: PROGRAMMED MATERIALS. Programmed materials have been used with various degrees of success in primary school. Like interactive radio, programmed teaching and learning systems closely correlate the materials and the means of delivering them and systematically apply various principles of instructional development and design. In general, they include step-by-step scripts for the teacher along with instructional materials for the children to use individually or in groups (Anzalone 1991). Major applications of programmed teaching and learning include Project UPE/IMPACT in Bangladesh, Project PAMONG in Indonesia, Project PRIMER in Jamaica, Project INSPIRE in Malaysia, Project IMPACT in the Philippines, Project IEL in Liberia (Cummings 1986), and Project RIT in Thailand (Potar 1984). Students in classes using programmed teaching and learning approaches have done better than students receiving conventional instruction in Liberia and Thailand, as well or better in the Philippines, and as well in Indonesia. Programmed instruction is particularly appropriate for multigrade classrooms.

Teachability: Children's Learning Capacity

The capacity to learn in school is determined in part by the prior learning experience and the health and nutritional status of each child. In general, deficits in nutrition and experience must be corrected during the preschool years. However, the effect of short-term hunger and some health deficits can be identified and corrected during primary school.

Box 3-5. Interactive Radio Instruction

Interactive radio instruction (IRI) is a successful means to improve the effectiveness of teaching time. Unlike traditional educational radio, which encourages passivity by having students receive instruction through lectures, interactive programs use radio creatively. IRI lessons are tightly controlled and systematically designed to engage the learner. The radio teacher interacts with students throughout the broadcast by leading them in question and answer sessions, songs, physical activities, and reading and writing exercises.

The Kenya Radio Language Arts Project, a good example of IRI, drew primary schoolchildren into a pedagogically sound dialogue covering the language curriculum. What might have looked like pandemonium was actually a well-designed, tightly structured lesson called English in Action that involved students actively in the learning process. The thirty-minute daily broadcasts, which were punctuated by music and short dramas, paused frequently so the children could respond and receive immediate reinforcement for their answers. Responses could be sung, spoken, written, or acted out. Typically children were given the chance to respond more than 100 times during each thirty-minute period.

Interactive radio can be adapted to the specific requirements of individual countries. The radio mathematics curriculum developed in Nicaragua has been adapted for use in Bolivia, the Dominican Republic, and Thailand. The interactive methodology has also provided the basis for new radio programs teaching language in the Dominican Republic, science in Papua New Guinea, mental arithmetic in Honduras, and English as a second language in Kenya and Lesotho. Some programs provide a comprehensive instructional package. Others complement existing textbooks and teacher training programs. Interactive radio has been successful in both formal and informal educational settings. In informal settings, interactive radio provides expert instruction and does not require the presence of trained teachers. In remote rural villages in the Dominican Republic, children without schools receive a basic primary education through interactive radio.

Sources: Friend 1989; Fryer 1989; Lockheed and Hanushek 1988.

Factors in Children's Teachability

HOME ENVIRONMENT. School learning is a joint process involving the home and school. This is nowhere more evident than in the early years of formal schooling. Family background affects the probability that children will enroll in, attend, and complete various levels of education. Data from the Philippines indicate that the occupational and educational level of the parents has shaped the school attainment of their children, with the same level of magnitude, since the early twentieth century (Smith and Cheung 1986).

Similar effects have been reported for Brazil, Indonesia, and Nepal (Chernichovsky and Meesook 1985; Jamison and Lockheed 1987; Psacharopoulos and Arriagada 1989). Family background also affects the learning of children in school. Children whose homes provide a stimulating environment, full of physical objects and learning materials, consistently learn more quickly in school than children from more deprived backgrounds. Schiefelbein and Simmons (1981) found that social class significantly predicted achievement in twenty-eight of thirty-seven studies examining the determinants of achievement in developing countries. The effect of family background on school achievement is most pronounced in subjects that are familiar or linked to parental knowledge (Lockheed, Fuller, and Nyirongo 1989).

NUTRITION. Several studies have explored the relationship between children's nutritional status and school indicators such as age at enrollment, grade attainment, absenteeism, achievement test scores, general intelligence, and performance on selected cognitive tasks, including concentration in the classroom. Three aspects of nutritional status affect achievement adversely: protein-energy malnutrition, temporary hunger, and micronutrient deprivation.

Protein-energy malnutrition is generally caused by a deficient diet, may be exacerbated by the child's parasite load, and is almost always accompanied by poverty. All nine of the studies reviewed by Pollitt (1990) reported a significant relationship between protein-energy nutritional status and cognitive test scores or school performance in China, Guatemala, India, Kenya, Nepal, the Philippines, and Thailand. Consistently, past and present nutritional status (as captured by height-for-age and weight-for-height data, respectively) was linked to higher cognitive test scores or better school performance. Taller children were also likely to be enrolled in school earlier than shorter ones.

One study found that Kenyan children who were comparatively well nourished had higher composite scores on tests of verbal comprehension and intelligence than children who were less well nourished (Sigman and others 1989). Furthermore, better-nourished females were more attentive during classroom observations than malnourished ones. For the children as a group, the best predictors of cognitive scores included food intake (current nutrition) and physical stature (nutritional history). Regardless of the family's social and economic resources, children with more adequate diets scored higher on the cognitive battery than those with poorer diets.

Similarly, in the Philippines, pupils with good nutritional status had significantly higher academic performance and mental ability than pupils with poor nutritional status, even when family income, school quality, teacher's ability, and mental ability were controlled (Florencio 1988). Although the relationship between health and nutritional status and academic achievement varied by grade level and subject matter, a significant positive relationship linked nutritional status to mental ability and academic achievement.

Children who are temporarily hungry—typically a result of not eating breakfast—are generally more easily distracted from their school work than those who have eaten (Pollitt and others 1983). Providing school breakfasts to Jamaican primary students significantly improved attendance and arithmetic scores, but not spelling (Powell, Grantham-McGregor, and Elston 1983). The reason might be that the two subjects require different problem-solving skills; spelling is learned largely by rote, while arithmetic requires the application of rules to novel situations. The presence or absence of breakfast affected achievement in schools with food programs, but not in the control group (Cotten 1985). This may be a consequence of proper targeting: students in the schools with food programs were perhaps at higher nutritional risk than those in the control schools and thus more susceptible to the demands imposed by temporary hunger. Differences in the causes and manifestations of temporary hunger and protein-energy malnutrition are important. Temporary hunger may be an educational problem for otherwise well-nourished children, but the effect of hunger is short-term and generally disappears when the hunger is satisfied.

Three micronutrients generally affect school performance: iodine, iron, and vitamin A. A study of school-age children with endemic goiter in Bolivia, and the effect of orally administered iodized oil on their intelligence and growth, indirectly supported the notion that correcting iodine deficiency improves mental performance (Bautista and others 1982). Another study examined the effect of iodine deficiency on mental and psychomotor abilities in Indonesian children and found significant cognitive performance differences among nine- to twelve-year-olds and similar but not significant differences among six- to eight-year-olds (Bleichrodt, Drenth, and Querido 1980). When educational background was controlled, however, few significant differences were reported. Research in Java showed that iodine-deficient children over the age of nine performed less well on tests of intelligence, motor skills, concentration, perception, dexterity, and response orientation than a matched iodine-replete population (Querido and others 1974).

Iron deficiency is likely to affect a child's alertness, which in turn affects attention and learning (Pollitt 1990). Iron deficiency also impairs the higher cognitive processes, such as conceptual learning, of preschoolers (Popkin and Lim-Ybáñez 1982). Pediatricians often describe iron-deficient children as irritable and uninterested in their surroundings, which inhibits their response to learning stimuli. Although apparently less attentive to environmental clues that facilitate problem solving, iron-deficient children can, once they learn a task, process the information as well as iron-replete children (Pollitt, Haas, and Levitsky 1989). Their motivation to persist in intellectually challenging tasks may be lowered, however, and their overall intellectual performance diminished (Pollitt, Haas, and Levitsky 1989). Iron deficiency also seems to produce behavioral changes that stem from altered brain functions, although the mechanisms related to this phenomenon are unknown.

Vitamin A deficiency has long been associated with nutritional blindness and severe cases of measles. Although total blindness generally precludes children from participating in primary school, lesser vitamin A deficiency can impair their academic performance by increasing night blindness and limiting their field of vision (especially peripheral vision). Vitamin A deficiency has recently been linked to morbidity and mortality caused by diarrheal and respiratory disease, even in children without clinical signs of the deficiency (Sommer, Katz, and Tarwotjo 1984; Tarwotjo and others 1987). Vitamin A deficiency also affects growth, including brain growth, which continues until the ages of seven to ten. Although studies of how vitamin A deficiency affects growth and morbidity have concentrated on preschool children, the same effects can be expected for schoolchildren.

OTHER HEALTH CONDITIONS. Persistent illnesses that contribute to repeated absence from school, heavy parasite loads (which contribute both to school absences and to malnutrition), and hearing and vision impairment adversely affect school learning. Children who are excessively absent from school tend to perform poorly and to drop out prematurely (Weitzman 1987). Absences can be caused by diarrhea and abdominal pain induced by parasitic infections; parasites can also adversely affect children's growth and other indicators of nutritional status. The impact of hookworm on school achievement and the mental development of children was studied in the United States and Australia at the turn of the century. Although their statistical methods do not meet current standards, the studies found that children infected with parasites did not perform as well as other children either in school or on a battery of psychological tests, including tests of general intelligence (Pollitt 1990). Infected children also suffered from inattention and limited persistence at school tasks. Children who had been dewormed made significant and large improvements in performance on all but IQ tests. Vision and hearing problems also place children at significant educational risk. Children whose sensory skills limit their exposure to classroom stimuli cannot be expected to receive optimal benefit from schooling.

The Pervasiveness of the Problems in Developing Countries

Unfortunately, the problems of inadequate preparation for school, malnutrition, and poor health persist throughout the developing world and limit the ability of children to learn in school.

DISADVANTAGED HOME ENVIRONMENT. Teachers often expect that children entering school know how to behave in a classroom setting and have been exposed to books, numbers, or complex verbal directions. Most children in developing countries lack such knowledge and exposure. In many countries, preschool children are carried by their mothers and have few chances to explore their environment. Moreover, their parents are often illiterate and unable to provide them with school-related or literacy-nurturing activities (Pollitt 1984a). Their homes lack books and magazines, maps, puzzles, crayons, and developmentally appropriate toys. While schools require abstract and representational use of language, the interaction between parents and preschool children in developing countries tends to be simple and concrete interchanges about the here and now (Haglund 1982). In short, what is provided by the home environment of most children in developing countries does not begin to match what is demanded by the primary school.

MALNUTRITION. In developing countries, malnutrition is often endemic. Conditions of particularly high prevalence include protein-energy malnutrition and micronutrient deficiencies (see table 3-8; Ashworth 1982; Pellett 1983). A synergistic reaction between protein-energy malnutrition and infection has been reported in several studies (Chen and Scrimshaw 1983). Temporary hunger is also undoubtedly widespread in both industrial and developing nations, although data on prevalence are not available. Essentially, temporary hunger is associated with short-term fasting, most typically when children do not eat breakfast.

Within the developing world, protein-energy malnutrition is the most prevalent nutritional problem. Epidemiological information on protein-energy malnutrition among school-age children is relatively scarce (Pollitt 1990). However, studies conducted in China, India, Kenya, Nepal, and the Philippines confirm that malnutrition is pervasive, particularly among poor, rural populations (Agarwal and others 1987; Jamison 1986; Moock and Leslie 1986; Sigman and others 1989; Tragler 1981). In Kenya, for example, more than 30 percent of children in school were stunted and underweight, while in Nepal the proportion of boys below the 75th percentile of weight for age (a measure reflecting current status and prior history) ranged from 59 percent (six-year-olds) to 84 percent (ten- to eleven-year-olds). Only 13.5 percent of the Uttar Pradesh, India, sample had normal heights and weights for their ages.

Micronutrient deficiencies (depletion of the body's store of various nutrients) are another widespread phenomenon in developing nations. Iron deficiency is the most common problem in many areas, afflicting an estimated 680 million people in Asia, 60 million in Africa, and 60 million in Latin America, and putting them at risk for related disorders (such as goiter, endemic cretinism, psychomotor and intellectual retardation, and impaired mental function) (Hetzel, Dunn, and Stanbury 1987). Among children between five and twelve, the prevalence of iron deficiency is estimated to be 49 percent in Africa, 6 percent in Latin America, 22 percent in East Asia, and 50 percent in South Asia (Pollitt 1990). Where it is

Table 3-8. Percentage of School-Age Children with Nutritional Deficiencies and Parasites, Various Years

Country	Chronic malnutrition[a]	Iron deficiency	Iodine deficiency	Parasites
Low-income countries				
Bangladesh	71.0	74.0	—	—
Bhutan	—	—	47.0	—
Burkina Faso	29.3	—	—	90.0
Burma	—	—	70.0	86.7
Burundi	—	4.2	0.6	—
China	5.0	86.9	—	2.4
Ethiopia	—	—	19.0	71.0
India	48.4	69.4	55.0	48.0
Kampuchea, Democratic	67.3	—	—	—
Kenya	25.0	13.0	38.5	35.0
Mauritania	—	—	—	22.6
Nepal	67.0		74.0	—
Niger	—	—	—	25.0
Senegal	—	—	—	22.0
Somalia	9.4	—	—	61.8
Sri Lanka	—	—	12.0	—
Sudan	—	—	70.0	45.5
Tanzania			48.0	55.0
Zaire	55.0	—	—	40.0
Zambia	28.1	13.2	—	45.0
Lower-middle-income countries				
Bolivia	26.2	11.2	65.3	—
Cameroon	—	—	75.0	41.4
Chile	49.7	6.7	19.0	—
Colombia	—	—	53.0	—
Costa Rica	11.3	—	3.5	30.5
Côte d'Ivoire	—	—	—	51.0
El Salvador	33.6	—	18.0	—
Egypt	13.5	—	70.0	—
Guatemala	37.4	—	10.6	—
Honduras	39.8	—	—	—
Indonesia	69.9	—	72.5	—
Jamaica	—	4.2	—	—
Lebanon	6.2	32.0	19.6	—
Nicaragua	22.0	—	—	—
Nigeria	—	—	58.0	50.0
Paraguay	—	—	26.5	—
Peru	49.0	—	22.0	43.0
Philippines	59.4	20.6	15.1	87.7
Syria	14.0	45.1	—	—
Thailand	8.0	11.4	23.5	—
Yemen, Arab Republic of[b]	—	—	—	28.0
Yemen, P.D.R.[b]	41.8	—	—	—
Zimbabwe	14.6	—	—	60.0
Upper-middle-income countries				
Argentina	—	—	3.2	—
Brazil	—	—	14.7	15.0
Korea, Republic of	18.1	—	—	—

Country	Chronic malnutrition[a]	Iron deficiency	Iodine deficiency	Parasites
Malaysia	52.7	—	—	89.0
Panama	21.9	—	—	—
South Africa	49.0	—	—	1.6
Trinidad and Tobago	—	17.6	—	—
Uruguay	—	—	10.0	—
Venezuela	6.6	—	13.4	—
High-income oil exporters				
Kuwait	6.1	19.3	—	—
Libya	—	—	46.0	18.0
Others[c]				
Antigua and Barbuda	—	17.9	—	—
Barbados	—	9.3	—	—
Bahrain	6.9	0.3	—	—
Cuba	—	—	30.0	—
Cape Verde	23.4	—	—	—
The Gambia	—	—	100.0	—
Grenada	23.0	—	—	—
Guyana	—	57.0	21.2	—
St. Kitts and Nevis	—	22.3	—	—
St. Lucia	—	36.4	—	—
Seychelles	10.7	—	—	—
Turks and Caicos	—	86.0	—	—

— Not available.

Note: This table lists available data on prevalence for school-age children. The studies refer to different times, employ different sample sizes, and use different standards. Care should be used when comparing among countries or among deficiencies.

a. Defined by height-for-age and anthropometric measurements.

b. At the time that data were gathered for the Republic of Yemen, the country was divided into the Arab Republic of Yemen and the People's Democratic Republic of Yemen.

c. "Others" include countries that do not report data or have populations of less than 1 million.

Sources: Florencio 1988 and 1989 for Philippines data; Galloway 1989 for all other data.

prevalent, iron deficiency anemia is generally attributed to the low intake of dietary iron and chronic blood loss due to hookworm, malaria, and schistosomiasis.

Although data on the prevalence of vitamin A deficiency have mostly been collected for preschool-age children, the scattered data on schoolchildren suggest that the problem is significant. For instance, in the Indian state of Uttar Pradesh 4.1 percent of schoolchildren had ocular signs of vitamin A deficiency, which indicates an advanced condition; in Ethiopia 4 percent of schoolchildren were affected by night blindness (Agarwal and others 1987; FAO 1988). Data from Tanzania show that two decades ago 17 percent of school-age children had dangerously low levels of serum vitamin A (FAO 1987).

OTHER HEALTH PROBLEMS. Two other important health problems plague schoolchildren in developing countries: parasites (see table 3-8) and hearing and sight impairment.

Helminths (Ascaris, Trichuris, and hookworm) are highly prevalent among school-age populations in developing countries, although precise estimates of pervasiveness have not been made. The profound effect of helminth infections on health and nutrition suggests that infected children are at serious educational risk because of unfavorable biochemical changes, altered immunological activity, and structural changes in organs such as the intestine and liver (Pollitt 1990).

Another parasitic infection that affects about 200 million individuals throughout the world, many of them school-age children, is schistosomiasis. This disease causes systemic changes, has pathological consequences, and adversely affects nutrition. Clinical features of the disease at various stages include fever, weakness, lassitude, muscular pain, nausea, vomiting, diarrhea, and fatigue (Pollitt 1990). Despite the lack of conclusive research, there can be no doubt that schistosomiasis impedes school attendance and achievement.

Many schoolchildren also have impaired sight and hearing, given the relationship between sensory function, on the one hand, and infection and nutrient intake, on the other. No comprehensive data are available for primary students, but a study in the Philippines revealed that 6–7 percent of those tested were visually impaired (Florencio 1988). This study also found the highest proportion of students with poor eyesight in the first grade. The percentage of children with normal vision increased as the grade level rose. This suggests that children with poor eyesight may drop out of school or repeat a grade at higher rates than their schoolmates with normal vision. Indeed, mental ability and visual ability were the two most robust predictors of academic achievement for the sample studied.

Hearing acuity was also examined in the Philippines; 13.5 percent of students had some degree of impairment. Hearing loss may be linked through subclinical hypothyroidism to mild iodine deficiency (Pollitt 1990). In China, Yan-you and Shu-hua (1985a, 1985b) found that the mean level of hearing for schoolchildren in an iodine-deficient, remote area was significantly lower than that of a comparison group of children. After three years of iodine supplementation, hearing differences between the groups disappeared.

Policy Responses

Preschools and nutrition and health interventions are designed to compensate for deficits in the experience, nutrition, and health of children.

PROMISING AVENUE: PRESCHOOLS. Children who come to school unprepared for the demands of formal education will not profit from instruction; many will leave school altogether. Preschool can prepare children for primary school and provide the necessary transition between home and formal schooling (Halpern and Myers 1985; Myers forthcoming).

Most evidence on the effect of preschool comes from the United States, and the most publicized results are from exemplary programs conducted under ideal conditions. More useful for consideration here are studies of the Head Start program (see box 3-6). Although the resources for Head Start are more abundant than would be the case in developing countries, Head Start is a useful example because it is a national program serving more than 450,000 children at 1,300 sites. The sustained effects of Head Start are encouraging, but not dramatic (Haskins 1989). Proponents point out that this is not surprising since "we simply cannot inoculate children in one year against the ravages of a life of deprivation" (Zigler 1987).

In developing countries, particularly middle-income countries, preschools have proliferated over the past two decades (see table 3-9; O'Connor 1988). In Mexico, for example, between 1970 and 1980 the number of preschools rose from about 3,000 to about 13,000 and then doubled to 26,000 by 1982 (Gorman, Holloway, and Fuller 1988). Despite this proliferation, the influence of preschool experience on school achievement has not been widely studied in developing countries (Myers forthcoming).

Halpern and Myers (1985) cite six studies of varying adequacy that were conducted in Latin America, as well as informal observations from Bangladesh, Brazil, Haiti, and Malaysia. They conclude that preschool has a modest but positive influence on initial adjustment to the demands of primary school in developing countries. The scattered studies from other developing countries confirm this picture (Kagitcibasi 1983). Brazil's PROAPE program illustrates the effect of combining preschool stimulation with supplementary feeding programs in a developing country: 73.5 percent of students who had been PROAPE participants successfully completed the first and second grades, compared with 59.5 percent of students who had not participated. Moreover, the academic performance of children involved with PROAPE for two years was consistently higher than that of children outside the program, ranging from 2 to 21 percent in three variations of the model (Didonet 1980; Halpern and Myers 1985).

One type of preschool generally neglected in discussions is the Islamic, or Koranic, school attended by millions of children in dozens of countries, often

Table 3-9. Median Preschool Enrollment Rate, Selected Years, 1965–80

Country income level	*1965*	*1970*	*1975*	*1980*
Low	0.0	0.0	0.0	1.0
Lower middle	1.0	2.0	3.0	4.5
Upper middle	5.0	6.0	14.0	15.0
High (excluding oil exporters)	13.0	19.0	26.5	25.5

Source: See table A-23 in the appendix.

Box 3-6. The Head Start Program in the United States

Begun in 1965 to combat poverty by promoting the development and school readiness of poor children, the Head Start program has endured into the 1990s and still serves nearly half a million children each year. Despite differences between conditions in the United States and developing countries, the size and longevity of Head Start makes the program relevant to the world community.

In the prototypical Head Start classroom, the teacher is trained in early childhood education; the aides are often parents who have been trained and are paid for their time. Children select from among activities the teacher has prepared and activities they initiate themselves. They spend most of the time working individually or in small groups. The teacher and aides move among the children and facilitate learning by asking questions, offering suggestions, and adding more complex materials, ideas, or language to the children's play. Children are allowed to acquire important skills at their own pace. These basic curriculum features are not unique to Head Start; they characterize many developmentally appropriate programs for preschool children.

The Head Start program goes beyond the classroom to provide health, nutrition, and social services for children and their families. Head Start stresses parental involvement in the belief that the most lasting effects are achieved by strengthening the parenting skills and self-esteem of parents.

There is evidence that Head Start has positive effects both in the short term and the long term. Children initially make large improvements on intelligence and achievement tests immediately after the program, but these gains tend to dissipate in a year or two. Nevertheless, Head Start has a lasting effect on school attendance, promotion, special education placement, and remaining in school. Since poor attendance and dropping out of school are prevalent among low-income students in the United States and even more common in most developing countries, improving attendance and school completion is crucial. These positive outcomes for Head Start participants, along with lower rates of grade repetition and placement in remedial classes, suggest that developmentally appropriate preschool experience can help prepare low-income children to meet the demands of school.

In fact, the children most at risk for educational failure may benefit the most from preschool programs like Head Start, at least in their school readiness. Children scoring below average before Head Start gained the most on two measures clearly relevant to school adjustment: a test of children's self-control or ability to control impulsive behavior, which is required in primary school, and a preschool inventory of social and cognitive skills that aid adjustment to school (see figure 3B-1). Providing disadvantaged children with the rudimentary social

before they enter public primary school (Wagner 1989). A five-year longitudinal study in Morocco determined that attending Koranic preschool significantly promoted literacy during the early grades of public primary school (Ezzaki, Spratt, and Wagner 1987).

competence and knowledge for coping with the school setting appears to be one of the useful contributions that preschool intervention can make.

Figure 3B-1. Gains in Children's Readiness for School as a Result of Head Start, 1969–70

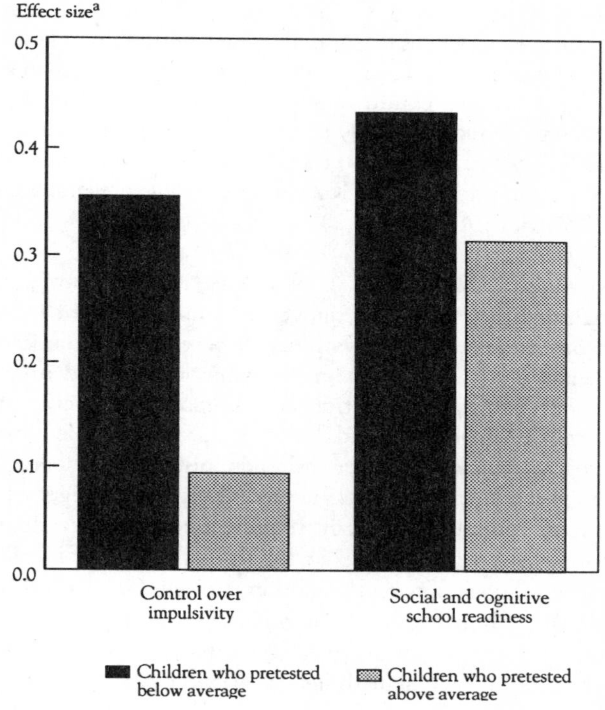

a. The effect size was calculated by taking the mean score of Head Start children, subtracting the mean score of children who did not attend Head Start, and dividing by the standard deviation of the latter group.

Sources: Haskins 1989; Lee, Brooks-Gunn, and Schnur 1989.

At the same time, the benefit of preschool often seems to be overridden by the primary school systems themselves: promotion quotas, negative teacher attitudes, poor instructional quality, and resource inadequacies all hinder learning (Halpern and Myers 1985). Preschool experience plays a role in the early progress of children in school, but only within a limited range. In many areas, the value of preschool intervention will be marginal unless primary schools are improved. The influence of preschool will be far more powerful when combined with upgraded primary education, education of the parents, and other efforts to improve the health and welfare of children, families, and communities (Hildebrand 1986).

Providing formal preschool experiences for all children is unlikely to be affordable for most developing countries. However, targeted preschool programs and alternatives to state-funded programs should be considered. Targeting preschool experiences to, for example, the children of families with particularly low incomes or children residing in poor rural or urban areas has been effective in developed countries and is being suggested in developing countries. Private preschools, possibly provided by nongovernmental organizations, should be encouraged in countries whose education budget cannot afford to support public preschools. Koranic preschools are one example of this approach (Wagner 1989).

PROMISING AVENUE: NUTRITIONAL INTERVENTIONS. Treatment of protein-energy malnutrition, temporary hunger, and micronutrient (specifically iron, iodine, and vitamin A) deprivation can be efficiently undertaken in school. Supplementary feeding is the most commonly applied intervention for treating protein-energy malnutrition and temporary hunger, while iron, iodine, and vitamin A supplements and deworming are the prevalent treatments for micronutrient deficiencies; all can be easily administered in schools.

Modest school snacks or breakfasts alleviate short-term hunger and its adverse impact on emotional behavior, arithmetic competence, reading ability, and physical work output (Pollitt 1984a; 1984b). The study of breakfast programs in Jamaican schools supports this notion (Powell, Grantham-McGregor, and Elston 1983). Although the cost is high, the benefits may be high as well, especially for nutritionally deprived students who are confronted with complex cognitive tasks (such as arithmetic) early in the day.

Supplementing iron, iodine, or vitamin A should be given a high priority where deficiencies of these micronutrients are prevalent. Regardless of the treatment (fortification, supplementation, or deworming), reducing iodine, iron, and vitamin A deficiencies is particularly cost-effective because the learning deficits related to them are both serious and reversible (Pollitt 1990). Supplementation reduces several associated conditions that impede learning. Cost-benefit ratios are favorable, benefits are sustainable, and targeting is relatively simple. Absorptive capacity and infrastructure requirements exist but are not onerous. Iodine can be supplemented inexpensively. A dosage can be administered once every two years at a cost of $0.12 per child with minimal risk of toxicity (Unesco 1989). As a long-term measure, however, fortification may be preferable to supplementation. Areas where iodine deficiency is prevalent can be readily targeted, which significantly increases the likelihood that children who are at risk will receive supplementation.

Iron supplementation for reducing anemia (which is, however, inappropriate in areas where malaria is endemic) is highly cost-effective, with a benefit-cost ratio substantially greater than 1. Levin (1985) studied three countries and found the benefit-cost ratio for dietary fortification to be between 7 and 70; the ratio for dietary supplementation ranged from 4 to 38. These benefits apply, however, not to educational outcomes, but to future earnings.

Vitamin A supplementation is effective, easy to administer, and inexpensive. A 75 to 80 percent reduction in the prevalence of the deficiency has been repeatedly associated with the universal distribution of vitamin A capsules (Berg and Brems 1986). Distributing vitamin A capsules in Indonesia twice a year reduced overall child mortality by one-fourth to one-third (Sommer, Katz, and Tarwotjo 1984). The cost of a massive dose of vitamin A in a capsule is low, about $0.05. Since students need to ingest them only twice a year, this intervention does not infringe on the teacher's other duties.

BLIND ALLEY: SCHOOL LUNCHES. Providing school lunches, rather than breakfasts or snacks, is of questionable value. Such feeding programs are rarely designed to meet nutritional goals; more often they are designed to offset the negative effect of hunger and undernutrition among preschool and school-age children, improve attendance, and transfer income to the poor. Only a few statistically sound studies have examined the effect of school lunch programs on enrollment and achievement, and the available analyses fail to confirm a clear relationship between school feeding and educational results. Ample evidence suggests that school feeding programs improve attendance (which may be beneficial to school performance), but this improvement may be a function of their impact on transferring income to poor families rather than of their nutritional value. Assessments of the long-term benefits of school lunch programs have, in general, failed to yield statistically meaningful findings on the improvement of nutrition. This may be because they operate a relatively small number of days (seldom more than 150 per year), food is substituted at home, rations are diluted at school, or the programs are poorly targeted (Levinger 1983). In particular, protein-energy malnutrition cannot be treated in the days available for school lunch programs, and those programs do not improve nutrition enough to warrant their relatively high cost.

The logistical demands of most school lunch programs create further problems. Although targeting is easy, ensuring that food reaches the children targeted and is not substituted by other food is frequently a problem, and the dilution of rations often mitigates the impact of targeting within schools. School lunches also require a well-developed infrastructure. Providing school lunches is a relatively expensive intervention even when the food and other commodities are donated; teaching time is lost as well.

PROMISING AVENUE: TREATING AND IDENTIFYING OTHER HEALTH PROBLEMS. Two other health problems are also appropriate for in-school treatment or assessment: parasitic infections and sensory (hearing and sight) impairment. Reducing a child's parasite load can increase school attendance and improve learning by reducing apathy and lethargy and may be a cost-effective approach to treating caloric, protein, and iron deficiencies in areas where parasitic infection is rampant. Reducing the parasite load through deworming costs only $0.25 to $0.40 per treatment (Unesco 1989). Deworming benefits both the treated individual and the community since environmental contamination (and thus the risk of reinfection) is reduced. Secondary effects of parasite treatment, however, must be taken into account. Maximum benefit-cost ratios are achieved when deworming is combined with sanitation, a clean water supply, and health education.

Visual and auditory screening, through simple eye charts and whisper tests that can be used at a negligible cost, can identify children who are sensory impaired. When problems are identified, teachers may apply classroom management techniques to compensate for the impairment, such as placing impaired students at the front of the room, monitoring the classroom performance of at-risk students, and being attentive to noise levels.

The relative priority is moderate. Given the limited resources available, schools will probably mediate rather than correct or treat the conditions identified through screening. Cost-benefit ratios are favorable only if the appropriate classroom management techniques are applied once a problem is discovered.

Summary

To determine which investments are most useful for improving learning achievement, this chapter presented what is known about curriculum, learning materials, instructional time, teaching, and children's teachability. For each of these areas, it described interventions that are cost-effective (promising avenues) and those that are ineffective, too costly, or not readily implemented (blind alleys). Table 3-10 summarizes these interventions.

Minor modifications that do no more than adjust the intended curriculum—what might be called tinkering about the edges—are considered to be blind alleys. The educational system in most developing countries needs a substantial overhaul of the curriculum: that is, a coherent, appropriately paced and sequenced instructional program for each subject taught. The most promising avenue for improving learning materials is to increase the quality and availability of textbooks and teacher guides. At this point, computers are not a sound investment in most developing countries given their high cost. Time for learning is critical and can be increased most effectively by setting and maintaining standards for instructional time. Although lowering class size below twenty students increases instructional time, reducing class size within the range that most developing countries can realistically attain will

Table 3-10. Interventions to Improve Learning Achievement and Estimated Annual Marginal Cost per Student
(constant 1985 dollars)

Variable	*Promising Avenue*	*Blind Alley*
Curriculum	Improve implemented curriculum ($?)	Adjust intended curriculum ($?)
Learning materials	Provide textbooks ($2–15) and teacher guides ($1)	Provide computers in the classroom ($120)
Teaching time	Require at least twenty-five hours of instruction per week for core subjects ($1)	Reduce class size from forty to twenty students ($32)
Teaching quality	Offer in-service training ($2) Use interactive radio instruction ($1–2) Supply programmed materials ($15)	Require lengthy preservice pedagogical training (see chapter 4)
Teachability	Target preschools at the disadvantaged ($32) Provide school snacks/breakfasts ($6) Supplement micronutrients ($0.50) Treat parasite infections ($2–4) Provide vision and auditory screening ($0)	Provide school lunches ($12)

Note: Cost figures for targeting preschools and reducing class size are based on average annual recurrent costs of $32 per student in developing countries (see table A-20 in the appendix).

Sources: Lockheed and Hanushek 1988; Levin and others forthcoming.

not have enough impact to be cost-effective. The most promising avenues for improving the quality of teaching received by primary-level children in developing countries are to offer in-service teacher training, supply programmed materials, and use interactive radio instruction.

Several avenues are useful for improving the teachability of children in developing countries. Because malnutrition and hunger limit children's learning, schools should offer breakfasts or snacks; iron, iodine, and vitamin A supplements; and treatment for parasites. However, no long-term advantage has been demonstrated for providing school lunches. Screening for visual and auditory problems is important for increasing the teachability of children, as are preschool programs, especially for poor children. Since the children with the greatest need seem to benefit most, preschool services should be targeted to the poorest children and those who live in rural and remote areas.

Are all these promising avenues equally important, or should priorities be set for these quality-enhancing interventions? Given limited resources, countries will probably have to choose among interventions or seek ways to mitigate a particular deficiency by increasing the amount of other inputs; countries will also need to identify the point at which returns begin to diminish. To set priorities, countries will need information on the full range of inputs—their effects on learning achievement, their cost, and their effectiveness relative to cost. Unfortunately, data on both effectiveness and cost are rarely available.

Conclusions based on cost or effectiveness alone are often misleading, since relatively inexpensive alternatives may be less effective and relatively more effective interventions may be more costly (Levin 1987; Levin, Glass, and Meister 1984; Lockheed and Hanushek 1988). Although some educational interventions can double or even quadruple annual recurrent costs per pupil (table 3-10), few educational interventions can double learning achievement; most increase achievement by 15–50 percent of a standard deviation. Thus interventions such as placing computers in the classroom and lowering the student-teacher ratio from 40:1 to 20:1 (400 and 200 percent of the annual recurrent cost per pupil, respectively) are unlikely to be as cost-effective as investing in textbooks, micronutrient supplementation, and interactive radio instruction (10–50, 1, and 5 percent of the average annual recurrent costs per pupil, respectively).

Notes

1. The correlation is given by the following equation, with *t*-statistics in parentheses below each coefficient:

$$\ln Y = 1.22 \;-\; \underset{(-3.19)}{0.085 \ln Z} \;-\; \underset{(-3.1)}{0.304 \ln \text{GNP}} \;-\; \underset{(-0.78)}{0.169 \ln \text{STR}}$$

$R^2 = 0.538$

$N = 22$

Y = number of school years per graduate relative to the length of the primary education cycle

Z = government expenditures on teaching materials per student per year

GNP = GNP per capita

STR = student-teacher ratio

2. External efficiency refers to the ratio of monetary outcomes, such as earnings, to monetary inputs (Lockheed and Hanushek 1991).

3. Benavot and Kamens (1989) differentiate two types of official languages: national indigenous languages that are spoken by more than half of the population and are the official language (for example, Chichewa in Malawi) and metropolitan or world languages that are the official language but not indigenous in origin (for example, French in the Central African Republic).

4. Two decades ago, the medium of modernity was television, which was introduced into schools in several developing countries. Although the potential for improving student achievement existed, implementation difficulties and failure to

sustain the intervention after external aid was removed brought about the demise of primary-level instructional television.

5. Learning time means the hours in which students are in contact with the teacher; time on task, which includes an element of student motivation, is included in learning time.

6. At the lower secondary level, actual class size can be compared with student-teacher ratios. In sixteen industrial and developing countries, the size of the eighth-grade class was, in all cases, larger than the secondary school's student-teacher ratio. In most cases, the observed class size was 50 percent greater than the student-teacher ratio.

7. These numbers are based on a teacher being able to help an individual student at a rate of 8 percent. That is, if a teacher can provide help 100 times in an hour, any one student will require help eight times in that hour. Hence in a class of 100 students, a teacher can offer help only one-eighth (12.5 percent) of the time.

8. Studies of the effect of homework on achievement in the United States fail to confirm consistent, positive effects (Rickards 1982). Even studies in which students are randomly assigned homework and no homework provide inconclusive results. Austin (1978) found sixteen achievement comparisons that significantly favored the homework group and thirteen that did not. Friesen (1979) reported twelve experiments demonstrating positive effects of homework on achievement and eleven experiments showing no effects or negative ones. Nonexperimental studies (Walberg, Paschal, and Weinstein 1985; Walberg 1986) often show positive correlations between a student's report of doing homework and student achievement, but the directionality of this effect is questionable.

4

Improving the Preparation and Motivation of Teachers

TEACHING QUALITY AND TEACHING TIME are key determinants of student achievement. Teaching time is largely determined by teacher motivation, while the fundamental prerequisites for proficient teaching are

> a broad grounding in the liberal arts and sciences; knowledge of the subjects to be taught, of the skills to be developed, and of the curricular arrangements and materials that organize and embody that content; knowledge of general and subject-specific methods for teaching and for evaluating student learning; [and] knowledge of students and human development. (National Board for Professional Teaching Standards 1989, p. 45)

Yet the teaching force in many developing countries is neither motivated nor trained. Most prospective teachers lack adequate general academic preparation, both new and experienced teachers lack many pedagogical skills, and motivation and professional commitment to teaching are low. Chapter 3 discussed strategies for improving the knowledge and teaching practices of incumbent teachers. This chapter presents strategies for improving the knowledge and skills of new teachers and the motivation of all teachers.

Governments and teacher training institutions face the daunting challenge of preparing, training, and retraining vast numbers of primary teachers. More than 19 million primary teachers are employed worldwide, with 8.9 million in low-income countries alone. Moreover, the teaching profession is likely to expand dramatically in the next decade, particularly in low-income countries. The projected growth of the school-age population will require the training and employment of hundreds of thousands more primary school teachers if the current level of enrollment is to be maintained. To give pri-

Table 4-1. New Teachers Required to Achieve Universal Access to Primary School by 2000
(thousands)

Country income level	*Estimated population age 6–11 in 2000*	*Teachers needed in 2000*	*Teaching force in 1985*	*Teachers lost through attrition by 2000*	*New teachers needed by 2000: Unadjusted for attrition*	*New teachers needed by 2000: Adjusted for attrition*
Low (excluding China and India)	154,448	3,965	1,727	1,076	2,238	3,314
China and India	268,314	8,335	7,244	3,473	1,091	4,564
Lower middle	136,570	4,421	3,543	1,756	878	2,635
Upper middle	97,642	3,956	3,778	1,725	178	1,902

Note: Data assume existing country-specific population growth rates, current student-teacher ratios, and nominal (3 percent) attrition of teachers.

Sources: See tables A-1 and A-9 in the appendix.

mary school–age children universal access to education by the year 2000, about 3.3 million new teachers will have to be trained in low-income countries (other than China and India), assuming that population growth and student-teacher ratios stay at their present levels and the teacher attrition rate is 3 percent. Each year the teaching force in these countries will have to grow at least 6 percent. Another 4.5 million teachers will be needed in middle-income countries (see table 4-1).[1] However, the present annual growth rate for new teachers is about 5 percent in middle-income countries, which compensates for an equally high attrition rate. As a result, middle-income countries are probably producing more teachers than they need.

To avoid producing new teachers with the same inadequate skills and professional commitment as many incumbent teachers, developing countries must design policies that (a) raise the level of knowledge of prospective teachers, (b) increase the pedagogical skills of new teachers, and (c) improve the motivation of all teachers. To improve the knowledge and skills of new teachers requires changing recruitment practices and preservice training; to improve teacher motivation requires restructuring the incentives for teachers to perform well. Low teacher competence and poor motivation are the result of the low status afforded the teaching profession in many countries. Status plays an important role in attracting academically prepared candidates and in encouraging them to remain teachers. Status depends on how society and prospective teachers perceive the extrinsic compensation and conditions of the workplace and the intrinsic rewards of professional accomplishment.

Historically, teaching was a highly regarded profession—in Korea, for example, teachers received the same honor as the king and as parents (APEID 1984b; IEES 1986). This is not the case in most developing countries today; the status of primary school teaching is low and has declined considerably in the past two decades (Warren 1990). This low status—manifest in low salaries, poor working conditions, and uncertain career paths—means that the most able students do not become primary school teachers; they either become teachers in secondary or higher education or they enter the private wage and public administration sectors, which offer more competitive salaries and better prospects for promotion.[2] Countries whose education systems are expanding rapidly recruit underqualified applicants to meet the growing demand for teachers, a practice that further lowers the prestige of teachers and teaching. As a result, teaching in primary schools neither attracts nor retains the best-qualified and most-motivated individuals.

How students rank a given occupation as a future career indicates their motivation to enter it. Students, even those enrolled in teacher training programs, rarely rank teaching as their first occupational choice. A survey in Turkey, for instance, found that only 4 percent of male and 7 percent of female students taking the university entrance examination in 1986 ranked teaching as one of their top three occupational choices (Murray 1988). This marked a decline from 9 percent and 16 percent, respectively, in 1982. In Liberia three-fourths of the teacher candidates surveyed had entered a teacher training program because they perceived no other option for a job (Liberia, Ministry of Education 1989). In Zimbabwe only 2 percent of the students surveyed in form four wanted to enter a primary teacher training program; the majority aspired to enter the university (Chivore 1986).

Although governments can rarely alter the national labor market in which they recruit candidates, they can improve the standing of the teaching profession within that market and raise its status. Raising the status of teaching would significantly strengthen the capacity of governments to recruit more competent teachers. By giving the teaching profession more attention, governments could increase public awareness of its importance at very little cost. Political leaders could emphasize the intrinsic benefits of serving society and growing professionally, which are benefits that teachers themselves associate with teaching in primary school. Governments in Zimbabwe and Nicaragua, for example, have made a concerted effort to highlight the value and importance of teaching (see box 4-1; Commonwealth Secretariat 1984; MacAdam 1984). Raising the status of teaching requires policies that strengthen the knowledge base of prospective teachers, enhance their teaching skills, and improve the conditions under which they work.

General Academic Background

A first determinant of teachers' effectiveness is their general academic preparation. In most developing countries, prospective teachers have only about nine years of general education and are often not the stongest students academically (Gimeno and Ibáñez 1981; Zymelman and DeStefano 1989; Cameron and others 1983).

Incomplete Academic Preparation in Secondary School

In developing countries prospective primary teachers have rarely completed a secondary education, although two-thirds of the developing countries that reported data in 1988 require candidates to have some upper secondary educa-

Box 4-1. Recruitment Incentives: Zimbabwe's Integrated Teacher Education Course

Governments must recruit enough teachers to meet their educational objectives. Zimbabwe's Integrated Teacher Education Course (ZINTEC) is a successful program that both attracts and retains new teachers.

In Zimbabwe the introduction of free primary education in the wake of independence created an unprecedented rise in the number of children attending primary school. Primary school enrollment increased from 800,000 in 1980 to more than 2 million by the end of 1982. To meet demand, the government increased the teaching corps from 21,000 to 54,000 by 1983. Because about 15,000 of these teachers had no training, the government also introduced a program to expand and train the teaching force rapidly. The result was a four-year sandwich program in which students attend a sixteen-week residential course, teach ten terms, and finish with another sixteen-week residential course. During the period of on-the-job training, students continue to study through a correspondence course and their teaching is assessed by visiting monitors. By 1986 more than 8,000 students had graduated from the program, which operated at four colleges.

The program emphasizes nonfinancial incentives. During the out-placement period, clusters of three or more students are deployed in each school. This encourages students to interact, maintains their enthusiasm, and facilitates school-based tutorials. The course curricula emphasize community projects—the construction of Blair toilets in rural villages is a notable example—and helps teachers play a pivotal role in community affairs. The ZINTEC project is considered a shining example of "Zimbabweanization" and a home-grown success. By emphasizing sacrifice and self-reliance and rejecting the concept of learning as a privileged commodity, the program benefits the teaching profession as a whole and inspires new applicants to become teachers.

Substantive material and professional inducements also encourage individuals to participate in the program. Tuition, food, and lodging are free during the residential courses, and correspondence materials are free during the period of out-placement. Students receive a stipend on joining the program, which increases as they complete successive parts of the cycle. On graduation, students receive the standard salary of a certified primary school teacher; in return, they must serve the government for three years.

Sources: Gatawa 1986; Sibanda 1982.

tion, and many have raised the minimum years of formal education that individuals must complete before entering teacher preparation programs (Unesco 1982). In Togo, for example, only 33 percent of all primary school teachers had completed lower secondary school plus three years of either general secondary school or a teacher preparation program in 1970; the proportion meeting this standard had risen to 77 percent by 1988. In India 75 percent of all primary teachers had completed lower secondary school plus three years in 1970, while 88 percent had by 1988 (Unesco 1982).

Countries with a high rate of population growth and a rapidly expanding primary education system have had to reduce the amount of general education required of individuals entering teacher training. One West African country in the early 1980s, for example, recruited children into teacher training courses at the age of eleven or twelve (Urwick 1987). Crash preservice programs were mounted, and unqualified, temporary teachers, often with no more than lower secondary schooling, were hired. The result of such an approach is that many teachers lack any sort of formal certification. In 1986, 46 percent of primary teachers in Zimbabwe were unqualified (Dorsey 1989). In Nigeria only 9 percent of the primary teachers in Kano State had formal teacher certification (Harber 1984).[3] In Haiti fewer than 10 percent of primary teachers were qualified and fewer than 30 percent were trained (IEES 1987).[4] In Liberia only 47 percent of teachers had received teacher training (Kemmerer and Thiagarajan 1989). Historical practices (for example, having upper primary school graduates teach in the lower primary school and junior secondary graduates teach in the upper primary school) have reinforced the problems. Such practices will not be discontinued until a high level of education has been attained. Some middle-income countries are reaching this stage.

The decline in the academic preparation of prospective teachers means that teacher trainees often lack the intellectual and academic skills they need to perform well in training; and, because the trainees are poorly prepared, teacher training must concentrate on ensuring that they gain a sound knowledge of the curriculum content, often at the expense of developing general and subject-specific pedagogical skills.

Poor Academic Accomplishment

Because prospective teachers have not completed a secondary education, they may lack the knowledge necessary to be effective teachers, and this problem may be growing. The general educational competence of prospective teachers appears to have fallen, even among secondary school graduates. One explanation may be that as access to secondary education has expanded, the academic results have been skewed downward. Prospective teachers generally have among the lowest grades in their graduating class. Rust and Dalin (1990) found that both the average secondary school rank and the average educational achievement of prospective teachers has declined. Murray (1988) reports that in Turkey in 1982, 56 percent of students graduating from secondary school and planning to become teachers had grade point averages between 41 and 60 (out of 100); in 1986, 81 percent did. Anecdotal evidence also suggests that those who choose to enter teacher training are among the least able of their classmates. For example, in China students accepted into postsecondary normal (teacher training) schools are among the worst students accepted for postsecondary education.

Part of the reason may be the inefficient selection process. Education ministries receive many applications for teacher training programs, but have few places to fill. The process of checking qualifications and conducting interviews can take months, and by the time selections are made, many applicants have dropped out of the process and many top applicants have entered universities or pursued other options.

Policy Options

Most training of primary teachers in developing countries takes place at the secondary level, lasts from two to four years, and replicates in large measure the curriculum content of general secondary school (Gimeno and Ibáñez 1981; Zymelman and DeStefano 1989; Cameron and others 1983). The cost of this type of general education is staggering. Requiring prospective teachers to obtain their secondary education in general secondary schools could save developing countries significant amounts of money.

According to recent Unesco statistics, in 42 percent of the countries surveyed, primary teacher training takes place only at the secondary level (Unesco 1982, table 3). Gimeno and Ibáñez (1981) noted that eleven of nineteen low- and middle-income countries had primary teacher training programs that required entering students to have only nine or fewer years of general education; these teacher training programs therefore began at the same level as upper secondary schools. Two-thirds of the programs offered courses that lasted three or more years, approximately the same as secondary schools, and as much as 86 percent of the curricular content replicated that of general secondary education (see table 4-2; Ghani 1990). In Somalia, Thailand, and northern Yemen, less than 15 percent of the curriculum was devoted to developing pedagogical skills and 5 percent or less to practice teaching; the remainder addressed general academic subjects (IEES 1984b, 1984c, 1984d). Substituting teacher training for general secondary education does not improve the preservice academic preparation of teachers. This approach fails to develop a wide range of pedagogical skills and is costly.

Teacher training costs as much as 35 times the annual cost per student of a general secondary education (see table 4-3). Although the difference in expenditures might be justifiable if the curricula were substantially different (teaching pedagogy, for example) or if particularly high levels of material in-

Table 4-2. Length and Content of Primary Teacher Training Programs in Selected Countries

Country	Length of program (years)	Content of curriculum (percent)		
		General education	Professional theory	Practice teaching
Ecuador	2	39	40	21
India	2	40	40	20
Lesotho	3	—	—	33
Malaysia	2	73	8	19
Morocco	1	80	10	10
Singapore	2	64	29	15
Somalia	2	86	11	3
Thailand	2	84	13	3
Yemen[a]	5	80	15	5

— Not available.

a. Data apply only to the former Arab Republic of Yemen.

Sources: Gimeno and Ibáñez 1981; IEES 1984b, 1984c, 1984d; Ghani 1990.

puts were required, it cannot be justified when the curriculum content is similar. The direct cost of teacher training programs is high because they tend to be residential and thus require funds to pay for students' food, lodging, and training salaries. The social costs of teacher training are also high when the students who enter the program decide not to become teachers and use the certificate instead as a key to further education or alternative employment.

PROVIDING GENERAL EDUCATION IN GENERAL SECONDARY SCHOOLS. One way to make primary teacher training less expensive is to introduce a cost-recovery system that would reduce the net costs of the stipends paid to teacher trainees. Such a system, however, might discourage students, especially the rural poor, from entering training. A better approach is to shift what is now a substantial part of teacher training programs—general education—to the secondary schools. The benefits would be multiple. First, a financial savings would be realized, because the cost of educating teachers in general subjects is significantly lower when done in secondary school. Second, scarce teacher training resources could then be concentrated on shorter, well-focused pedagogical training. Third, the social costs of training teachers would also be lowered, because teachers would be less likely to use their certification as a stepping stone to different employment or further education. Fourth, students who receive a solid general education in secondary school are likely to perform well in teacher training.

SHORTENING TEACHER TRAINING. Teacher training programs could also be shortened and focused more narrowly on providing appropriate pedagogi-

Table 4-3. The Annual Cost of Teacher Training as a Multiple of the Annual Cost of General Secondary Education in Selected Countries

Country	*Cost per student*
Low-income countries	
Bangladesh	1.64
Cape Verde	9.07
Central African Republic	9.07
China	8.51
The Gambia	10.43
Ghana	2.96
Guinea-Bissau	6.55
Haiti	6.31
Madagascar	8.60
Mali	12.82
Malawi	3.07
Mauritania	4.61
Nepal	2.65
Pakistan	25.53
Seychelles	0.53
Somalia	0.87
Swaziland	4.28
Tanzania	4.11
Tonga	34.67
Zambia	3.25
Lower-middle-income countries	
Botswana	2.83
Dominican Republic	8.68
Guatemala	1.36
Indonesia	1.10
Liberia	10.12
Nicaragua	3.81
Nigeria[a]	3.21
Average ratio	7.06

Note: Calculations are based on costs in current domestic currencies.

a. Data represent state and not federal institutions.

Sources: Unesco 1982, tables 3.5 and 4.3; Tilak 1989b; Benson 1985.

cal skills which would further reduce the cost of producing a teacher. Since less than 25 percent of the curriculum in many countries is devoted to teaching prospective teachers how to teach, the length of these programs could presumably be cut 75 percent without losing instructional time in this area.

Pedagogical Skill Development

Incumbent teachers lack pedagogical skills in part because the attention paid to general academic education during teacher training detracts significantly from the time available for developing pedagogical skills. The long-term impact is that teachers are unable to deal creatively with the pedagogical challenges of the classroom.

Objectives of Skill Development

Teachers with a wide repertoire of teaching skills are clearly able to teach better than those with a limited repertoire (Fuller 1987; Haddad 1985). Effective preservice training needs to build on a sound knowledge of the curriculum, teach pedagogical methods, and encourage practice teaching under the supervision of an experienced and capable teacher.

Proficiency means that a teacher is competent in the subject matter and understands how to transmit knowledge effectively. Teachers must understand the subject matter for themselves and be able to elucidate that knowledge in new ways, recognize and partition it, and clothe it in activities, emotions, metaphors, exercises, examples, and demonstrations so that it can be grasped by students (Shulman 1987). The goal of teacher education is not to indoctrinate teachers to behave in rigid, prescribed ways, but to encourage teachers to think about how they teach and why they are teaching that way. Teachers must comprehend both the subject and the pedagogical skills needed to promote the exchange of ideas. These pedagogical skills include classroom management and organization, appreciation of each student's characteristics and preconceptions, formal and informal evaluation of students, personal reflection, and critical self-analysis (Shulman 1987).

The Content of Teacher Training

Unfortunately, most preservice teacher training does not develop these skills enough. Most time is spent on general academic courses, and the remaining time is poorly used. Either classes focus on broad theoretical issues rather than on specific strategies to enhance the students' comprehension, or they provide a limited range of pedagogical skills (see table 4-4). In Haiti 23 percent of preservice training was devoted to pedagogical training, but only 8 percent of course time directly taught pedagogical skills; the remainder was devoted to general topics such as psychology or the history of education (IEES 1987). The skills that enhance student achievement, such as asking questions, checking comprehension, and providing feedback, are neglected. Moreover, the educators themselves are poorly prepared and lack the skills to train teachers effectively. A recent internal World Bank report found that in twelve states in Nigeria, only 30 percent of 4,500 instructors in primary teacher training institutions had an undergraduate degree in training primary school teachers. In Zambia only 20 percent of the instructors were more qualified than their students, and 9 percent were actually less qualified (Kelly and others 1986).

Table 4-4. The Curriculum in Teacher Training Colleges in Haiti, Nepal, and Yemen

	Haiti [a]		*Nepal* [b]		*Yemen* [c]	
Subject	*Average hours per week*	*Percentage of curriculum*	*Total credit hours*	*Percentage of curriculum*	*Average hours per week*	*Percentage of curriculum*
Academic training	17.7	53	1,550.0	86	25.4	70
General	—	—	500.0	28	—	—
Special	—	—	1,050.0	58	—	—
Pedagogical training	7.6	23	150.0	8	5.4	15
Introduction to education	0.0	0	50.0	3	0.0	0
Philosophy of education	1.0	3	0.0	0	0.0	0
Professional ethics	0.0	0	0.0	0	0.0	0
General pedagogy	1.0	3	0.0	0	0.0	0
History of education	0.3	1	0.0	0	1.2	3
Theories of education	0.3	1	0.0	0	0.0	0
Psychology	2.3	7	100.0	6	1.2	3
General didactics	1.0	3	0.0	0	0.0	0
School administration	1.0	3	0.0	0	0.0	0
Special pedagogy	0.7	2	0.0	0	0.0	0
Methods of teaching	0.0	0	0.0	0	1.8	5
Instructional materials	0.0	0	0.0	0	1.2	3
Student teaching	0.0	0	100.0	6	2.0	5
Social education	3.3	10	0.0	0	0.0	0
Practical training[d]	4.7	14	0.0	0	3.8	10
Total	33.3	100	1,800.0	100	36.6	100

— Not available.

a. Three-year course.

b. Two-year course.

c. Five-year course. Data apply only to the former Arab Republic of Yemen.

d. Includes physical education, arts and crafts, music appreciation, and agricultural training.

Sources: IEES 1984d, 1987, 1988.

Policy Options

In many countries, teacher training transmits few teaching skills. During the two to three years of training, less than 25 percent of the time is devoted to developing teaching skills, and graduates often perpetuate the poor practices of their instructors when they begin teaching primary school themselves. Three policy options build on a good secondary education to improve pedagogical skills: revising the standards for admittance, emphasizing the development of pedagogical skills, and expanding practice teaching. For incumbent teachers, providing pedagogical support by improving the supervision and advice of the principal is also important.

RAISING THE CRITERIA FOR ADMITTANCE. The chief mechanism for changing the nature of teacher training is to revise the criteria for admission so that teacher training institutions are not obliged to provide a general secondary education to prospective teachers. Thus, wherever possible, admittance to a teacher training program should be restricted to students with a complete secondary education. As an interim measure, some systems may have to admit secondary school completers who have not passed their terminal examinations, but they should only admit individuals who are highly motivated to stay in teaching and prepared to accept a lower starting salary than secondary graduates.

EMPHASIZING PEDAGOGICAL SKILLS. Pedagogical skills can also be improved by emphasizing courses that develop the teacher's ability to reason about the content of instruction. As already discussed, teachers must have a sound knowledge of the curriculum and be able to transfer it to their students. They must be able to analyze critically the material they wish to present, exploit analogies and examples to convey the information, and adapt the material to the interests and abilities of their students. Teachers must be able to organize and manage the classroom and to evaluate, discipline, and encourage students in a manner that promotes learning (Shulman 1987).

INSTITUTING PRACTICE TEACHING. Practice teaching, in which a novice teacher leads a class under the direct supervision of an experienced educator, inculcates the instructional skills that enhance student achievement and prepares prospective teachers to cope with unexpected events in the classroom. The effectiveness of practice teaching depends in large part on the capabilities of the mentor teacher and the management of the program.

Practice teaching can be incorporated into preservice training in three ways: (a) by integrating observation and practice teaching with course work over a period of months or years, (b) by concentrating practice teaching in a block lasting from two to five weeks (in Ethiopia, for example, practice teaching takes place in regular schools for two to three weeks under the supervision of subject-matter specialists; similarly, student teachers in Jordan, Malaysia, Morocco, and Portugal spend five full weeks in practice teaching), and (c) by spending a full year as an intern or supervised teacher. Year-long internships characterize practice teaching in El Salvador, Gabon, Haiti, Jamaica, Lesotho, and Tunisia. In some countries practice teaching follows classroom instruction, while in others it is sandwiched between years of formal courses (Haddad 1985). However it is structured, practice teaching must be taken seriously: prospective teachers must be evaluated on their performance, and successful performance must be a prerequisite for certification.

Motivation

Even competent teachers who are well prepared cannot teach effectively under adverse conditions. Poor motivation, which translates into teacher absences, indifferent classroom practices, and early departure from the profession, impedes a teacher's ability to teach. Governments can improve motivation by improving the professional and environmental conditions under which teachers work.

Teacher Absenteeism

Many countries do not get the best performance from either their incumbent or their new teachers. Lack of motivation and professional commitment produce poor attendance and unprofessional attitudes toward students. Teacher absenteeism and tardiness are prevalent in many developing countries. In Nigeria, for example, government officers complained that during their visits to supervise schools teachers were absent or late for no apparent reason (Harber 1984). Other surveys report high absenteeism among teachers in Mexico, New Guinea, and Sri Lanka; absenteeism is especially acute in rural areas (Baker 1988b). Students obviously cannot learn from a teacher who is not present, and absenteeism among teachers encourages similar behavior among students. In some countries, such as Mali and Somalia, parents react to high rates of teacher absenteeism by refusing to enroll their children in school.

Teacher Attrition

In many countries the best primary school teachers leave to enter the private sector or to teach at higher levels of education. The annual attrition rate of primary teachers is 10 percent in Haiti, where teaching is viewed as a transitory occupation. Many primary school teachers aspire to become tailors or chauffeurs because the pay is better (IEES 1987). In Liberia teachers have left to become executive management officers and medical and legal workers (Liberia, Ministry of Education 1989). In Korea the attrition rate has climbed above 5 percent and is a cause for concern (APEID 1984b).

To meet the increasing demand for teachers, countries must reduce attrition. Reducing attrition would yield a number of benefits. First, countries would receive an economic and social return on the investment they make in training teachers. The longer teachers remain in the profession, the greater the country's return to its initial investment. Second, countries would not have to train replacements for the teachers who leave. If the attrition rate were to increase at the same time as the demand for teachers, countries would have to build more teacher training colleges to meet the growing demand.

The recurrent cost of teacher training (salaries, administrative placement costs, and stipends) would also increase. A nominal attrition rate of 3 percent would triple the number of teachers who need to be trained just to maintain the current number of teachers (table 4-1). Third, if the teachers who leave have benefited from experience and in-service training, their productive capacity will be higher than that of new teachers. The effect of experienced teaching on student achievement is, however, mixed: it was positive in fewer than half of the studies reviewed (Fuller 1987; Mwamwenda and Mwamwenda 1989). Finally, communities in which the turnover of teachers is high are likely to respond by refusing to support new teachers and removing children from school.

The Factors Affecting Attendance and Attrition

Several factors promote absenteeism among primary school teachers, discourage new teachers from remaining in the profession, and drive out experienced teachers. These include (a) absolute salaries that are so low teachers must hold other jobs to supplement their income, (b) lower salaries than those of workers in the private sector or secondary teachers, (c) poor working conditions, (d) scarce opportunities for professional advancement, and (e) deficient local supervision, authority, and administrative procedures. For example, in Africa teachers identified the following as incentives to enter or remain in teaching: salaries, nonsalary benefits, working conditions, and professional status, as indicated in table 4-5 (Chivore 1988; Kemmerer and Thiagarajan 1989; Liberia, Ministry of Education 1989). Many of these incentives are scarce or inadequate in developing countries.

LOW ABSOLUTE SALARIES. Salaries and benefits are important factors that motivate individuals in any profession. In developing countries, teacher sala-

Table 4-5. Factors in the Attractiveness of Teaching Most Often Cited by Teachers in Liberia and Zimbabwe

Liberia, 1987

1. Payment of salaries on time
2. Government provision of housing
3. Adequate instructional materials in classroom
4. Sufficient opportunity for in-service training

Zimbabwe, 1986

1. Salaries that are competitive with the civil service and private sector
2. Opportunity to upgrade training and education
3. Adequate housing
4. Opportunity for promotion

Sources: Kemmerer and Thiagarajan 1989; Chivore 1988.

Table 4-6. The Average Salary of Primary School Teachers in Relation to the Average Nonagricultural Wage and to Gross National Product per Capita in Selected Countries
(1985 constant dollars)

Country	*Year*	*Average primary teacher salary (1)*	*Average nonagricultural wage (2)*	*Ratio of (1) to (2)*	*Ratio of (1) to gross national product per capita*
Low-income countries					
Burundi	1985	1,704.00	1,660.35	1.03	7.2
Ghana	1986	787.90	723.04	1.09	2.0
India	1986	1,041.90	771.95	1.35	3.6
Kenya	1984	1,295.40	1,512.15	0.86	4.6
Malawi	1986	849.70	662.05	1.28	5.8
Tanzania	1979	1,072.00	790.34	1.36	3.0
Zambia	1982	2,483.90	2,668.22	0.93	6.1
Lower-middle-income countries					
Chile	1981	2,845.70	3,837.65	0.74	2.0
Costa Rica	1986	4,760.30	2,550.39	1.87	3.2
Guatemala	1979	2,489.60	929.36	2.68	1.7
Honduras	1982	3,307.30	4,104.00	0.81	4.2
Mauritius	1986	1,161.50	809.96	1.43	0.9
Peru	1985	1,145.40	633.29	1.81	1.3
Thailand	1986	1,861.60	1,616.00	1.25	2.3
Zimbabwe	1984	3,549.90	3,407.92	1.04	6.6
Upper-middle-income countries					
Greece	1985	7,219.60	4,731.65	1.53	2.2
Mexico	1986	1,733.30	2,240.29	0.77	0.9
Portugal	1985	5,055.70	2,597.71	1.95	2.6
Venezuela	1984	5,125.10	9,671.47	0.53	1.5

Sources: See table A-22 in the appendix; ILO 1988.

ries and emoluments account for as much as 95 percent of government recurrent expenditures for primary education (see table A-19 in the appendix). In some countries the average salary of teachers is significantly higher than the GNP per capita (see table 4-6), but in other countries, teacher salaries are too low to provide basic necessities. Absenteeism is then likely to be high because teachers must supplement their income with other jobs. A survey in Liberia found that about 30 percent of all full-time teachers also worked their land or held other wage-earning jobs (Liberia, Ministry of Education 1989). In Indonesia, according to data from the Ministry of Education, teachers work five to ten hours a week at other jobs. In Somalia 36 percent of primary teachers hold second jobs, and in Haiti, too, teachers generally supplement their income (IEES 1984c; Kemmerer and Thiagarajan 1989).

Irregular payment of salaries is also a problem in many countries, which encourages teachers to take on additional work or to abandon the profession altogether. It also encourages teachers to travel to the central administration to collect their salary, often leaving their classes unattended for up to a week. In Liberia salaries were often paid three to five months late; trainees and incumbents cited this as the least attractive aspect of teaching, and former teachers ranked it as a major reason for resigning (Liberia, Ministry of Education 1989). Teachers surveyed in Haiti and Yemen were also discouraged by the late payment of salaries (Kemmerer and Thiagarajan 1989).

Teacher salaries have clearly declined in both real and relative terms in low-income countries. In the few countries with a weakening private sector the position of teachers appears to have improved relative to other occupations, even though their salaries may have declined in real terms. In middle-income countries the teaching profession has fared slightly better absolutely, although the morale of teachers is still low because teacher salaries remain lower than those of their peers in the private and public sectors.

As the teaching force in many developing countries has expanded over the past two decades, teacher salaries have eroded. Figure 4-1 displays total

Figure 4-1. Index of Recurrent Spending per Primary Teacher, Selected Years, 1970–85

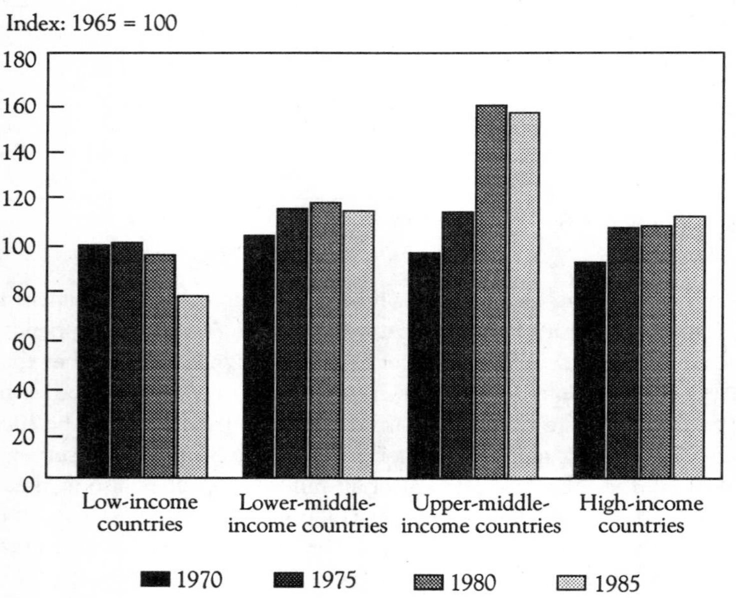

Note: The median percentage of public recurrent expenditure for teacher emoluments, measured in constant 1985 dollars, was divided by the number of primary school teachers in each country and indexed to show recurrent spending per teacher.

Source: See tables A-9, A-17, and A-19 in the appendix.

recurrent spending per teacher, expressed as an index, with salaries in 1965 equal to 100.[5] In the low-income countries for which data are available, real expenditures per teacher fell between 1965 and 1985. Lower-middle-income countries held teacher salaries relatively constant in real terms, while real spending per teacher increased sharply in upper-middle-income countries and gradually in high-income countries. The figures for all low-income countries taken together mask even greater declines in some regions and countries. For instance, spending per teacher has fallen 30 percent on average among West African countries and 20 percent among East African countries since 1970. A similar decline also occurred among the Central American countries for which data are available. Schultz (1985) notes that rapid population growth has reduced the unit cost of education, particularly the cost of teacher wages.

A recent World Bank study found that real teacher earnings declined 20 percent in Francophone Africa and 13 percent in eastern and southern Africa. Figure 4-2 shows that the change in real earnings varied among African countries. The real salaries of teachers in countries with relatively low GNP per capita eroded the most. Only in Niger did real teacher salaries increase between 1980 and 1985; the remaining seventeen countries surveyed re-

Figure 4-2. Change in Real Average Salary of Teachers in Seven Sub-Saharan African Countries, 1980–85

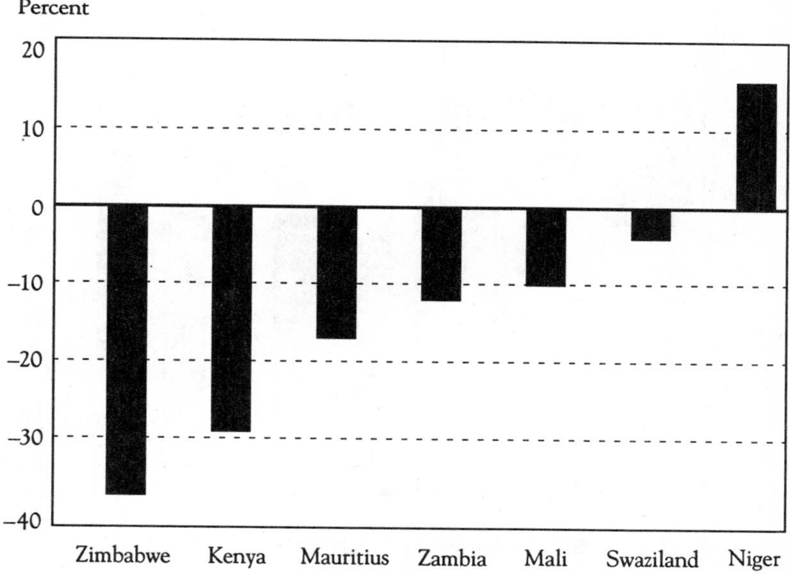

Source: Zymelman and DeStefano 1989.

ported declines (Zymelman and DeStefano 1989). Another study in seven developing countries noted that in 1985 the median salary of the best qualified primary school teachers had fallen to between 30 and 60 percent of the median salary in 1970 (Tibi 1990). In some countries, teacher salaries are simply too low. In Somalia the average teacher earns about one-half the average cost of living, even though wages were doubled recently (in current terms); teachers now earn the equivalent of US$6 per month, which is 25 percent of GNP per capita (Kemmerer and Thiagarajan 1989).

LOW RELATIVE SALARIES. When teacher salaries decline relative to those in other sectors, the opportunity costs of remaining in teaching increase as well, and the more qualified prospective and incumbent teachers are likely to leave the profession (Murnane and Olsen 1990). Figure 4-3 presents the real salaries of teachers and real average earnings in the manufacturing sector for Sub-Saharan African countries between 1980 and 1985. The figure, which shows only relative changes and not starting points, presents a mixed picture.

Figure 4-3. Change in Real Teacher Salaries and Average Manufacturer Salaries, 1980–85

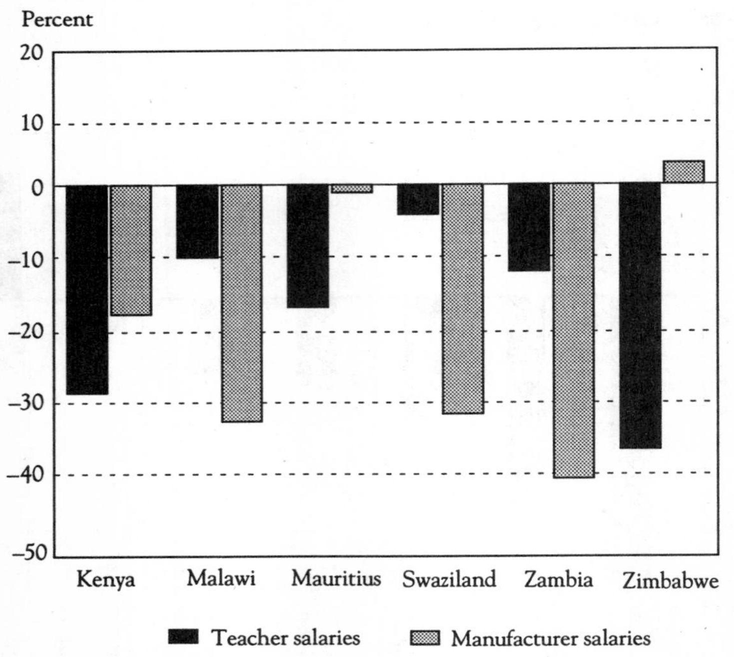

Source: Zymelman and DeStefano 1989.

In Zambia, for instance, the average teacher salary equaled the average manufacturing wage in 1980. By 1985, teachers made 50 percent more, on average, than manufacturing workers did, despite a 15 percent decline in real earnings. In other countries, however, teacher salaries declined relative to manufacturing wages. Teachers in Kenya earned just 85 percent of the average manufacturing worker's wage in 1985. In Zimbabwe, teachers earned 70 percent more than manufacturing workers in 1980, but only 5 percent more in 1985.

The data on teacher salaries are less complete for other regions. One study conducted in three Latin American countries (Argentina, El Salvador, and Panama) found that teachers earned the same as or somewhat less than the average civil service worker, but more than the average worker in manufacturing (Cox-Edwards 1989). However, a study from Brazil painted a more pessimistic picture. In 1980 male primary school teachers earned 8 percent less than clerks and female teachers earned 12 percent less, even though they had an average of two more years of schooling (Psacharopoulos 1987a). As early as 1970, teachers earned, on average, 95 percent of the average household (cash) income. In urban centers, teachers earned much less than the average household (Birdsall and Fox 1985). In Haiti teacher salaries in nongovernment (usually church-related) schools, which serve about 70 percent of all primary school students, ranged from US$25 to $80 per month; GNP per capita is $30 a month (IEES 1987).

In Asia teacher salaries were quite competitive in the Republic of Korea and Singapore in the early 1980s; they have always been low in Indonesia (Beeby 1979; Cox-Edwards 1989; IEES 1984a). In Pakistan only top-ranked primary school teachers earn the average salary in manufacturing (APEID 1984a). In countries with a strong wage sector, teacher salaries are falling behind. However, in countries with a small and expanding wage sector, teacher salaries remain relatively strong. In some cases, teachers earn relatively high salaries but are not paid on time.

Even within the education sector, primary school teachers receive low salaries. Salary schedules provide significant differentials based on each teacher's level of preparation and the educational level at which he or she teaches. In Botswana, for example, a university graduate teaching in an upper secondary school earns fifteen times more than a lower secondary school graduate teaching in a primary school (Botswana, Ministry of Education and Culture 1986). This differential reflects a priority that may have been sensible in the past but is dysfunctional for countries that need to emphasize primary education and basic literacy.

NONSALARY BENEFITS. Teacher compensation includes salary and nonsalary benefits. Nonsalary incentives offer a selective way to augment individual incomes. For instance, in Africa, housing, transportation, and other benefits augment teachers' base salaries by 20 percent on average. Monthly supplements to primary teacher salaries as a percentage of average monthly salaries range from 45 percent in Burkina Faso and 33 percent in Senegal to as low as 8 percent in Madagascar and Zaire (Zymelman and DeStefano 1989). Benefits can help offset the low salaries received by primary school teachers and improve the relative salaries, although teachers do not receive the in-kind payments such as bonuses and commissions that many private sector employees do. In Côte d'Ivoire, for example, teachers earn higher base salaries, on average, than private sector employees with equivalent educational qualifications, but this premium vanishes when benefits are included in the respective total compensation packages (Komenan and Grootaert 1988).

Hardship allowances and in-kind supplements are other important incentives for attracting teachers to areas in need. Rural schools, in particular, have a shortage of high-quality teachers, and many governments offer a panoply of special benefits for rural assignments: higher pay, more rapid promotion, enriched housing benefits, and subsidized access to public services. In Nepal, for example, the salary premium for rural teachers can exceed 100 percent of the base salary paid to urban teachers (IEES 1988).

Although such extrinsic incentives are needed, intrinsic rewards are also important. Mexico, for example, has experimented with a program guaranteeing advanced training based on length of service in rural areas (Dove 1986). The provision and encouragement of in-service training to improve the professional skill of rural teachers, combined with efforts to encourage greater communication within schools, can mitigate the perception that teaching in rural schools is a hardship. In addition, teacher candidates from rural areas can be recruited and deployed locally. Returning authority over the recruitment and deployment of teachers to local communities may reinforce the connection between schools and the community and may promote teacher retention. Steps can also be taken to lessen the professional isolation of teachers posted in rural areas. In Pakistan, for example, the government built housing clusters for female teachers to reduce rural isolation, encourage professional interaction, and increase safety.

Some South Asian countries pay teachers in urban areas higher salaries than they pay rural teachers, to compensate for the high cost of living. This may be inefficient: subsidies to urban teachers further distort the already skewed distribution of teachers by encouraging teachers to remain in urban areas. Moreover, family networks and support systems tend to be stronger in urban areas, and the nonmaterial rewards associated with urban dwelling—better services and improved quality of life—are enough to attract potential teachers.

Comparatively short work weeks, coupled with extended vacations, are traditional benefits of working in primary education and motivate individuals to enter the teaching profession. On average, teachers work 120 hours per month (while wage earners work 185 hours) and receive twelve weeks of vacation a year (Unesco 1982).

POOR WORKING CONDITIONS. In general, conditions within the classroom discourage potential candidates from becoming teachers and force incumbent teachers to leave the profession. Teachers cannot do their job efficiently without basic instructional materials. In developing countries, school buildings are often poorly maintained and facilities are inadequate. At one site in Nigeria, according to an internal World Bank report, three schools serving more than 3,000 students shared two pit latrines. The Philippines has a perennial shortage of classrooms. Often more than fifty-five students are crammed into classes built for half that number (APEID 1984c). In Haiti many classes have to be taught outdoors (IEES 1987). These conditions demoralize teachers, weaken their professional commitment, and, as indicated in chapter 3, affect student performance.

ADMINISTRATIVE INADEQUACIES. Local authority and supervision are important for ensuring teacher productivity. Principals currently have little authority to sanction teachers who do not turn up for work. Centralized bureaucracies undermine the capacity of principals to manage their staff by imposing complicated procedures for redress. In Nigeria principals are frequently transferred among schools, and the Ministry of Education does not pursue complaints (Harber 1984). Central authorities in countries such as Indonesia, Mexico, and Papua New Guinea often fail to provide support services, and the lack of experienced and dedicated supervisors or inspectors discourages improved teaching practices and encourages absenteeism (Baker 1988b).

CAREER ADVANCEMENT AND PROMOTION. Organizations traditionally accept employees only after careful screening, and promote them based on their abilities. New staff are hired on a probationary status while their performance and fit with the organization are assessed. If they pass probation, they are awarded more secure status, and they spend their early years learning the performance standards of the firm or institution, mastering their craft, and, ideally, working closely with skilled mentors. Advancement and salary increments are linked to actual performance.

Neither of these practices—careful induction and performance-based promotion—is as commonly applied in teacher personnel systems as it should be. Temporary or part-time teachers are commonly hired to fill vacancies in developing countries, but they are usually unqualified, and opportunities for further training and advancement are limited. Regular new teachers, in effect, enter the teaching service when they begin a teacher training program. Teachers automatically receive annual salary increments and promotions, even though their length of service is not consistently related to the achievement of their students (Fuller 1987). Moreover, salary increments increase steeply in the first years of employment, and raises thereafter depend more on tenure than on training (Zymelman and DeStefano 1989). Thus salary scales provide incentives to remain in the profession for a few years, but not to become better trained or to make teaching a career.

Policy Options

The pressure to expand educational capacity has stretched resources between the need to employ new teachers and the need to compensate incumbent teachers adequately. Expanding the teaching force often means that fewer resources are available for the training and economic incentives needed to improve the existing teaching force. Motivating and improving the existing cadre of teachers may occur at the expense of expanding children's access to basic education.

The differences among countries greatly affect priorities for allocating resources. For example, Korea and Indonesia, both middle-income countries with declining enrollment growth, have enough teachers to cover future enrollment (APEID 1984b; IEES 1986). Their educational resources can thus be directed toward training and motivating incumbent teachers and toward improving the managerial and administrative support for the active teaching force. However, in many low-income countries, particularly in Africa, teacher shortages persist as the population continues to grow. These countries must direct their resources primarily toward filling shortages. Nevertheless, developing countries should consider various policy options for addressing the issue of teacher motivation and productivity.

PROVIDING SUFFICIENT COMPENSATION. The absolute decline in real wages undoubtedly affects the motivation, attendance, and performance of teachers. Paying salaries that cover basic needs is essential to ensure that teachers attend school regularly. Governments in countries where the base salary of teachers is equal to or greater than salaries in other sectors should invest their resources in incentives that promote teacher performance. These include direct fiscal incentives, nonsalary benefits, and opportunities for promotion.

REDEFINING SALARY SCHEDULES. Salary schedules provide the underlying structure of opportunities and incentives for advancement. Current schedules are often tied to the uniform scale of the civil service. They are based on certification and experience, but they tend to reward years of formal education (degrees earned) more highly than years on the job. This increases the teacher's incentive to pursue preservice training, but reduces his or her willingness to remain in the profession.

In many developing countries, pay scales tend to be front-loaded, with salaries based on certification rather than on experience. In Togo, for example, new teachers holding the highest certificate earn 2.78 times more than new teachers in the lowest (unqualified) entry category (Zymelman and DeStefano 1989). In Nepal the highest-ranked primary school teacher (who has two years of university training) earns about 60 percent more than the lowest-ranked teacher (who has not gone beyond lower secondary school). Among teachers with similar educational qualifications but different amounts of on-the-job experience, a salary gap may also exist. In Sudan, for example, the most senior teacher earns 3.25 times as much as a first-year teacher with equivalent education. Education and certification, however, can move teachers up the salary scale much faster than years of teaching experience can. In twenty of the twenty-five Sub-Saharan African countries studied, individuals must teach for fifteen years or more to earn the maximum salary; they need far fewer years of advanced schooling to attain that same salary (Zymelman and DeStefano 1989).

To increase the general education prerequisites for teacher training (and hence for teaching) and to reward performance as well as experience can have significant consequences for the budget. Teacher salary schedules may have to be separated from the uniform scale of the civil service. Specific elements to be delinked could include (a) increasing the general education prerequisites for teaching without increasing salaries, (b) decelerating the rate of salary increase in the first few years of teaching, and (c) rewarding teachers on the basis of their performance, as well as qualifications and experience.

Linking salaries to the civil service may sustain the professional identity of teachers in some countries, but uniform salary schedules and automatic pay increases are dysfunctional. They neither reward superior performance nor penalize poor performance. Changing the structure of compensation will, however, require tradeoffs to be made in the recruitment of high-quality new teachers and the retention of existing teachers. Increasing the starting pay while reducing the rate of increase over time might boost the supply of qualified, new teachers but deter experienced teachers from staying in the field. Reducing starting salaries and increasing the rate of salary increment might discourage prospective teachers but motivate those already in the profession.

Countries that cannot separate the salary scales for teachers from those for civil servants might have to recruit less well-educated (but more affordable) teachers. They would then have to invest in interventions such as interactive radio instruction, programmed learning, and distance education programs to upgrade the skills of existing teachers. Changes in compensation packages should not, however, be used in isolation. Instead they should be among the array of incentives available to ministries.

REWARDING TEACHER PERFORMANCE. Under a system of merit pay—a common method of encouraging performance—a significant portion of each teacher's salary is based on her or his performance, as assessed by supervisors. This form of remuneration has several problems. Unlike factory production, teacher supervisors cannot monitor a consistent set of activities. Establishing an evaluation system that justifies the supervisor's decisions is costly in both time and money. The system tends to promote divisiveness within schools, as teachers attempt to bias supervisors by detracting from the performance of their colleagues. Furthermore, teachers seem to respond negatively when they are not awarded merit pay. Instead of working harder, they become bitter and refuse to cooperate (Murnane and Cohen 1986). Finally, in small rural schools with only one or two teachers, evaluation is made harder still by the scarcity of external supervisors. Merit pay can, however, be successfully implemented in one-teacher schools; at schools in the Seti Zone Project in Nepal, teachers were paid directly by supervisors but only if they attained observable performance goals (Bennett 1991).

One alternative to awarding merit pay is to provide grants that support innovative teaching projects in the classroom. Grants could be targeted toward specific goals (literacy, for example), and the teachers who implement such projects could receive a portion of the grant when the project is successfully completed. The teachers would be recognized nationally, and their innovations would be adopted by other schools. Unlike merit awards, teachers would be given clearly established objectives, and energetic teachers would be recognized and rewarded, which would raise the visibility of the teaching profession (Murnane and Cohen 1986).

In a variation on the grant approach, Pakistan gives salary increases to teachers, principals, and assistant education officers who produce a first-grade enrollment of 98 percent in boys' schools and 50 percent in girls' schools. If they maintain these targets for grades two and three, they receive even larger raises. Another option is to reward teachers collectively rather than individually. Schools that significantly improve the performance of their graduating students would receive across-the-board salary increases for all teachers. This could, however, have serious equity consequences.

Financial incentives appear to improve teacher attendance. One North American study found that absences declined significantly and perfect attendance increased dramatically, from 8 to 34 percent, during the first year of an attendance incentive plan. In this program teachers drew one share from a pari-mutuel for each absence below the mean number of absences the previous year (Jacobson 1988).[6] Appropriate supervision and delegation of authority to the school also improve attendance. As will be discussed in chapter 5, empowering principals helps to maintain school discipline and keep teachers accountable for their performance.

IMPROVING ADMINISTRATION AND COMMUNICATION LINKS. The administrative support system in developing countries is inadequate. Teachers are poorly deployed, either to regions where they do not want to teach or to subjects that they are not prepared to teach. In Botswana, for instance, the government was recruiting costly expatriate mathematics teachers when a new management information system revealed that there were already enough mathematics specialists to fill the demand, but they were teaching other subjects.[7] The government canceled the expatriate program and saved a substantial amount in salaries.

Regional deployment does, however, pose many problems. Because the opportunity costs (such as health risks and isolation) are high, incentives to work in rural areas must also be high; they are rarely high enough to attract teachers (Dove 1982). As long as working conditions in rural areas remain worse than those in cities, and incentives do not compensate for the hardship, teachers who are not quickly redeployed will become dissatisfied. Governments must improve working conditions in rural areas, as well as provide adequate incentives to attract teachers. Central bureaucracies must also accelerate their response to demands from local authorities. The timely payment of teacher salaries, the rapid hiring of replacement staff, and consistent supervision all increase the time that teachers spend teaching.

ENCOURAGING TEACHER PROFESSIONALISM. Organizational devices that motivate teachers to become more professional—to examine their own knowledge and pedagogical practices and to strengthen their weaknesses—are urgently needed. In-service training programs encourage teachers to engage students more actively. In a series of studies conducted by Bristol University, teachers said that they would like more opportunities to improve their abilities through in-service training (97 percent of those questioned in Botswana, 92 percent in Ghana), although most (52 percent overall) admitted that they would attend such programs more readily if their qualifications or salaries would improve as a result (INSET Project 1982). Until teaching is seen to require professional growth and responsibility, the effect that in-service workshops have on the behavior of teachers will be short-lived.

One modest step would be to link in-service teacher training directly to salary increments and promotions. This would require some standardization of the specific competencies taught and the length of instruction provided by in-service workshops. The Gambia has successfully equated a series of in-service courses with more conventional (and costly) preservice programs. The less costly in-service sequence is tightly linked to promotions (Commonwealth Secretariat 1984). In Ghana the administrative costs of selecting candidates for training and promotion have been reduced by advertising vacancies, which also ensures that applicants want the post (Thompson 1990). Governments should carefully assess the knowledge and skills that are effectively delivered by teacher training programs. Linking status and promotional opportunities to training is sensible only if that training reflects the increased investment.

Career ladder plans have been proposed as one way to retain able and talented teaching staff. Although these plans must be adapted to the particular circumstances of each country, most career ladder plans share certain features: predetermined criteria for advancement (including training and certification), objective evaluation procedures, and opportunities for teachers to accept new roles (such as senior teacher, department head, and vice or assistant or deputy principal).

For forty years Great Britain has been operating a career ladder that has significantly improved education. In that system, teachers progress up a five-rung ladder through competitive promotions. British schools have a limited number of positions, which are nationally advertised as they become vacant. As teachers climb the ladder, they assume more and more responsibility. In the United States, Tennessee has a statewide career ladder in which senior teachers (rung three of four) are encouraged to teach remedial students in the summer, develop curricula, and counsel apprentice teachers (the lowest rung). Wisconsin has a similar program in which teacher specialists nurture staff and guide apprentices. The benefits and challenges of implementing ladder programs are discussed in box 4-2.

Another means of rejuvenating professionalism among teachers involves the formation of unions or professional associations with the right to negoti-

Box 4-2. The Use of Career Ladders to Promote and Retain Talented Teachers

Career ladders have a number of benefits: they produce good role models by placing seasoned, enthusiastic teachers where they can influence younger teachers; they develop leadership by preparing teachers to be principals and assistant principals; they give teachers more authority and allow them to become involved in decisionmaking at the school level; and they offer teachers the opportunity to develop new skills and accept new challenges. Moreover, career ladders can be designed to suit particular needs. Levels can exist to accommodate teachers who deserve to be promoted but are content to remain in the classroom, and teachers who receive promotions can be placed in rural schools.

Introducing a system of career ladders is not easy, however. Transition plans must be developed, teachers must be informed, and a system of performance accountability must be designed. Finding cost-effective ways to evaluate teachers' performance, providing enough opportunities for advancement, and maintaining good relations between administrators and teachers—these tasks pose other challenges. And funding can be a fundamental difficulty, particularly when teachers' salaries are tied to a rigid civil service salary scale. Although attaining new skills and meeting new challenges are important rewards in and of themselves, career ladders motivate teachers best when each higher rung carries the incentive of higher pay. Separating teachers' pay scales from the civil service scales, as Malawi has done by establishing a separate teaching service, can make career ladders easier to implement.

Sources: Burden 1984; Murphy and others 1984.

ate salaries, benefits, and working conditions. These organizations would allow teachers to improve their lot, participate in decisionmaking, and develop curricula. Such self-determination of the role, function, and condition of service is essential to all professional groups. Active teachers associations, such as the Zambia Science Teachers Education Project (ZAMSTEP), which has improved teaching in science and mathematics, or Ghana's EDSAC 1, which encourages community involvement in junior secondary education, have introduced successful innovations.

IMPROVING WORKING CONDITIONS. The conditions under which many of the world's primary school teachers work are clearly untenable. Research has linked student performance to the provision of adequate facilities. In Botswana a survey of fifty-one primary schools revealed that students performed significantly better on academic tests when they had adequate classrooms, desks, and books (Mwamwenda and Mwamwenda 1987). In Malta teachers were motivated by their working conditions and by factors intrinsic to pedagogy. Thus teachers were more motivated when their students performed well (Farrugia 1986). When students must share a textbook or when they have no paper on which to write, they do not perform well and more complex forms of instruction (such as working in student groups) become more difficult. Technologies that help structure the delivery of lessons without discouraging teacher or student initiative are especially important. Textbooks and instruction manuals, for instance, structure the curriculum and ensure that specific material is covered during the school year (Lockheed, Vail, and Fuller 1986). Improving working conditions enables teachers to function better and students to perform better. When students perform better, the teacher's motivation is reinforced, as is classroom practice.

Summary

A key determinant of student achievement is the quality of teaching. An effective teacher should possess at least a thorough knowledge of the subject matter being taught, an appropriate repertoire of pedagogical skills, and motivation. The teaching force in many developing countries fails to meet these standards. Governments that are serious about achieving an effective primary education system will have to commit themselves and their resources to solving this problem. Governments must design policies and programs aimed specifically at improving the academic and pedagogical preparation of teachers and providing incentives to strengthen their motivation and professional commitment. The challenge is particularly difficult for low-income countries, which must not only improve the quality of the current teaching force but also expand its size if they are to achieve universal primary education.

To address the problem of inadequate academic background, countries will have to shift the general education component of teacher training to secondary schools, shorten preservice teacher training, and improve the process of recruiting students for teacher training institutes. Strategies for developing good pedagogical skills should include revising the admission requirements, emphasizing pedagogical methods, and incorporating practice teaching into preservice training. Improving teacher motivation is perhaps the trickiest task that governments face in their effort to upgrade the teaching force. Doing so will require a variety of measures, such as paying adequate salaries and providing nonsalary benefits, improving working conditions, offering opportunities for professional advancement and incentives for good performance, and strengthening supervision and support.

Notes

1. The projected supply of teachers was estimated using the aggregate growth rate of the teaching force between 1980 and 1985 and extending it to 2000 (see table A-9 in the appendix). The current production of teachers is assumed to be maintained until 2000. The aggregate output rate was 2 percent for low-income countries (7 percent without China and India), 5 percent for lower-middle-income countries, and 3 percent for upper-middle-income countries.

2. This is not true of highly centrally planned and administered economies such as China.

3. According to World Bank data, this proportion had increased to 14 percent by 1989.

4. In Haiti qualified teachers must have completed a teacher training program or its equivalent. Teacher training upgrades the skills of current teachers through in-service programs.

5. This rough indicator represents total recurrent spending on primary schooling divided by the number of teachers. Different governments, of course, spend different shares of their recurrent budget on administration, instructional materials, and maintenance, and some teacher benefits, such as housing subsidies, may not come from the education ministry's budget. The bulk of recurrent spending at the primary level is, nevertheless, for teacher salaries. Thus the measure provides a reasonable indication of changes in teacher compensation as the teaching force expands and government resources change over time.

6. Hypothetically, if only one teacher had fewer than the mean number of absences for the previous year, that teacher would win the entire pool. Two teachers, one with one fewer absence and the other with two fewer absences, would split the pool in the ratio of 1:2, and so on.

7. Management information systems collate appropriate educational indicators to track changes over time and develop cost and financing plans.

5

Strengthening Institutional Capabilities

THE PREVIOUS CHAPTERS stressed that to improve student achievement, the level and effectiveness of inputs must be increased and the quality of teaching must be improved. Establishing managerial and institutional strength throughout the education system is essential to the success of these efforts. Effective schools require effective school management, and effective school management requires the support of well-developed national and intermediary organizations. The leadership and resources of these organizations enable schools to translate policies into teaching and learning.

Strengthening the managerial and institutional capability of the education system in developing countries generally confronts three problems: (a) ineffective organizational structures, (b) undeveloped managerial capacity, and (c) poor information systems. The bureaucratic, social, and political context in which education systems function compounds these problems. The administrative structure of education reflects and is intricately linked to the wider system of public administration, which complicates attempts to reform the education system. Moreover, education is, by nature, provided by a multitude of small, dispersed, and disparate units. In virtually all countries, schools must respond to interests and goals that are often in conflict. Parent associations, teacher unions, and locally elected school boards often exert a powerful influence on both the objectives schools are expected to reach and the resources available to them. Local priorities of the system often do not coincide with central priorities. Horizontal and vertical lines of communication among the agencies that provide educational services are often blocked. Responsibility and authority are rarely supported by adequate resources. In addition, primary schools must contend with the pressures generated by the competitive examinations that guard the gates to secondary and tertiary levels of education. Reforming such a complex system is a major challenge for most governments.

Primary schooling is delivered through organizational structures that are fundamentally similar throughout the world. At the base are groups of schools organized in geopolitical districts. Almost all schools are managed by a head teacher or principal. In many countries, individual schools or groups of schools are overseen by school boards or parent-teacher organizations, some of which have considerable influence. An intermediary administrative organization at the district level provides some supervision and technical support. At the top of the structure is a national ministry or department of education, which plans and administers the education system as a whole and assigns one department principal responsibility for primary schooling. In larger countries this structure extends over four levels, including a state education agency or provincial ministry that falls between the district and the national ministry. Although this three- or four-level structure describes virtually all education systems, education systems are, in fact, diverse because the distributional patterns of decisionmaking are diverse (see table 5-1).

Constraints on Managerial and Institutional Effectiveness

Real authority accompanies the responsibility for mobilizing and allocating resources. Some systems are highly decentralized and authorized to mobilize and use resources at the district, municipal, or state levels. This approach has long been practiced in large countries such as Brazil and China. Other systems are relatively centralized. Financing and authority are concentrated at the national, or in large countries at the state, level, and districts and schools carry out a centrally mandated and financed program of schooling. The degree of centralization affects how the education system functions at the central, intermediate, and school level.

Ineffective Organizational Structures

A recent study of efforts to reform the administration of education in two Latin American countries illustrates the advantages and disadvantages of highly centralized and decentralized systems (Hanson 1986). In Venezuela the education system suffered from excessively centralized power. So extreme was the consolidation of authority that teachers had to request excused absences directly from the Ministry of Education. Such excessive centralization completely precluded regional autonomy and produced rigid standardization and tremendous inefficiency. In contrast, the Colombian educational system was so decentralized that schools often ignored mandates from the central administration. In order to impose order and uniformity and to compel national and state authorities to share decisionmaking authority, the ministry established a resource distribution program and state-level administrative units. Although systems in most countries are not so extreme, many struggle with similar problems.

Table 5-1. The Locus of Decisionmaking Authority in Primary Education Systems in the Republic of Korea, Nigeria, and the Philippines
(percent)

Decision and decisionmaker	*Republic of Korea*	*Philippines*	*Nigeria*
Authorizing major expenditures			
Central or regional body	9	66	42
School board	78	7	33
School principal	11	5	1
Teachers	0	0	0
Selecting principals			
Central or regional body	40	83	38
School board	39	3	45
School principal	0	0	0
Teachers	0	0	1
Selecting teachers			
Central or regional body	6	63	37
School board	71	7	49
School principal	3	14	1
Teachers	0	1	0
Determining the range or type of science courses			
Central or regional body	61	82	82
School board	3	0	5
School principal	5	5	5
Teachers	28	5	5
Choosing science texts			
Central or regional body	89	76	59
School board	5	1	12
School principal	1	2	9
Teachers	1	0	6

Note: Percentages are based on the responses of teachers and school administrators to questions about decisionmaking authority. The four categories of decisionmaker do not include all the options, so percentages may not total 100.

Source: Unpublished data from the Second International Science Study, provided by the International Association for the Evaluation of Educational Achievement, Hamburg.

Arguments support centralization in newly independent countries where a strong sense of nationhood has yet to be developed or in small countries where such control is feasible. Centralized control may also be more efficient for some purposes, such as producing textbooks and training teachers, because it achieves economies of scale, which are particularly important when financial and managerial resources are scarce. Most important, even when local governments can provide education more efficiently than the central government can, the involvement of the central government is necessary to ensure that the level of educational services provided is appropriate and equitable (Winkler 1989).

Highly centralized control can, however, bottleneck information and resource flows and limit the ability of schools to respond to local needs. When decisionmaking authority is concentrated at the center, higher-level staff devote their time to such tasks as appointing and assigning teachers, reviewing and accounting for thousands of expenditure requests, and planning budgets and allocating resources for individual schools. When their time is consumed with administrative duties that could be delegated to lower levels, high-level administrators cannot effectively carry out their principal functions: planning broad policies, designing strategies for implementing those policies, monitoring the consequences of policy implementation through testing and evaluation, providing financial resources and technical expertise, and obtaining and distributing educational materials.

Centralized systems have proven to be most effective in countries that are politically and economically stable and have strong administrative systems, good infrastructure, comparatively well-educated and well-compensated teachers, and a relatively homogeneous context for schooling (for example, the Republic of Korea). In ethnically and linguistically diverse countries, in those where schools are not centrally located, and in those with poor transportation and communication systems, rigid centralization blocks the flow of resources and information. In such circumstances, education systems are likely to be more efficient and effective if certain functions and responsibilities are devolved to lower levels.

THE CENTRAL LEVEL. Administrative weaknesses at the central level generally arise when managers do not have the authority or resources to do their job effectively, when lines of communication are blocked, when roles and responsibilities are unclear, or when managers are consumed by routine tasks.

Reports on education projects supported by the World Bank between 1984 and 1988 repeatedly underscore these administrative weaknesses. In Bangladesh functions and responsibilities overlap levels of administration. In Nepal imprecise job descriptions, no delegation of authority, and overlapping functions adversely affect the delivery of primary education. In Honduras management deficiencies arise from an outdated organizational structure, cumbersome procedures, weak coordination among ministries, and haphazard processes for formulating policy. In Guinea Bissau the organizational structure is unnecessarily complex, administrative units barely function, and operational policies and procedures are ill defined and ad hoc. Highly centralized control is a problem in countries such as Bhutan, Burundi, and Senegal, and severe structural deficiencies characterize Brazil, the Central African Republic, Ghana, Indonesia, Jamaica, Pakistan, Peru, Solomon Islands, and Togo.

THE INTERMEDIATE LEVEL. Offices at the intermediate level tend to be poorly financed, inadequately staffed, and not authorized to act. Their principal responsibility is to provide professional support and technical assistance to individual schools. Typically, however, the district office operates solely as an inspectorate, linking the central administration with schools and neglecting its own role as a source of professional support.

In most countries even the role of inspector tends to be weak and ineffective because of severe resource constraints. To inspect and supervise schools effectively requires regular school visits. Only a relatively small share of resources, other than teacher salaries, are allocated to intermediate-level support for schools (see table 5-2). In Honduras only 53 percent of the schools were visited, once each, in 1986. In Senegal school visits were limited in 1985, despite a favorable staff to school ratio, by a lack of transportation: only 28 vehicles served more than 600 staff in 41 regional directorates. The recent situation in Kenya is somewhat better, with 225 vehicles for some 600 supervisory staff, but inspectors rarely visit remote schools because of limited transportation, bad roads, and bad weather (Eisemon and Schwille 1991). In contrast, the allocation of funds to lower levels in Ethiopia enables schools to receive close supervision and technical and training support from a network of 106 Awraja (district) pedagogical centers, which are in turn supervised by 15 regional offices.

Resource constraints adversely affect the ability of intermediate organizations to operate effectively, the delegation of authority rarely matches responsibility, and the lines of authority are often unclear, mirroring weaknesses at the central level. In Tanzania, for example, the regional education officer is accountable administratively to the regional development director, politically to the regional commissioner, and professionally to the director of primary education. The district education officer is accountable administratively to the district executive director and professionally to the regional education officer. At the ward level, education coordinators, operating through powerful ward executive officers, provide pedagogical and supervisory support for schools in their wards and mobilize community resources, but they have no role in disbursing allocations from the central budget. It is remarkable that

Table 5-2. The Distribution of Primary Education Budgets, Selected Countries
(percent)

	Recipient of support			
Country	*Central services*	*Intermediate-level offices*	*Local schools*	*Ratio of schools to supervisors*
Ethiopia	39	19	42	22:1
Honduras	61	30	9	45:1
Kenya	77	23	0	21:1
Senegal	60	28	11	4:1

Note: Budget allocations for teacher salaries are excluded.
Source: World Bank data.

these administrators can even identify their responsibilities, much less carry them out effectively.

Another familiar problem at the intermediate level is the conflict between the functions of inspection and evaluation and those of professional guidance and support. Not surprisingly, intermediate-level staff who are responsible for inspection and evaluation often have trouble developing the effective working relationships they need to support school and staff development. Combining these functions in one position or one unit often hampers the effective performance of both.

THE SCHOOL LEVEL. At the school level, authority and responsibilities are acutely mismatched. Principals are largely excluded from decisions that affect their ability to improve student achievement. Curricula are designed centrally, and the diverse capacities and interests of schools and students are often ignored. Teachers are appointed, assigned, and evaluated centrally, leaving principals little control over the choice or discipline of their teachers. In many countries, teacher deployment policies fail to take into account regional (including instructional language), subject matter, or grade-level needs, which undermines the ability of principals to build and maintain an effective school environment. Nor do principals have the authority or resources to organize staff development programs that address the problems and challenges faced by teachers in their school. At best, they have access to more general programs created nationally. The dearth of authority at the school level is most prevalent in highly centralized systems, but even in decentralized systems authority may not be delegated below the intermediate level.

Learning should be the product of both the school and the community. To achieve this unity of purpose, school managers must have the authority to mobilize and use local resources for school improvement, particularly when central-level resources are scarce. School leaders are severely constrained, however, by centrally administered regulations that require funds raised locally to be submitted to the central ministry or that restrict the purposes to which they may be put. In the absence of testing and monitoring systems that assess performance, schools are generally held accountable for their use of basic inputs: enrollment allocations, student-teacher ratios, schedules, time allocations, and reporting requirements. The rules that govern basic inputs may be necessary, but they do not create incentives for schools to focus on the interventions that improve student achievement.

In extreme cases, of course, schools are highly autonomous because financially constrained central ministries and intermediary organizations find effective levels of input delivery and supervision difficult or impossible to establish. Textbooks are not distributed, teacher salaries are months in arrears, inspectors do not visit schools, teachers are often absent, schools are frequently closed, and no one takes responsibility. Systems are fragmented and fail to delineate who is responsible for what and to whom. Teachers blame parents for not reinforcing the value of education at home and administrators for setting and enforcing ill-conceived rules and for lacking resources. Parents blame teachers for teaching poorly. Administrators blame politicians for providing inadequate funding and teachers for not doing their job.

Undeveloped Managerial Skills

Many education managers throughout the system have a wide range of managerial skills and competencies, but many do not. Managerial and administrative capacity has weakened in almost all developing countries, where the rapidly expanding education systems require more skilled education managers and administrators than have been trained. The primary reason for such deficiencies is simple: training (whether preservice or in-service) is unavailable, inadequate, or inappropriate. Moreover, opportunities and incentives for advancement, clearly defined career paths, and systems for assessing performance are absent. The lack of such inputs not only hinders the professional development of managers but also dampens their motivation to perform well.

THE CENTRAL LEVEL. At high levels of school administration, frequent staff turnover and expansion are common, contributing to unstable policies and practices. Key positions may be filled by individuals who have little experience in educational policy and planning, often because they have been rapidly promoted or appointed to fill vacancies. Institutes that specifically train education administrators are limited in number and often handicapped by shortages of staff, teaching materials, and in many cases funds. Additionally, educators rarely enter institutes of public administration, which could teach the relevant administrative and managerial skills. The result is that many ministries are staffed by unqualified and inexperienced individuals who are unable to manage the overall education system, to formulate and enhance policies, to plan and carry out routine developmental tasks, and to exercise adequate control over resources.

Not only are general administrative and managerial skills lacking, but technical skills are scarce as well. In a number of countries, staff in textbook divisions have no commercial or publishing expertise; staff in curriculum divisions have little experience or background in child development or curriculum design; staff in examination agencies are inexperienced in test development, psychometrics, monitoring, and evaluation.

THE INTERMEDIATE LEVEL. Intermediate-level staff are responsible for supervising schools, which includes providing pedagogical and technical support. Yet they are often the least-qualified to advise others how to teach or manage a school more effectively or how to implement a new program

(Chapman 1990). Inspectors and district education officers in many countries are appointed on a political basis. Frequently they have little or no experience at the school level and little or no expertise in curriculum development and pedagogical methods. Moreover, when intermediate-level staff supervise the implementation of a new educational policy, program, or practice, they have rarely been involved in the decisionmaking process and thus care little about the project. Such shortcomings in staff competence and motivation, which limit the adequacy of supervision, are further compounded by the financial and logistical barriers to supervision discussed above.

Beyond being weak supervisors, intermediate-level staff frequently lack solid managerial skills. Low managerial competence at the intermediate, as at the higher, level stems from the absence of systematic staff development programs; supervisory training tends to be haphazard at best. Entrepreneurial skills are especially critical for intermediate-level administrators who play a role in mobilizing resources, and once again, those skills are usually developed by trial and error and not specialized training.

Resolving the conflict between the dual functions of inspection and professional support challenges even experienced and competent supervisors. As at other levels, the absence of monetary and status incentives, professional opportunities, and systematic feedback further diminishes the potential of supervisors to be effective agents for improving schools.

THE SCHOOL LEVEL. The school principal, who is in the best position to observe and influence teachers, is the best source of instructional supervision. The support, recognition, and approval of principals are key factors in changing teaching practices (Chapman 1983; Fullan and Pomfret 1987; Waugh and Punch 1987). A recent study of primary school effectiveness in Burundi documents a strong and significant relationship between the frequency of teacher supervision and student achievement: student test scores rose as the number of times the school director visited the classroom increased. Frequent teacher supervision improved the punctuality of teachers and their adherence to the curriculum, which in turn produced higher scores (Eisemon, Schwille, and Prouty 1989). Furthermore, the active support of principals consistently increases the chances that school improvement programs will be implemented successfully. At the same time, principals may suffer from isolation and lack of communication, just as teachers do. To play an effective role in school improvement, principals need regular mechanisms for communicating their questions and concerns to supervisors.

The absence of strong managerial skills, which is evident at all levels, is particularly glaring at the school level. Principals perform multifaceted and complex tasks. They maintain relations among the school, community, and parents, supervise teachers, oversee the maintenance of facilities and equipment, manage a range of reporting and record keeping duties, and, in small schools, teach as well. Yet they perform all these tasks under the chronic shortage of materials, clerical support, operating funds, and resources for staff development. In addition, ceilings on rank and salary diminish the attraction of being a principal.

In almost all developing countries, principals are selected from among teachers on the basis of seniority and then trained, although systematic training is limited. Training before the appointment is virtually nonexistent, except when a teacher has served as a deputy or assistant principal. Studies in Egypt, Indonesia, and Paraguay have found that a principal's teaching experience and training (number of courses taken) are related to higher student achievement (Fuller 1987; Heyneman and Loxley 1983; Sembiring and Livingstone 1981).

Only a handful of countries, including China, Ethiopia, Kenya, Malaysia, Papua New Guinea, the Philippines, and Thailand, have addressed the need to improve school management, primarily by establishing institutions to train principals. Such institutes face three problems. First, they cannot accommodate the number of new principals needed to run the burgeoning number of schools. Second, no consensus has been reached about what the curriculum should reflect and who should provide the training. Staff often transplant curricula and methodologies derived from their overseas training without adapting them to the sociocultural context and needs of their country and community. Third, the national policies for training administrators are not coherent, which hinders the effectiveness of these institutes. Malaysia and Thailand, however, are leading the way in developing coherent policies (Den Hartog Georgiades and Jones 1989).

Poor Information Systems

At all levels strong management is based on good information systems composed of achievement testing, monitoring, and research. Testing, monitoring, and research programs have been vital to educational reform over the past quarter century in developed countries. They provide the information needed to improve educational policy and programs at the national, district, and school levels. In developing countries, good information systems are particularly important for providing information on the cost and effectiveness of inputs because they allow educational policymakers to decide how to allocate resources most efficiently (Lockheed and Hanushek 1988; Middleton, Terry, and Bloch 1989). Few countries have developed the information systems that are adequate to the task.

ACHIEVEMENT TESTING. To inform education policy and provide common indicators of student performance, achievement tests that measure student performance with respect to national curriculum objectives must be standardized. This means that tests administered for purposes of national, regional, or local monitoring must cover the same content (topics and levels of difficulty) and must be administered and scored exactly the same way for all students taking the test.[1] To measure trends in achievement over time, it must be possible to "equate" test scores from one test administration to the next.

Standardized test data can be combined with monitoring systems that capture information on costs, student flows, material inputs, and staff to enable managers to identify which schools and districts are performing well and which are not. These data can also be provided directly to districts and schools, to help managers at the local and intermediate levels identify areas of the curriculum that need improvement and to help them monitor change.

Two types of standardized tests, serving different functions, are most common: (a) end-of-cycle certification and selection tests and (b) periodic tests of student achievement. At the primary level, certification and selection examinations are administered to all students seeking either a credential or admission to secondary school; they serve to limit access and are more reliable and equitable criteria than others that do not measure academic merit, such as political loyalty, family alumnus status, personal wealth, ethnicity, or gender. Periodic tests of student achievement are administered to all students enrolled in any or all grades; they inform parents, teachers, administrators and national authorities about the learning progress of individual children. Because they are administered to all students, both end-of-cycle and periodic achievement tests are expensive. To save on costs, most developing countries administer only certification/selection examinations.

About half of all countries test students on their mastery of the primary curriculum in end-of-cycle examinations. These examinations rarely provide information about how well the education system imparts learning skills. They fail to do so for four reasons. First, the examinations are rarely standardized (Kellaghan and Greaney 1990). Second, the number of students who pass a national examination is often a predetermined fraction of the individuals taking the test, rather than the number who meet an explicit standard of achievement. For each of the past ten years in Malawi, for example, 70 percent of the students taking the primary school graduation examination have passed, since the passing score was consistently chosen so that 30 percent of the students would fail. Third, because the examinations serve a selection, rather than an assessment, function, they are often very difficult. They are designed to discriminate among students with the highest level of proficiency, not to accurately measure various levels of achievement across the full range of skills, from beginning skills to advanced ones. Fourth, the examinations test the recollection of facts rather than problem-solving skills. Students are often asked to repeat whole sections from their textbook rather than to apply what they know to new problems. They are rarely asked to interpret what they have read. Thus students who complete their primary education may not be able to solve problems that require higher-order thinking.

MONITORING. The monitoring of high-quality indicators is essential for generating a flow of information about the status and evolution of education systems. Monitoring involves the use of surveys, routine reporting, observation, and other techniques to gather relevant data.

Weak monitoring systems are a significant liability for effective education strategies in many developing countries (Lockheed and Hanushek 1988; Middleton, Terry, and Bloch 1989; Oakes 1986, 1989; Verspoor 1989; Windham 1988). Learning indicators are seldom examined. Other basic statistics (on, for example, teacher qualifications and deployment, student characteristics, school attendance, dropout and repetition rates, and costs) are not routinely collected and reported. In general, the absence of information, most notably information on educational outcomes and costs, severely hampers efforts to improve education.

RESEARCH. Weak research capacity is a problem in almost all developing countries. The components of a strong research capacity include trained and motivated personnel; an indigenous literature; a base of general information; institutionalized procedures for collecting, processing, analyzing, storing, and retrieving data (both quantitative and qualitative); a technical support staff; organizational arrangements for conducting research and disseminating results; and adequate funds. Few nonindustrialized countries have developed this capacity; as a result, little educational research is carried out in the developing world, even when resources from donors are made available for this purpose (Lockheed and Rodd 1990). The ability to design, develop, and adapt effective educational policies and programs is constrained as a consequence.

Strategies for Addressing Managerial and Institutional Weaknesses

Organizational reform, or the strengthening of managerial and institutional capacity, must be the cornerstone of all school improvement strategies. Successful efforts require both time and resources, and national education laws and charters often have to be modified. Indeed, legal barriers, including those established in national constitutions, have proved to be formidable obstacles to organizational reform efforts (Hanson 1986). Particularly important to the success of institutional change are the support and commitment of the various stakeholders, including politicians, administrators, teachers, and parents.

Improving the Organizational Structure

Improving the organizational structure of education requires giving managers at all levels the authority and resources to do their jobs effectively, opening lines of communication, and clearly defining roles and responsibilities. In many countries this entails realigning authority and functions among central ministries, intermediary organizations, and schools. In principle, the goal of restructuring is to authorize school managers to manage and improve the process of instruction and to mobilize local resources, including community participation—in short, to authorize them to run effective schools. Effective schools require strong leadership and support from the central and intermediate levels.

DEVELOPING NATIONAL INSTITUTIONS AND SUPPORT AGENCIES. At the central level key institutions and support agencies must be developed and mechanisms to encourage school initiatives must be established. The planning and policy institutions are the professional and analytical heart of the education system and the key to improving the organizational structure. These institutions propose and analyze policy options, prepare investment plans, develop new curricula and materials, and design and monitor standards of performance. Support agencies are responsible for much of the logistics associated with delivering educational services, such as paying teachers, building schools, and producing and distributing educational materials. Some countries will have to create new units altogether. Others will have to restructure existing units.

Experience in a large number of countries suggests the importance of establishing well-defined, functional units for delivering key national support services.[2] These may be line units within the ministry of education or separate organizations funded by the ministry budget. Many functions can often be combined in a single organization with broad responsibilities for managing educational reform (Rondinelli, Middleton, and Verspoor 1989).

National production and national distribution of textbooks are central to improving learning. These are not one-time activities. Developing sustainable capacity for creating, testing, producing, and distributing student learning materials—as well as teacher guides and in-service training materials—requires a significant commitment of the central ministry's resources. Experience in Bangladesh, China, Ethiopia, and the Philippines suggests that textbook agencies, whether autonomous or divisions of the (national or federal) education ministry, are essential to establishing and sustaining a program of materials development. Box 5-1 describes important considerations in setting up textbook agencies. Even small countries, where publishing industries are relatively weak because the market for textbooks is small, may have to set up a clearly defined unit for developing instructional materials, carrying out the design, development, and testing necessary to achieve high quality, and managing the storage and distribution of materials. Such a unit has been successfully established in Lesotho.

Material interventions other than textbooks, such as interactive radio and programmed learning (see chapter 3), can play a significant role in improving the quality of teaching. These interventions require highly specialized skills for designing, testing, and producing materials. Interactive radio, for example, requires additional skills in broadcasting production and use. Successful applications of these technologies have established specialized units to handle specialized tasks. In many cases such units are most effective when they share

Box 5-1. Textbook Agencies in Indonesia and the Philippines

Two early textbook projects supported by the World Bank demonstrate the importance of establishing textbook agencies as separate entities with appropriate authority and sources. A project in Indonesia, approved in fiscal 1973, was inteded to improve the quality of primary education through an integrated program of instructional materials and teacher training courses. A project approved in the Philippines three years later aimed to develop the institutional capacity for continuously developing and supplying educational materials. The difference in objectives is significant, and the project in the Philippines was much more successful than the project in Indonesia.

Indonesia. A Project Implementation Unit, which formed a separate unit within the Ministry of Education, managed the publishing activities financed by the project. The unit was staffed at the senior levels with ministry personnel who had other responsibilities and thus could not give consistent attention to the production of textbooks. Almost all the remaining staff were technical. The unit suffered from a lack of staff, of full-time management at the top, and of middle-level management to direct the day-to-day work of the technical staff. The committee that was established to approve manuscripts and program plans was disbanded in the first year of the project, and its approval functions were subsequently performed by ad hoc committees. The failure to establish an institutional framework that clearly delineated responsibilities and coordinated the various components of the unit created a project plagued with difficulties.

The Philippines. In contrast to the Indonesia project, the Philippines project supported a relatively integrated organizational arrangement for producing textbooks, in part because a coherent structure already existed and in part because the project benefited from the experience gained by the Bank staff in Indonesia. The existing Textbook Board was responsible for managing all activities related to developing textbooks, from the initial planning through the manufacturing and distribution of books to schools. The government recognized the board as the only entity responsible for textbooks in the Philippines and gave it enough resources to carry out its work.

Organizational differences affected the performance of each project. Both projects encountered similar problems with scheduling, quality control, and so on. The Textbook Board in the Philippines had, however, more independence, more staff, and full-time management and was thus able to diagnose and respond to problems better than the Indonesian unit.

Source: Searle 1985.

facilities and professional expertise with agencies that develop textbooks or curricula.

ENCOURAGING LOCAL INITIATIVES TO IMPROVE SCHOOLS. In addition to developing and strengthening key institutions and support agencies, the central ministry should establish mechanisms to encourage school initiatives. One such mechanism is the school improvement fund. This is a financing program that supplements regular budgetary support for schools by authorizing intermediate-level administrators to make grants for projects proposed by individual schools. The central ministry sets specific guidelines for projects, sometimes targeting national or regional goals such as giving disadvantaged groups greater access to education or improving a particular component of the curriculum, such as science. However, resources support local initiatives designed to be effective within local contexts. In this respect, school improvement funds depart significantly from traditional, centrally mandated and centrally controlled programs that improve education from the top down.

School improvement funds have been established with success in a number of countries, including Colombia and Yemen. Experience suggests that guidelines for the use of funds must be carefully designed, monitored, and changed as needed. Support should be provided for programs that complement national inputs, notably textbooks, teachers, and time allocations. Grants might thus be made for supplementary reference materials, teaching aids (such as maps and charts), in-service training, teaching assistants drawn from the community, before- and after-school tutoring, and outreach programs that encourage disadvantaged students to enroll. Grants could also include requirements and financial support for monitoring, thereby stimulating improvements in local evaluation practice.

PROVIDING INTERMEDIATE-LEVEL SUPPORT. The most important role that intermediate organizations can assume is to support schools by providing technical assistance, staff development, and extra resources. For many countries, this means that intermediary units will have to expand their role from that of an inspection agency alone to that of a service agency. Clearly, school inspection is critical to identifying schools that do not meet the minimum physical and administrative standards and criteria for student and teacher attendance. Once intermediate-level staff identify problems, they must have the authority and resources to help schools resolve those problems.

In World Bank education projects, intermediary supervision and support have been essential to the success of efforts to improve schools (Verspoor 1989). Supervision can be strengthened through regular school visits by central office staff and through the establishment of supervisory staff positions at the intermediate level (where they do not already exist). In Bangladesh and Pakistan, for example, new supervisory arrangements contributed substantially to improving school performance (see box 5-2).

Box 5-2. Improving School Supervision in Pakistan and Bangladesh

Pakistan and Bangladesh have sought to improve the quality of school supervision by creating new tiers of officials. Pakistan established the post of learning coordinator in its 1979 Primary Education Project. One year later, Bangladesh introduced the category of assistant Upazilla education officer.

Pakistan. The learning coordinators in Pakistan provide pedagogical support to teachers and conduct recurrent in-service training. Each coordinator visits between ten and twenty schools at least once a month. Although their tasks vary, learning coordinators generally observe teachers in the classroom, inspect lesson plans, and seek to improve the quality of teaching. A learning coordinator in Sind, for example, gave demonstration lessons and prepared audiovisual aids.

The major benefits of the learning coordinators' work have been to reduce teacher absenteeism significantly, to improve the quality of teaching, to increase student enrollment and attendance, to promote a greater sense of professionalism among teachers, and to improve communication between schools and the district management. The program is not without problems, however, which stem from the process through which learning coordinators are brought into the district educational offices. Initially, the learning coordinators were hired and paid by units set up under the Primary Education Project. This created jealousy and resentment among incumbent district education officers, who felt that their authority was being undermined, and among supervisors, who did not receive the special allowances and motorbikes that the learning coordinators received.

Bangladesh. The new supervision system in Bangladesh recruited 225 assistant Upazilla education officers to supervise teachers and principals. The creation of these supervisory positions dramatically improved the ratio of supervisors to schools from 1:100 to 1:20. Officers visit the schools at least twice a month, once for school-based teacher training and once for general supervision. The quality of these officers is particularly high since they were selected from a field of 5,000 applicants, mostly head teachers who had degrees and taught in primary schools.

Supervision focuses on classroom management, teacher absenteeism, and the condition of the physical facilities. The result of scheduled and surprise visits has been to increase teacher attendance. Little emphasis has been placed on subject teaching. Having recognized this gap, however, the ministry plans to include supervision of subject teaching in the training program in the future. Another major problem is the lack of transportation, since officers visit schools throughout a wide geographical area. Finally, few women have been recruited, which does not support efforts to increase the enrollment of girls.

Sources: Verspoor 1989; Warwick, Reimers, and McGinn 1989.

Supervision has also been improved in countries that have implemented school cluster systems. A school cluster groups schools for administrative and educational purposes. In many cluster systems, smaller schools are attached to a core school, which exercises a degree of authority over them. All schools in the cluster benefit to the extent that they share educational resources, collaborate in staff development and curriculum improvement programs, and involve principals and teachers in local efforts to improve school effectiveness. In rural areas where a school may have no more than one or two teachers, the cluster reduces the teachers' sense of isolation and gives them the opportunity to discuss problems and learn from each other.

Clusters have been documented to improve the communication among schools, the sharing of resources, the frequency of in-service workshops, and the sense of purpose and morale of school personnel. The cluster system also decreases teacher absenteeism, reduces repetition rates, and increases the number of students passing scholarship examinations. Clusters have the greatest impact when the district-level administration is relatively decentralized: educational specialists can interact more frequently with clusters, and financial officers can respond quickly to their requests. A study of school clusters in Thailand found that the influence of clusters depends on both the efficient use of resources and the degree of school receptivity. The degree of school receptivity, in turn, depends on school-level management, particularly the leadership of the principal (Tsang and Wheeler 1991). Clusters are prevalent in Latin America, where they are called *nucleos*; they are also found in Burma, India, Nigeria, Papua New Guinea, Sri Lanka, and Thailand (see box 5-3).

Increasing the authority of intermediate agencies over budget and personnel decisions strengthens their ability to respond rapidly to local needs. In consultation with principals, intermediate organizations can develop an annual budget for the schools and for their own activities. At a minimum, a discretionary line item can be included in the budget to enable district or regional offices to reallocate small amounts during the year to meet unanticipated problems or to encourage specific initiatives. Authorizing schools to purchase materials and services locally or in bulk can improve the efficiency of schools. Where resources are available, the intermediate level can administer school improvement funds.

Finding practical ways for principals and the staff of intermediary organizations to participate in selecting and assigning school staff is also important. In small countries, teachers need to be appointed and assigned at the national level to ensure that rural schools are staffed. In larger countries, this function can be decentralized and assigned to intermediary organizations at the state or provincial level. In both cases, consulting principals before assigning teachers to specific schools allows them to articulate their needs and strengthens their authority over teachers and teaching. Intermediate organizations influence school quality more when they have a voice in appointing principals from a list of qualified candidates.

Box 5-3. School Clusters in Sri Lanka, Thailand, and Papua New Guinea

School cluster systems—which group schools for administrative and educational purposes—can be placed on a continuum reflecting how tightly they are organized and how much authority is vested in the designated head.

Sri Lanka occupies one extreme of this continuum: each cluster is organized as a partial federation or amalgamation of neighboring schools, and the cluster's head is authorized to move staff and resources among schools. Typically, a large multilevel school that has a proven record of preparing students for higher-level education is linked with ten or more smaller schools that have fewer resources. The largest school is usually designated the core school; its principal heads the cluster and steers a coordinating council. This council plans common activities, including joint development projects, fund-raisers, educational programs, and extracurricular activities. To strengthen weaker schools, teachers move among schools in the cluster on a short- or long-term basis. Clusters do not receive additional external resources, so they typically share scarce educational materials, such as library books, science laboratories, and audiovisual equipment, and plan joint in-service teacher training.

The system is similar in Thailand, where clusters are tightly organized and schools share many activities and resources. An example is the tin-box library project in which leaders in one cluster purchased seven tin boxes and filled them with library books. Each school had one box for fifteen days and then passed it to the next school, enabling each to receive a far wider selection of books than they would have had on their own.

In contrast, clusters in Papua New Guinea are loosely organized, and leaders have little formal authority. Schools are only grouped for the specific purpose of in-service teacher training and sharing a common resource center. These centers, besides conducting in-service training, house an education library, provide facilities and resources that teachers can use to develop teaching aids, offer loans to schools for specific educational equipment, participate in curriculum development and implementation, provide a forum for discussing school problems, and conduct research and evaluation.

Source: Bray 1987b.

DELEGATING AUTHORITY TO SCHOOL MANAGERS. At the school level, principals need the authority to manage their schools effectively. To do so, they require a high degree of control over matters such as allocating instructional time, evaluating and disciplining teachers, using material inputs, developing staff, mobilizing and using community resources within general guidelines, and evaluating the progress and problems of their students and school. Without the authority to make or influence these decisions, principals cannot improve schools. Restrictions that limit the decisionmaking ability of principals should be reexamined and modified as their responsibilities and functions are realigned. In many cases, delegated authority must be accompanied by appropriate resources. In all cases, granting authority to principals must be coupled with giving them adequate and appropriate training and resources.

STRENGTHENING COMMUNITY INVOLVEMENT. Parent-teacher associations or councils can play an important role in overseeing district or school programs. School management committees, for example, that are composed of community representatives, parents, and teachers, have helped to improve school enrollment and maintenance in Bangladesh. Giving community groups a voice in efforts to improve schools may increase the flow of community resources to schools. Community support may also increase the demand for education and ensure that school and district administrators are held accountable for their decisions.

Communities support their schools when the perceived benefits outweigh the costs of education. The benefits of improved schooling heavily outweigh the costs in the medium term. In the short term, the nature of community support will depend on three factors: the ability of each school to respond to community interests and needs, public perception that school leadership is actively seeking to improve, and the establishment of mechanisms that give community organizations a voice in school affairs.

Developing Managerial Competence

Strengthening managerial competence is essential at all levels and requires substantial resources and a long-term perspective. The lack of skilled managers and the low level of morale should be tackled by providing systematic staff development programs, increasing professional opportunities and incentives, clearly defining career paths, and establishing systems for assessing performance. Training must be linked to clear, long-term strategies for organizational development. To maintain managerial competence, countries must develop specialized institutes for training educational managers at all levels (see box 5-4 for an example of such an institute in Malaysia). Training lower- and higher-level managers together is important because it facilitates communication, establishes networks, and perhaps most important, imparts a sense of common purpose. Managers at various levels may use different skills, but all managers should have the training to do their jobs effectively, a clear idea of the opportunities for advancement, and the incentives to perform well and advance.

DEVELOPING CENTRAL-LEVEL MANAGERS. Refining and testing the criteria for selecting managers and improving the assessment of individuals as potential managers are important activities that deserve more attention. Although

Box 5-4. Management Training in Malaysia

In Malaysia, as in many developing countries, the great majority of educational managers and other professional staff lack the training and experience they need to fulfill their professional roles effectively. To remedy this, the government established the National Institute of Educational Management, a training institute designed to improve the planning, implementation, and management capability of public administration staff. The government also limited the fellowships available for obtaining a master's degree overseas, in order to encourage students to seek training locally. The cost of sending one student overseas for training was estimated to equal the cost of training seven students in Malaysia.

The Institute moved into its new 121-acre site at Sri Layang in mid-1985. From 1985 to 1987 the number of participants taking the Institute's 100 courses each year rose steadily from 6,235 to 8,665—more than double the number originally projected. School and higher-level managers were trained together. Each year more than 3,500 of the trainees were educational managers, including principals, while the remainder were professional staff from the Ministry of Education, other government agencies, and voluntary and private organizations. Course participants saw the Institute as an effective organization for training and upgrading educational managers and other professional staff.

personal ambition, academic qualifications, and length of service all contribute to successful management, they do not in themselves produce good managers. The practice of promoting talented technical or professional staff (teachers and principals) to the position of central-level manager should be replaced by recruiting or identifying individuals with specific managerial skills.[3] Most central-level managers should have substantial analytic skills that can be sharpened by training programs that provide experience in a broad range of essential government functions. Standard criteria should be used to evaluate managerial performance, and poor performers should receive prompt feedback and coaching; moreover, arrangements should be made for removing individuals whose performance does not improve after a reasonable period. Competition based on performance as well as training can ensure that the caliber of managers is high.

All managers should have a basic level of education and training in the fundamentals of managerial practice. Training should focus on the following topics, particularly as they apply to education policy and practice: managing human, financial, material, and information resources; setting objectives and strategies; motivating, training, counseling, and developing the careers of staff; supervising and controlling the unit's work and evaluating its results;

managing financial systems and procedures; and defining the nature of accountability, delegation, communication, and decisionmaking.

DEVELOPING INTERMEDIATE-LEVEL MANAGERS. Intermediary supervision is crucial for improving school effectiveness. Individuals promoted to the intermediate level should be chosen not for their academic qualifications and length of service alone but also for their past performance. In most cases, intermediate-level managers should be recruited from the ranks of successful principals or teachers.

In many cases, intermediate-level managers fulfill an inspection function that may be separated from a support function, depending on the number of staff and the amount of budget resources. They still must be trained, however, and much of that training should echo the basic management training courses suggested for higher-level managers. Particularly important is training focused on curriculum objectives and teaching methods, the creation of timetables and the use of school facilities, personnel management and evaluation, maintenance of facilities, student record keeping, and accounting and statistical data collection. Ideally, staff at the intermediate level would be sufficient for supervisors to spend one day in each school every month and to organize at least five days of professional development for all teachers within the area every year.

DEVELOPING SCHOOL MANAGERS. Research on the role of principals in improving schools is compelling. Virtually every line of inquiry identifies school managers as a key ingredient for success (Fullan 1982). Yet little is known about what makes a successful school manager or how to train one.

In most cases, the process for selecting school managers should be more systematic. Selection could be improved by establishing a larger pool of candidates than the current needs. Candidates would be teachers who combine seniority and a record of achievement. Final candidates could then be chosen based on their performance in an intensive preinduction training program. Preservice training could also take the form of an on-the-job apprenticeship with a good principal. Such apprenticeships are particularly effective in the early stage of career development (Leithwood 1989). The experienced principal acts as a mentor to the assistant or apprentice principal and often receives a stipend. Vacation workshops reinforce on-the-job learning by bringing apprentices together for additional training.

Although principals in both preservice and in-service training programs would study many of the topics covered in programs for supervisors and higher-level managers, they would concentrate more specifically on developing instructional leadership skills. A training course for primary school principals and district inspectors in Malawi includes courses in educational psychology, curriculum development, testing and evaluation, study skills, educational administration, inspection and supervision, trends in primary education, audiovisual education, in-service teacher education, school management and organization, educational leadership, topics in primary school administration, and curriculum studies in a number of subjects, including Chichewa, arithmetic, science, and English. This curriculum is taught over the span of three years—eight weeks per summer—and combined with a seven-month field assignment each year.

An in-service training program for principals in Thailand was particularly successful in changing the behavior of principals. The program, administered from 1985 to 1988 for all principals in the country, consisted of three phases. The first phase included a one-day orientation in which all participants were tested on their knowledge of six areas of administrative responsibility: academic development, personnel, general clerical duties and finance, student affairs, building and facilities, and the relationship between the school and the community. Participants then worked on exercises, which they completed during an intensive study period. The second phase, which began two weeks later, consisted of five days of group activities, including simulations of typical administrative problems that principals face. At the end of this phase, another test was administered. Candidates who scored below 60 percent were asked to retake the test a month later. The third phase consisted of a year of follow-up supervision by district and provincial supervisors. Principals who passed the second test and received good evaluations during the follow-up year were awarded certificates (Wheeler and others 1989).

This in-service training program, coupled with a national test for sixth-grade students, led 2,000 principals to resign or be voluntarily reassigned to teaching. According to officials in Thailand's Office of the National Primary Education Commission, principals understood and accepted the message that being an effective school administrator requires leadership, hard work, and attention to academics. In fact, the service training program was deliberately more stressful and formal than a comparatively relaxed companion in-service program for teachers. Education officials wanted principals to recognize that they were accountable for what occurred in their schools. While this program was being administered, the requirements for becoming a principal were also changing. Minimum qualifications were set so that appointments require the approval of district and provincial administrators, and all candidates must complete a training program in educational administration. Candidates are now chosen by how they score on the tests that complete the training program and the quality of the research project they undertake in educational administration (Wheeler and others 1989).

Continued professional development requires a sustained program of support. Associations of principals, with newsletters and other professional activities, can be effective (Leithwood 1989). Periodic in-service seminars, organized both within and across districts, also encourage principals to discuss problems and seek solutions. Feedback on performance and recognition for accomplishment are important incentives for continued improvement.

Strengthening Information Systems

Good management requires the systematic collection and use of information that supports efforts to improve school effectiveness. Information must be useful for and made available to policymakers whose decisions affect the delivery of educational services at all levels: this includes teachers, principals, and district staff. Information systems should be strengthened in three areas: testing, monitoring, and research.

MEASURING LEARNING. Mechanisms to routinely measure student learning are essential for improving the quality of education. Strengthening tests, including achievement tests and examinations for certification or selection, can yield positive effects. First, the content of tests can be changed to improve pedagogy. For example, Kenya used to give graduating primary students a certification examination on which 74 percent of the questions required the recall of facts; none tested reasoning skills. At the time, teaching practices stressed memorization. After the examination was reformed in 1974, only 23 percent of the questions tested recall and 28 percent tested reasoning ability, and teachers changed their teaching practices to meet the requirements of the examination (Somerset 1987).

Second, both certification/selection examinations and other kinds of tests can be standardized with respect to content, administration, and scoring. Standardizing the content requires that the same or equivalent questions be asked of all students. Standardizing test administration requires uniformity in the written and verbal instructions given to examinees, in the length of time afforded them, in the materials provided to them, and in the physical testing environment. Standardized scoring requires explicit, impartial procedures for correcting tests. Certain formats for questions—multiple choice, true-false, and matching—minimize subjectivity and are less expensive to score than other formats, such as as essay or open-ended questions, or "performance assessments" such as portfolios or samples of behavior. For test scores to be meaningful, student performance must be measured against an inelastic yardstick of achievement, such as a norm reference group or criteria references (Anastasi 1988; Berk 1984). Such standardization enables comparisons to be made across regions within a country and over time to monitor progress.

Third, sample-based assessments can be substituted for periodic tests of achievement administered to all students. A number of industrial and developing countries evaluate the effectiveness of their education systems in this manner—by testing representative groups of students in certain grades on selected subjects (see R. Ibrahim, Satoshi Kawanobe, M. M. Miguel, and Winai Kasemsestha in NIER 1986). For purposes of international comparison in selected subjects, sample-based assessments have been carried out recently in the Republic of Korea, two states in Brazil, twenty provinces and cities in

China, and two cities in Mozambique (Educational Testing Service 1991); a number of other developing countries are planning to follow suit (IEA 1991).

Sample-based assessments have several advantages over testing the entire population of students, especially as regards total (not unit) costs and the quality of the data. Sample-based assessments typically require the employment of special test administrators; teachers are not allowed to test their own classes. This ensures that the administration conforms to standard procedures and that the data are treated appropriately. One important consequence is improved data quality and completeness. Although sample-based assessments may cost more for each student tested (because of sampling and administrative expenses), the total costs are lower (because only a fraction of all students are tested). This means that the direct costs of printing, transporting, storing securely, and scoring tests are far less than those incurred when testing an entire grade or age cohort.

Fourth, standardized achievement testing can also improve the quality of education by producing tangible criteria that hold the system accountable for education. A standardized achievement testing program that permits scores to be disaggregated by district enables central staff to monitor the progress of districts over time, as a means both of controlling quality and of identifying districts that need additional help. Beginning in 1978 in Kenya, the Certificate of Primary Education (CPE) mean scores for districts and schools were widely publicized. Field officers in low-scoring districts and teachers in low-scoring schools came under immediate public scrutiny: some districts demoted their principals, while others took positive steps, such as forming parent-teacher groups. Since then, the annual issuance of the CPE results has created much public discussion, and "almost certainly, it has been this public concern which has been responsible for the upturn in the performance of the weakest districts in 1980 and 1981" (Somerset 1987, p. 106). Similarly, the initiation of a national test for sixth-grade students in Thailand has increased the accountability of teachers, principals, schools, and districts and thus increased student achievement (see box 5-5).

Finally, standardized achievement testing is critical for monitoring and assessing the impact of policies and programs. Major reforms to improve the quality and efficiency of education rarely analyze how a particular reform affects both costs and learning. School finance analysts concentrate on the cost dimension, while ignoring learning outcomes; educators focus on student achievement, while ignoring costs. Questions of efficiency can be addressed only by combining measures of learning achievement with cost information.

A central testing agency, with enough staff to provide testing extension services to schools and districts, can lend substantial support to school improvement programs. By developing tests for national assessments and providing technical support at the local level, a testing agency can help local educators evaluate the effect of school improvement initiatives. While few

Box 5-5. Establishing Accountability by Testing Achievement in Thailand

Responding to the lack of administrative control being exercised over the thousands of primary schools in Thailand, the Office of the National Primary Education Commission initiated national testing for sixth-grade students in 1984. The program began with a pilot text in 15 percent of the districts in every province, expanded the following year to a sample of students in every district, and has continued that way ever since.

From 1984 to 1988 the Commission invited the educational directors of all provinces to annual meetings where mean scores, standard deviations, and rankings based on scores and degree of improvement were announced. Provinces that had outstanding scores or that had made dramatic gains presented their strategies at these meetings. Directors from provinces ranking in the bottom third met privately with the secretary general and other key staff of the Commission to discuss specific problems, plans for improving test scores, and special needs that might justify additional resources.

Provinces began to rank districts and to meet regularly with the heads of the district offices to discuss their progress. In turn, districts initiated their own testing systems and began to rank schools and individual teachers by how their students performed on the tests. Now school principals often use test results as a criterion for recommending specific teachers for merit promotions.

Schools, principals, and classroom teachers have indeed begun to pay greater attention to the academic task of schooling. Except in 1987, the achievement scores and percentage of students who have mastered a given area have increased substantially (see table 5B-1). Although the reason that scores dropped in 1987 is unclear, scores remained, nonetheless, significantly higher in 1987 than in 1984.

Table 5-B1. Student Achievement on the National Sixth-Grade Examination in Thailand, 1984–87

	Mean score				*Percentage of students with satisfactory achievement*			
Subject	*1984*	*1985*	*1986*	*1987*	*1984*	*1985*	*1986*	*1987*
Thai language	49.1	56.8	58.4	56.3	47	69	76	64
Mathematics	33.1	36.5	47.8	46.2	11	18	41	40
Life experience	44.0	45.7	54.5	50.8	30	37	65	52
Work-oriented study	55.3	57.6	65.8	62.1	55	60	81	67

Sources: Thailand, Ministry of Education 1987; Wheeler and others 1989.

countries at present use standardized tests for monitoring achievement, Chile, Ethiopia, Indonesia, Jamaica, Kenya, the Republic of Korea, and Malawi have national testing authorities that set and mark examinations. These authorities provide a base for broadening the role of professional bodies in assessing educational outcomes. Other countries should consider establishing similar authorities.

BUILDING MONITORING SYSTEMS. Strong monitoring systems depend on the routine collection and analysis of basic education statistics and indicators. Such data can provide descriptive "snapshot" information on a school system, trace anticipated changes from implemented reforms, and enable diagnostic investigation of their relationship. The most valid and effective indicators, able to identify new problems as well as address old questions, "must be based on a model of the educational system that is sufficiently general, yet sufficiently complete, that its validity will continue over time" (Porter 1991, p. 15). This can be achieved by carefully selecting and designing indicators to measure four aspects of an education system: the inputs, the process, the outputs, and the outcomes.

Input indicators are those that capture quantitative, qualitative, and cost data on such things as physical facilities, students, teachers, administrative staff, and materials and equipment (textbooks, writing materials, workshops, library books, and so on). Process indicators might analyze aspects of the curriculum, school organization, or teaching style. Output indicators include years of schooling attained by students, dropout and repetition rates, number of graduating students, and measures of student achievement. Outcome indicators cover spinoff effects of education relating to employment, earnings, health, and nutrition.

In many cases a testing agency can undertake the monitoring function and thereby take advantage of its own technical expertise. Collecting and reporting data from schools and aggregating them at the intermediate level allows comparisons of student achievement to be made across districts, which enhances the ability of districts to monitor indicators of school quality. By relating input indicators to output indicators, ratios measuring the effectiveness of the system can be derived. By incorporating cost data, the relative efficiency of these ratios can be ascertained. Of course, if the basic statistics for indicators are not assembled on a regular basis in a timely manner, then no reliable measures of effectiveness or efficiency can be obtained.

DEVELOPING RESEARCH CAPACITY. Systems of testing and monitoring will give the ministry of education a rich flow of data and information on schools. In turn, the ministry must have the capacity to use these data effectively to identify good practices and to spot problems. Both economic and social research will be needed to complement assessments based on routine information. This research should address issues such as the effectiveness of learning materials, the impact of training on the performance of teachers and principals, the educational outcome and equity effect of strategies to improve schools, the effectiveness of supervision, and the impact of alternative instructional practices. Studies of cost are needed at different levels. Once the research is complete, information about good practice must be disseminated among schools. Publicizing successful school-based initiatives not only provides models of locally developed innovations but also reinforces and rewards local accomplishment.

Combining these elements (testing, monitoring, and research) into a strong information system requires comprehensive and long-term institutional development strategies. Experience suggests that these functions should be located in specialized institutions, which require substantial resources and professional development. Korea, in particular, has created an array of institutes that significantly improve educational policymaking (see box 5-6).

Summary

Increasing the effectiveness of educational inputs and improving the quality of teaching can improve student achievement, but the success of such efforts depends on the managerial and institutional strength of the entire education system. Three major problems severely weaken the managerial and institutional capacity of the education system in developing countries: (a) ineffective organizational structures, (b) lack of managerial competence, and (c) poor information systems.

Addressing these problems requires both time and resources. Efforts to improve organizational structures should focus on developing national institutions and support agencies, encouraging local initiatives to improve schools, providing intermediate-level support to schools, delegating authority to school managers, and strengthening community involvement in education. Managerial competence can be fostered by providing systematic programs to develop staff, increasing professional opportunities and incentives, clearly defining career paths, and establishing systems for assessing performance. To strengthen information systems, national systems for testing achievement should be created, monitoring systems built, and research capacity expanded. Only by developing strong managerial and institutional capabilities will countries be able to achieve and sustain high-quality primary education systems.

Box 5-6. Educational Research Institutes in the Republic of Korea

Testing, monitoring, and research are essential components in the success of management reforms. Korea's research centers assess and provide information about the educational system and disseminate examples of good practice among schools.

The Korean Educational Development Institute is an independent, government-funded educational research and development organization. Founded in 1972, it conducts comprehensive and systematic research and development on educational goals, content, and methodology; helps the government formulate educational policies; produces educational programs for television and radio; improves the effectiveness of the teaching and learning process by exploring the potential advantages of using broadcast media in the educational process; publishes and disseminates significant findings; and promotes cross-cultural exchange in the field of education.

The National Institute of Educational Evaluation has exclusive responsibility for evaluating education at the national level. Established in 1985, the Institute attempts to raise the social credibility of educational evaluation by managing various types of national examinations. It assumes a central role in systematic research on educational evaluation. Its major functions include managing various types of national examinations, evaluating the academic achievement of students nationally, researching methods of evaluation, supporting the evaluation activities conducted by individual schools and other educational organizations, and compiling and analyzing statistical data and materials on education.

The Korean Institute for Research in Behavioral Sciences was founded in 1967 and charged with conducting scientific research on human problems in order to further the nation's social development. In the field of education, the Institute conducts research, develops materials, and provides field guidance to raise the overall level of Korean education and advance educational development. The Institute also produces testing instruments to measure the psychological characteristics and capacities of students.

City and provincial boards of education also sponsor educational research. These boards establish and operate independent research organizations that conduct research, theoretical inquiries, and research-based development activities. They also design and produce educational materials. As of 1986 fourteen such educational research institutes were functioning in Korea. Moreover, institutes attached to colleges and universities carry out academic research on educational issues.

Source: Bolvin and others 1978.

Notes

1. Exceptions are permitted for students with physical impairments, such as blindness.

2. For example, primary schooling in Kenya is supported by a National Examinations Council, a Teachers Service Commission, the National Council for Science and Technology, and the Kenya Institute of Education. Ethiopia has established organizations for developing and testing curriculum, producing science equipment, publishing textbooks, maintaining equipment, conducting educational research, and monitoring performance. The Korean Educational Development Institute, established to provide professional leadership for educational reform, offers a high level of curriculum development and research support to both primary and secondary education.

3. Evidence on the effectiveness of early identification programs is not available for developing countries.

6

Improving Equitable Access

DESPITE IMPRESSIVE INCREASES in enrollments during the past two decades, many governments have not achieved universal and equitable access to education. As of 1985, between 114 million and 145 million school-age children, many from traditionally disadvantaged segments of society, were out of school in developing countries. As enrollment begins to slow and the population continues to grow, extending access to out-of-school children becomes harder and more expensive. This represents a major challenge for developing countries, especially the poorest ones.

Equitable access is only one aspect of educational equity, which also includes equity in process and in outcome (Campbell and Klein 1982). Equitable access has been defined as equal opportunity to enter school and as equal access to learning; both imply differential treatment in admissions. Equity in process means education that is fair and just but does not necessarily treat everyone the same. Equity in outcome implies that all students are provided with educational experiences that ensure the achievement of uniform goals; this may well imply differential process. This discussion of equitable access focuses both on access to school and on access to learning.

Disparities in Schooling

Broadening access to schools is not just a matter of increasing the number of school places. School participation is an interaction of supply, demand, and the learning process. Supply refers both to the availability and to the quality of school facilities, materials, and teachers. Demand is created by the decisions that parents make based largely on the opportunity cost of schooling, but also on the influence of cultural and religious factors. The learning process involves the experience that children have in school. Educational supply, educational demand, and the learning process are not consistent across the entire primary school population. Certain groups of children are educationally disadvantaged in virtually all societies; this is reflected in their enrollment, tendency to stay in school, and educational attainment. In countries with high enrollment and dropout, efforts to keep children in school can have as much impact on the overall enrollment rate as efforts to increase opportunities for enrollment (McGinn 1988). Dropout is particularly alarming at the primary level because children who drop out of school early soon lapse back into illiteracy.

Four characteristics significantly affect the access children have to school, their demand for education, and their treatment within school: their residence, gender, poverty, and minority status. These characteristics often occur together, and the effect of this interaction is significant.

Residence

The single most important determinant of primary school enrollment is the proximity of a school to primary school–age children. Since schools are readily available and accessible in urban areas, urban children are more likely to attend school than rural children. The International Council for Educational Development estimates that less than half of all rural children in most countries and as few as 10 percent in many countries complete four or more grades in schools (Anderson 1988). Major impediments to education in rural areas include (a) a general lack of resources, including teachers, materials, facilities, and equipment; (b) a lack of reinforcement for education; (c) language problems that arise when the curriculum is taught in a national (usually urban) language that is not used in rural areas; (d) household and production chores that compete with the school schedule; and (e) primary schools that offer, for example, only three or four grades (ICED 1974).

Studies have repeatedly demonstrated that distance from school is a critical factor in determining whether or not children, especially girls, attend school. In some cases parents have to pay for transportation when the school is far from home. Even when children walk to school, families incur indirect costs: the time that children spend in transit and the fatigue they feel after making the trip. In Nepal, for every kilometer that a child walks to school, the likelihood of attendance drops 2.5 percent (CERID/WEI 1984). In Egypt 94 percent of the boys and 72 percent of the girls enrolled when the school was located within 1 kilometer of their homes; when the school was 2 or more kilometers away, the percentages dropped to 90 and 64 percent, respectively (Robinson and others 1984). In Côte d'Ivoire 40 percent of school-age children, nearly all of them in rural areas, have no access to educational facilities of any kind (Glewwe 1988).

Enrollment and dropout rates are decidedly worse in rural than in urban areas. Table 6-1 shows the regional enrollment rates in selected countries. These figures reflect differences not only in residence (that is, urban or rural), but also in socioeconomic level and often in ethnic or racial composition.

Table 6-1. Regional Disparities in Gross Enrollment in Primary Schools, Selected Countries, Various Years
(percent)

		Gross enrollment		
Country	*Year*	*In region with highest enrollment (1)*	*In region with lowest enrollment (2)*	*Difference between (1) and (2)*
Bangladesh	1987	40	20	20
Brazil	1982	101	74	27
Burundi	1979	59	15	44
Ethiopia	1982	95	15	80
Guatemala	1985	76	53	23
Kenya	1984	119	22	97
Malawi	1987	72	25	47
Malaysia	1976	91	76	15
Niger	1980	65	1	64
Pakistan	1986	64	44	20
Senegal	1981	86	24	62
Sierra Leone	1977	95	45	50

Source: World Bank data.

Such figures underscore the interrelationship of factors that affect participation in schooling.

In Guatemala the enrollment and dropout rates sharply differ between urban and rural schooling. In urban areas 56 percent of seven-year-olds enter school and 8 percent drop out compared with 25 and 19 percent, respectively, in rural areas (Lourié 1982). In Brazil the disparity in dropout rates is even greater: the probability of dropping out of school is about three times higher in rural than in urban areas within the same region (Psacharopoulos and Arriagada 1987).

Coleman and Clark (1983) also found a discrepancy in the curriculum of urban and rural schools. They reported that 73 percent of the urban elementary schools in Liberia followed the prescribed curriculum compared with 39 percent of the rural schools. This discrepancy probably reflected a shortage of staff and materials in the rural schools.

The achievement of urban and rural students also differs: urban students scored significantly higher than rural students on tests in Peru, the Philippines, and Thailand (Arriagada 1983; Lockheed, Fonacier, and Bianchi 1989; Thailand, Office of the Prime Minister 1981). In El Salvador, however, rural students scored higher than urban students, perhaps because they were older and many were repeater-passers—that is, students who had passed but were repeating the year because no higher grade was offered (Robinson 1977).

While rural children face significant barriers to enrolling and staying in school, urban students in highly dense areas attend school in overcrowded and often dilapidated classrooms. A 1986 study of school quality in Malawi

Table 6-2. Student-Classroom Ratio in Four Cities in Malawi, 1986

City	Classrooms	Students	Student-classroom ratio
Blantyre	547	32,217	59:1
Lilongwe	279	43,792	157:1
Mzuzu City	94	11,824	125:1
Zomba	134	7,565	56:1

Source: Malawi, Ministry of Education and Culture 1986.

found, for example, that the average class size was eighty-eight students for standard three; sixty-four students for standard four; and seventy-three students for standard seven (Malawi, Ministry of Education and Culture 1986). In some cities the student to classroom ratio exceeded 100:1; in Lilongwe it reached 157:1 (see table 6-2). Urban schools in Brazil are severely overcrowded, even though most operate with three and sometimes four shifts each day. An internal World Bank report states that in urban areas in Pakistan, as many as 120 children are routinely crammed into a 12' x 17' room; each child thus has 1.7 square feet of space, while the international standard is approximately 14 square feet per child. Such overcrowding is clearly not conducive to learning.

Gender

Although the disparity between the enrollment of girls and boys in primary school has narrowed since 1960, the percentage of girls who enroll continues to lag behind that of boys throughout most of the developing world; Latin America is the exception. In developing countries about 25 percent of all school-age children, and 40 percent of all girls, are not in school. In Africa the enrollment rate of six- to eleven-year-olds is 69 percent for boys compared with 57 percent for girls. In Asia the disparity is even greater: 77 percent for boys and 60 percent for girls (Kelly 1987). Regional averages mask, however, serious gender disparities in certain countries, as indicated in table 6-3. Moreover, not every country has reduced the educational disparity between the sexes. Since 1970 the gender gap in school attendance has actually widened in Afghanistan, Nepal, and Pakistan.

The obstacles to female education stem from many factors: national educational policies that affect boys and girls differently; uneven distribution of primary schools, especially in rural areas; lack of schools for girls in systems segregated by sex; shortage of female teachers and general reluctance among females who have their certification to teach in isolated rural areas or in urban slums; perceived irrelevance of primary school curricula to women's employment possibilities; demand for the household labor of girls; late entry of girls in school and increased likelihood of pregnancy or preparation for marriage; and restrictions placed on the physical mobility of older girls (Chamie 1983).

Table 6-3. Gross Enrollment by Gender in Primary Schools, Selected Countries, Various Years
(percent)

Country	*Year*	*Total*	*Male*	*Female*	*Male-female difference*
Afghanistan	1988	25	33	17	16
Bangladesh	1989	70	76	64	12
Benin	1988	63	83	43	40
Bhutan	1988	26	31	20	11
Bolivia	1987	91	97	85	12
Burkina Faso	1987	32	41	24	17
Burundi	1986	59	68	50	18
Central African Rep.	1988	67	83	51	32
Chad	1987	51	73	29	44
Côte d'Ivoire	1985	70	82	58	24
Egypt	1987	90	100	79	21
Ethiopia	1988	36	44	28	16
The Gambia	1988	61	76	47	29
Ghana	1988	73	81	66	15
Guatemala	1986	76	82	70	12
Guinea	1988	30	42	19	23
Guinea-Bissau	1987	53	69	37	32
India	1988	99	114	83	31
Liberia	1984	40	51	28	23
Malawi	1988	72	79	65	14
Mali	1987	23	29	17	12
Mauritania	1987	52	61	43	18
Morocco	1988	67	80	53	27
Mozambique	1987	68	76	59	17
Nepal	1988	86	112	57	55
Niger	1988	30	38	21	17
Pakistan	1987	40	51	28	23
Saudi Arabia	1986	71	78	65	13
Senegal	1988	59	70	49	21
Sierra Leone	1988	53	65	40	25
Somalia	1985	15	20	10	10
Sudan	1985	50	58	41	17
Yemen, Arab Rep.[a]	1983	67	112	22	90
Yemen, P.D.R.[a]	1983	66	96	35	61

a. At the time that data were gathered, the Republic of Yemen was divided into the Arab Republic of Yemen and the People's Democratic Republic of Yemen.

Source: Unesco 1990.

Girls are much less likely to attend school than boys and, once in school, particularly coeducational institutions, they are often discriminated against by teachers who believe that females are incompetent. This is particularly true in Latin American and Islamic cultures. Tinker and Bramsen (1975), in their study of Islamic schools in Nigeria, noted that girls did not ask questions in class, were not asked questions by their teachers, and generally had to sit in the back of the class, away from the boys. Moreover, textbooks often portray women in passive and powerless roles, thus reinforcing negative stereotypes (Michel 1986). A primary text in Swaziland, for example, depicted a boy happily playing ball and a girl shying away from a snake. A study of the twenty-nine most widely used primary textbooks in Peru found that three-quarters of all references to and illustrations of people were about men and that 80 of the 104 occupations mentioned were reserved for men (Anderson and Herencia 1983). Similar discrepancies characterize the representation of men and women in textbooks from Costa Rica, Egypt, Kuwait, Lebanon, Qatar, Saudi Arabia, Tunisia, Yemen, and Zambia (see box 6-1; Michel 1986). Although there is scant empirical evidence that sex stereotypes in textbooks affect the participation or scholastic achievement of girls in primary schools, such biases presumably affect their aspirations and expectations for the future.

Differences in the achievement of boys and girls become more pronounced at higher levels of the education system; disparities exist, however, even at the primary level. Data from the International Association for the Evaluation of Educational Achievement show that boys exceed girls in mathematics and science achievement at all ages in most parts of the world (Finn, Dulberg, and Reis 1979; IEA 1988). In Malawi, according to unpublished data from the National Examinations Board, boys outscored girls on the Primary School Leaving Examination in all subjects except house crafts and needlework, for which boys were not tested. The sex differences were quite large (two-thirds of a standard deviation for science and the general essay and one-third of a standard deviation for English and arithmetic) and far exceeded those reported for primary school students in other developing countries.

Poverty

In all countries, children of poor families are less apt to enroll in school and more apt to drop out than children of better-off families (Anderson 1988). Families pay for the education of their children both directly and indirectly. Direct outlays include school fees, activity fees, examination fees, supplies, uniforms, transportation, and lunches. Indirect or opportunity costs include the household labor not done or the income not earned by children in school. Parents decide to bear the cost of educating their children if they perceive that the returns from education (such as higher income in the future, a more productive household overall, or greater prestige) justify the expense. Because poverty is often linked to the limited educational attainment and low occupational status of the parents, poor families do not reinforce the value of education.

One study in India and Nepal found that in the richest 10 percent of the families, the rates of enrollment exceeded those of the poorest 10 percent by 50 to 100 percent (Evans 1981). In Côte d'Ivoire the enrollment rate of the poorest 20 percent of the school-age population was 33 percent, whereas that

Box 6-1. Gender Discrimination in Costa Rican Textbooks

Comparing the textbooks used in Costa Rican primary schools in 1975 and in 1985 reveals that gender bias persists despite efforts to promote a gender-neutral education. In both years, the references to females were less frequent than the references to males, and they reflected sex stereotypes. The twenty-eight books analyzed included books that were commercially distributed and books that were distributed by the Ministry of Public Education. Both types were used extensively.

Commercial books. Sexist images were as prevalent in 1985 as in 1975, when three-fourths of the images portrayed were of males and one-quarter were of females. The percentages had not changed by 1985. The most common male images were of historical figures, men pursuing intellectual activity, and men working in agriculture and ranching. In contrast, females were most frequently depicted doing domestic tasks, pursuing intellectual activity, and caring for children. The category of intellectual activity included any behavior suggesting that the person was a student, which probably explains why it was so common. The textbooks did not show females involved in professional or scientific pursuits.

One of the few stories about women is notable for its attempt to stifle female creativity. The story is about a poor street vendor who, while making plans for the future, drops her basket of wares. The text asks, "What should the woman have been doing instead of imagining future possibilities?" The implication is that the consequences of a woman's imagination are negative.

Government books. The textbooks distributed by the Ministry of Public Education in 1975 contained 1,285 images of males and 551 images of females (72 and 28 percent of the total, respectively). The most common male characters were historical figures and men or boys farming, ranching, or doing other work. The most common female images were of women or girls in domestic roles, teaching, playing games with dolls, and playing house. Again, women were not portrayed as professionals, scientists, or police officials.

In 1984 and 1985, with financing from the United States Agency for International Development, the Costa Rican Book Council rewrote and published these government-distributed books in the series Towards the Light. This series shows promise. Of the people portrayed, 69 percent were males and 31 percent were females, a slight move toward achieving equal representation. The type of images also changed. First, although girls did not appear strong or dominant, neither did they appear dependent. Second, girls were represented more often with characteristics that are stereotypically male and less often in the traditional roles of teacher and housewife. Third, the texts sometimes depicted males involved in child care and domestic activities.

Source: González-Suárez 1987.

of the wealthiest 20 percent was 64 percent (Glewwe 1988). In Egypt parents most frequently cited the cost of schooling as the reason they did not send their children to school (Robinson and others 1984). In Nigeria 98 percent of the students surveyed said that the inability of parents to support their children in school was the most significant cause of dropout (Odebunmirey 1983). Studies in Liberia, Nepal, and the Philippines found that the parents' educational attainment, income level, and attitude toward schooling were the most important determinants of children's attendance (Brenner 1982; Jamison and Lockheed 1987; King 1981; Smith and Cheung 1981). A study in Brazil found that after controlling for regional differences, the two main factors determining whether a child attended school were household resources, which had a positive effect, and the demand for child labor, which had a negative effect (Psacharopoulos and Arriagada 1987).

In poor families, children's labor is often critical to the income or survival of the household, especially in rural areas. Children who work have little or no time to attend school. In the Philippines 15 percent of boys and 9 percent of girls in rural areas must work in the paid labor force and therefore cannot attend school. Another 63 percent work in unpaid agricultural labor, which also affects school attendance (King 1981). In India and several African countries poor, rural girls seldom participate in school because they must draw water, prepare food, gather wood, tend younger children, and help with farm activities (Kelly 1987). The following quotation describes the situation in Mali:

> The reason why a lot of children don't go to school in Mali is economic. School is not free. Primary pupils have to buy desks, chalk, chairs, and make a monthly contribution. This cost is very high for the average family in Mali. Many parents who have limited resources only want to invest in boys' education and not in girls'. In the towns, girls stay at home to mind other children or to sell things from roadside stalls. (Dall 1989, p. 7)

When working children do attend school, they have little time to study, which weakens their academic performance. Poor children are also apt to be malnourished, which lowers their achievement level even further.

Child labor is not confined to rural areas. Throughout the developing world, millions of urban children work in industrial and related activities (Bequele and Boyden 1988). In 1981 in India alone, according to the national census, 12.6 million children were working in the industrial sector (Kanbargi 1988). The problems encountered by rural working children are exacerbated in urban settings. Whereas rural child labor is traditionally carried out within the context of the household, urban child labor takes place within an employer-employee structure. When children are incorporated in this structure, the parental protection that exists in domestic and agrarian activities is generally absent. Urban child laborers work long hours at strenuous and often dangerous tasks. The effect on their schooling is considerable: those who do manage to attend school are less able, less industrious, and less regular in their attendance, putting them at a disadvantage throughout their school years and beyond (MacLennan, Fitz, and Sullivan 1985). As is true of rural workers, they are unable to attend school regularly because they would lose essential income and their parents perceive that the cost far outweighs the benefit of education.

Religious, Ethnic, and Handicapped Minorities

Although few data describe their participation in primary schooling, religious and ethnic minorities appear to suffer from inequities in the education system. In Nepal in 1973, caste or ethnic background, together with the economic status and level of literacy of parents, was found to be a significant determinant of the educational attainment of girls. Within the Hindu caste hierarchy, the female literacy rate ranged from 0 percent in one group (Sarkis) to 24 percent in another (Newars). The Newars sent 53 percent of their daughters to school, while the Sarkis sent none (Unesco 1984).

Vast disparities in schooling are also seen among ethnic groups in Guatemala, compounding the contrasts between urban and rural life discussed above. Guatemala's rural world is split into two subpopulations—Indian and non-Indian—which are often estranged by cultural and linguistic barriers. While 48 percent of non-Indian children in rural areas attend school, only 26 percent of Indian children do (Lourié 1982). Dropout rates are also higher for social minorities, as seen in India, Sri Lanka, and Viet Nam (Unesco 1984).

Discrimination is sometimes explicit, as when schools are designated along religious, ethnic, or racial lines, and sometimes implicit, as when the language of instruction favors one group over another. Children who speak a language other than the language of instruction confront a substantial barrier to learning. In the crucial, early grades when children are trying to acquire basic literacy as well as adjust to the demands of the school setting, not speaking the language of instruction can make the difference between succeeding and failing in school, between remaining in school and dropping out.

Morocco, for example, has a language policy that creates a disturbing element of social bias in the education process.[1] In Morocco, French, which is not an indigenous language, is introduced in the third grade. As the unofficial working language of the government and the media, it is widely spoken and written among the educated minority. In rural areas and among the urban poor, however, children have little exposure to French outside school, and its early introduction in primary school and increasing emphasis thereafter are among the main causes of student failure and hence of high repetition and dropout rates. A survey conducted by the Ministry of Education in

1978 established that knowledge of French was the single most important determinant of successfully completing primary school examinations (Salmi 1987).

The mentally and physically handicapped also remain on the fringe of the educational system. Although the extent of their participation in education in developing countries has rarely been studied, anecdotal evidence suggests that they are often excluded from schools. Teachers are not trained to teach them, and schools are not equipped to deal with their special needs. Most systems lack the infrastructure, capacity, and resources needed to provide them with a basic education.

Strategies to Promote Equitable Access

Policies for promoting equitable access must address issues of supply, demand, and process. Policymakers must recognize and address the inequities that disadvantage certain groups of children and design policies that target those groups. In areas where the demand for education is high, the priority will have to be providing more and better facilities. In other areas, policies and programs will have to stimulate demand. In many cases, attention will also have to be paid to school and classroom processes that inhibit the achievement of disadvantaged students. The issues of supply, demand, and process overlap and interact with one another in complex ways, so categorizing strategies under these three headings is more convenient than realistic.

Increasing Supply

The push to expand access to schooling by increasing the supply of school places has dominated the agenda for education development since the 1960s. Enrollment has increased impressively since then, but access to education is still limited, and certain groups of children are completely excluded, especially in low-income countries. Efforts to increase supply must continue and should seek creative approaches for extending access to disadvantaged groups.

SCHOOL CONSTRUCTION. Building more schools is an obvious and necessary means to increase the number of school places. Yet the persistent disparity in school attendance among groups of children means that the location of new schools should be carefully mapped before construction begins. Since distance is a significant factor determining school attendance, particularly for girls and rural children, a tradeoff exists between building large schools that benefit from economies of scale but are hard to reach and small schools that are accessible but possibly more expensive.

School construction is not cheap and may require more resources than many countries can afford. Yet many countries could develop and use new school designs that meet minimum standards but are much less expensive than those typically used at present. Greater reliance on local materials could reduce the cost of school construction substantially. Furthermore, using local materials often improves the quality of construction. In Niger, for example, a classroom made of concrete costs five times more than one made of *banco*, the most common construction material in rural areas, and *banco* keeps the classroom cooler in the summer and warmer in the winter than concrete. Thus the cheaper alternative is also the most conducive to learning. In Senegal a recent pilot project has maximized the use of local materials and reduced the cost of school construction from US$300 to $155 per student place. Similar projects are under way in Burkina Faso, the Central African Republic, and Mali. Besides reducing costs and improving quality, such projects help transfer the responsibility for construction and maintenance from the central government to the local community (World Bank 1988b).

SCHOOL EXPANSION AND RENOVATION. Before undertaking the expense of building new schools, governments (or communities) should examine the possibility of upgrading existing school facilities to create additional places. Since schools in rural areas do not always offer every grade and thus limit the progress of successful students, adding more classrooms is one way of encouraging students to continue their education. Constructing buildings to replace shelterless schools with no physical facilities is another successful strategy to improve enrollment. In Pakistan, a country in which shelterless schools are widespread, the government has accorded these efforts a high priority. In many cases, using existing buildings such as churches and community centers as schools is feasible and cost-effective. Building latrines is another strategy that might increase enrollment, particularly of girls. According to Qasem (1983), 71 percent of rural schools and 53 percent of urban schools in Bangladesh had no latrines, which discouraged girls from attending.

TEACHER RECRUITMENT AND DEPLOYMENT. Schools cannot operate without teachers, and shortages of teachers are common in rural areas. Possible solutions include multiple shifts and multigrade teaching, which are discussed below. In addition, incentives are needed that encourage teachers, especially females, to work in isolated areas. Increasing the supply of female teachers is an important strategy for increasing the access of girls to schools. In India a rural community recruited female secondary school graduates to teach classes, which significantly decreased the dropout rate of children, especially girls (Chamie 1983). In Tanzania efforts to recruit and train female graduates from local primary school, combined with arrangements for certified teachers to supervise and assist them, substantially increased the number of teachers available to work in isolated areas. Other incentives might include the provision of boarding facilities, increased training, or even additional pay. A special program in Nepal to increase the attendance of girls provided free tuition, stipends, residential facilities, books, stationery, and medical allowances that permitted rural women to qualify and function as teachers (Unesco 1984).

PRIVATE SCHOOLS. The government provides primary school education in most developing countries, and public schools enroll approximately 90 percent of all primary school students. Most public schools are free, or almost free, to students. In many countries, however, tightening fiscal constraints have kept the public sector from expanding free public education. This has created a particularly serious problem for the poorest countries, where the demand for schooling is expected to increase dramatically over the next decades. One option available to these countries is to lift the restrictions on private schooling. Private schools can benefit the urban poor indirectly by relieving some of the pressure on severely crowded urban schools. Private primary schools do introduce a serious problem, however, by encouraging wealthy families to withdraw their children from public schools and reduce their commitment to public education.

Koranic schools and mosque schools also supplement the public school system.[2] From 1978 to 1983 Pakistan had 8,200 mosque schools taking in grades one, two, and three; an additional 20,000 mosque schools have opened since 1986 (Unesco 1984; Warwick, Reimers, and McGinn 1989). The major benefit of these schools is to increase enrollment opportunities for all primary school students at low cost and to encourage parents to enroll their daughters. In one of the largest provinces (Sind), mosque schools serve scattered rural settlements with populations under 200 and sometimes even under 100. Given the financial and logistical difficulties of building government schools in such communities, the use of existing mosques for primary schooling is a convenient solution. Unfortunately, these schools may not provide a high-quality education (Warwick, Reimers, and McGinn 1989). Koranic preschools or after-school classes are effective, however, and improve access particularly when their enrollment is coordinated with that of government primary schools (Wagner 1989).

MULTIPLE SHIFTS. Multiple shifts both increase enrollment and reduce unit costs. By organizing classes into separate sessions (for example, morning and afternoon shifts) and having teachers share facilities (classrooms, desks, texts, and equipment), a multiple-shift system can accommodate double or sometimes triple the number of students that a single-shift system can, as well as reduce certain capital and teacher costs. When the staff teach in more than one session, multiple shifts generally allow teachers to increase their income. In most cases, multiple shifts also lower school fees and make more working hours available to child laborers, thus benefiting poor children. A multiple-shift system can also reduce overcrowding in urban classrooms.

Multiple shifts have been used effectively and extensively in urban areas in Brazil, Burundi, Egypt, Korea, Malaysia, Turkey, Yemen, and to some ex-

Table 6-4. Official Instructional Time in Single- and Multiple-Shift Primary Schools in Selected Countries
(hours per week)

Country	*Single session*	*Double session*	*Triple session*
Ghana	22.9	19.6	n.a.
Jamaica	25.0	22.5	n.a.
Laos			
Lower school	19.0	19.0	n.a.
Upper school	22.0	22.0	n.a.
Malaysia			
Lower school	22.5	22.5	n.a.
Upper school	24.0	24.0	n.a.
Nigeria, Imo State	22.1	22.1	n.a.
Philippines			
Lower school	25.0	23.3	n.a.
Upper school	30.0	24.2	n.a.
Senegal	28.0	20.0[a]	n.a.
Singapore			
Lower school	22.5	22.5	n.a.
Upper school	24.5	24.5	n.a.
Zambia			
Lower school	17.5	17.5	17.5
Upper school	26.7	25.0	n.a.

n.a. Not applicable.
a. The school year is extended by ten days.
Source: Bray 1989.

tent Bangladesh. At the same time, governments in Korea, Malaysia, and Singapore hope to abolish multiple shifts in the near future. They view single shifts, when affordable, as much more desirable.

A problem often associated with multiple shifts is that they reduce teaching time and make covering the curriculum difficult. Multiple shifts at the primary level do not, however, necessarily shorten instructional time (see table 6-4). In fact, even the shortened double-shift allocation of instructional time is longer in some countries than the single-shift allocation in others. In Zambia, for instance, the twenty-five hours of time allocated to instruction in upper primary schools with double shifts is longer than that of schools with full single shifts in Ghana, Laos, and Nigeria (Bray 1989). Even when the school day is shortened, the school year can often be extended to compensate for the reduction in classroom time.

The savings realized by establishing multiple-shift systems can be dramatic. In Zambia the extensive use of double and triple sessions reduced by 46 percent the estimated cost of achieving universal primary enrollment by 2000 (Bray 1989). The savings realized by multiple shifts is sometimes less than anticipated, however, because (a) maintenance costs are higher as a consequence of increased wear and tear on the facilities, (b) extra space is needed for storage and for students who come early or stay late, (c) specially designed and thus more expensive classrooms are needed in some countries to withstand the afternoon sun, (d) custodians who work during very early or very late hours require extra pay, and (e) teachers must be paid more. Taking these factors into account, studies in Jamaica and Malaysia calculated a savings of 32 and 25 percent, respectively. Although substantial, these savings are lower than the estimated savings of 46 percent in Zambia (Bray 1989).

The results of studies examining the effect of multiple shifts on student achievement, although somewhat inconsistent, are generally positive. Academic achievement often varies between multiple- and single-shift schools because multiple-shift schools serve a lower-income population. When multiple shifts reduce class size and the curriculum is streamlined, student achievement does not seem to suffer. Senegal has experimented with double shifts since 1982. The initial reaction of parents was negative because their home schedules were disrupted and teachers were not given sufficient training and materials. Nevertheless, student scores on writing, reading, and mathematics are generally higher in multiple-shift than in single-shift schools. The positive results are attributed to lower student-teacher ratios.

Although generally instituted in urban areas, multiple-shift systems can also target girls and rural children. Where schools are segregated by sex and no institutional facilities have been built for girls, the establishment of multiple shifts could give female teachers and students access to school facilities. Village school systems that only offer lower primary classes could institute multiple shifts to use school facilities for upper primary classes (Chamie 1983). In fact, a shorter school day might be more welcome in rural areas, where the opportunity costs associated with school attendance are generally higher than they are in urban areas.

MULTIGRADE CLASSES. Multigrade classes, in which one person teaches several grades, also improve access in rural communities. Multigrade teaching addresses the problem of uneconomically small classes as well as that of incomplete schools. Settlement patterns in Sierra Leone demonstrate the need for multigrade classes in rural areas. A 1979 sector study, examining demographic data and using a recommended pupil-teacher ratio of 35:1, concluded that the smallest community that could support a full-sized primary school (grades one through seven) would have 980 persons—provided enrollment was 100 percent. In Sierra Leone, however, more than 65 percent of the population lived in settlements of fewer than 900 persons, and only 1.3 percent of all settlements were capable of supporting a complete primary school (Sierra Leone 1979).

Effective multigrade teaching, which requires special instructional materials and teacher training, has been successfully implemented in a number of countries (APEID 1989). The small schools experiment in Indonesia developed materials that enabled teachers to instruct all grades in one room by using conventional texts and programmed techniques. The project spread from 5 schools in central Kalimantan in 1978 to 100 schools in central Kalimantan and southern Sulawesi, 25 schools in Madura, and 132 schools in other isolated areas of East Java five years later (Cummings 1986). A detailed evaluation in 1984 showed that students who were in the project performed better and were more self-reliant than students who were not (Bray 1987a). Another example of successful multigrade teaching is the Escuela Nueva program in Colombia, which has become a critical element in the government's strategy to combat the deficiencies of its rural basic education (see box 6-2).

BIENNIAL INTAKES. An alternative strategy for rural areas where classes are uneconomically small is to admit students every other year instead of every year. This system allows communities to have their own schools without instituting multigrade teaching. A disadvantage is that some children must wait an extra year before starting school. Since rural children often have high dropout rates, a policy that delays their entry may, in fact, also reduce their time in school. Therefore, biennial intakes may not increase educational participation.

SINGLE-SEX SCHOOLS. In countries where schooling is segregated by sex, providing schools for girls is essential to increasing female enrollment and attendance. Single-sex schools are generally more important and prevalent at the secondary level, when boys and girls reach puberty, but countries such as Pakistan and Saudi Arabia maintain sex-segregated schools at all levels. Various mechanisms exist for establishing single-sex schools: constructing separate buildings for boys and girls, using the same facilities for boys and girls at different times (double shifts), and conducting single-sex classes in coeducational schools. The quality of schools and the achievement level of students do not appear to be directly affected by the coeducational or sex-segregated nature of school systems at the level of primary school.[3] Nevertheless, historical evidence from developed and developing countries indicates that girls' schools typically receive fewer material resources than do boys' schools (Tyack and Hansot 1988).

NONTRADITIONAL SCHOOLING. In general, formal primary schooling is the preferred means of teaching basic literacy and numeracy skills. Since many disadvantaged children live and work in areas where schools are nonexistent or do not fit their circumstances, other strategies for providing education may be necessary. Nontraditional schemes are particularly important for children who work during formal school hours.

Box 6-3 describes multifaceted programs in India and Kenya that increase the access of working children to schooling. Another nontraditional approach is interactive radio, discussed in detail in chapter 3. The Dominican Republic effectively uses radio learning centers in communities without schools. Young school-age children meet every afternoon under the supervision of an adult from the community; they listen to one hour of radio lessons in mathematics and language, plus some social studies and science. Achievement tests demonstrate that the children in the radio groups learn as much language and more mathematics than children attending traditional schools in comparable communities (see table 3-7). Furthermore, these radio schools cost about half what traditional schools cost.

Increasing Demand

Strategies to mobilize demand are as important as, if not more important than, strategies to increase the supply of school places. The social, economic,

Box 6-2. Improving Rural Education in Colombia through Escuela Nueva

Based on its 1978 study detailing major weaknesses in the provision of rural basic education, the government of Colombia accorded top educational priority to improving schooling in rural areas. The result was a comprehensive ten-year Rural Primary Education Plan, which set forth a national strategy for attacking key problems and built heavily on the Escuela Nueva program.

Escuela Nueva was launched in 1975 after a decade of experimentation. From its official inception in 1976, the program's coverage grew from 500 to 3,000 schools by 1983 and to 8,000 schools, or approximately 30 percent of the schools targeted throughout the country, by 1987.

The program is designed for multigrade classrooms in rural areas and seeks to achieve both educational and social goals. The program aims to (a) provide the full five-year primary cycle, (b) improve the relevance and quality of education, (c) improve student achievement, (d) improve educational efficiency and productivity, and (e) integrate the school and the community.

The strategies for implementing the program focus on curriculum, teacher training, administration, and the relationship between the school and the community. The content of the curriculum, which can be readily adapted to the circumstances of a particular community, is simple and sequential, with an emphasis on problem-solving skills. Presented as semiprogrammed learning guides, the curriculum permits a flexible promotion system. To complement the curricular materials and to meet the challenges of multigrade teaching techniques, Escuela Nueva has developed a special classroom design featuring resource corners, simple furniture, and a library with 100 books.

Escuela Nueva places special emphasis on teacher training and continuous in-service upgrading of skills. An initial ten-day workshop introduces teachers to the program's philosophy and content, teaching strategies, school organization, and student evaluation. Two follow-up workshops given during the year

and cultural factors discussed in the first section of this chapter have a powerful, adverse effect on the demand for schooling. Special efforts are needed to address those constraints. In general, strategies include reducing direct costs, reducing indirect costs, and mobilizing community support.

REDUCING DIRECT COSTS. The most obvious way to increase the demand for education is to reduce the direct costs of sending children to school. These costs are often significant in developing countries. In Nepal, for example, the total household income in rural areas is estimated to be Rs1,500–2,000 annually. The direct cost of attending primary school is estimated to be between Rs90 and Rs300 annually, which pays for copybooks, pencils, exami-

provide further training, particularly on how to adapt curricular materials to the needs of the students and the characteristics of the community. Supervisors undergo similar training, with greater emphasis on the pedagogical aspects of supervision. Regular and frequent supervisory visits form an integral part of the program and provide pedagogical rather than administrative support.

An administrative structure has been developed that delegates authority. At the central level, a national committee ensures that the program's policies are in accordance with national goals and objectives. At the departmental level, committees are responsible for planning and implementing the program and for training personnel in the department. At the school level, a student council operates the program.

Escuela Nueva considers integration of the student, school, and community as crucial not only to the school's effectiveness as a learning institution, but also to the community's assumption of responsibility for its own development. A committee of parents works with the student council to develop joint projects for the community and the school.

Sustained evaluation of Escuela Nueva has found sound, positive results. Escuela Nueva students have significantly higher levels of achievement than students in traditional schools. Their social self-esteem and civic behavior are higher, and parents are more satisfied with their children's schooling. Moreover, due to the flexible promotion system, repetition rates are lower. Teachers have more positive attitudes toward teaching and are more involved in the community than teachers in traditional rural or urban schools. Supervisors see their jobs more positively and emphasize providing support rather than merely overseeing school operations.

Sources: Colbert de Arboleda 1987; Rojas and Castillo 1988.

Box 6-3. Addressing the Needs of Child Workers in India and Kenya

To increase the access to education in developing countries, the needs of working children in both rural and urban settings must be addressed. As long as the earnings of children are essential to the survival of families, educators will have to devise creative and practical approaches to educating children. Effective ways of dealing with child workers in India and Kenya are described below.

India. Responding to the persistent social and economic problems created by millions of working children, the government of India has launched pilot projects in areas where child workers are concentrated.

One of these areas is Sivakasi, where 45,000 children are employed in the match and fireworks industry. With the ultimate objective of eliminating child labor in this industry, the government sponsored a pilot project to provide better health care, improve working conditions, raise the community's general awareness of the problems of child labor, and provide informal education. Because most children working in this industry have either dropped out of or never attended school, the educational programs concentrate on imparting basic literacy skills. These programs are organized through a registered society whose members belong to voluntary organizations involved in informal education. The Central Board for Workers' Education also organizes informal education and consciousness-raising classes. It holds ten weekend classes, with 30 pupils in a class, and involves 1,800 children each year.

Kenya. In 1964 children accounted for 21 percent of Kenya's paid labor force, and although the number of children working for wages appears to have declined, child labor is still prevalent. Since the 1977 Employment (Children)

nation fees, and school uniforms (Jamison and Lockheed 1987). Thus families could spend up to 20 percent of their income by sending just one child to school.

Several developing countries have attempted to reduce the cost of education for rural children and girls. These efforts include lowering or eliminating school fees, providing instructional materials and uniforms, offering free or subsidized transportation, directly subsidizing households for the cost of materials and uniforms, and providing school feeding programs, boarding facilities, and scholarships.

Indonesia, Kenya, and Tanzania increased their enrollment significantly after abolishing primary school fees. In Pakistan the *mohalla* school project reduced the cost of education by holding classes in homes and abandoning the requirement that students wear uniforms and even shoes; the enrollment of girls and rural children increased dramatically. Abolishing school fees can, however, have adverse results if unintended consequences are not carefully explored. In Kenya the abolition of fees increased the enrollment of poor

Rules prohibited industries from employing children, child workers are no longer officially counted. Only two programs address the needs of working children in Kenya, and both are nongovernmental. One is the Undugu Society of Kenya.

Established in 1973, the Undugu Society targets street children, mainly parking boys and girls, who direct motorists to parking bays and are between the ages of four and sixteen, and child prostitutes, who are mostly girls between the ages of twelve and sixteen. The Society sponsors schools, provides loans, and offers a number of programs in basic education, vocational training, and income generation.

The Undugu Basic Education Program consists of informal education, with an emphasis on practical skills and literacy in Kiswahili and English. The program is divided into three phases, and individuals move from one to the next according to aptitude. The more capable students are channeled into primary schools, and the majority are prepared for self-employment. The course lasts four years. The combination of this and other programs has enabled the Undugu Society to become one of the most successful private organizations providing supportive, preventive, and developmental programs for street children in Kenya. It has devised an alternative approach to primary education, shown that parking boys and girls can become useful members of society, promoted the informal sector, upgraded about 1,000 slum dwellings, helped single mothers earn decent incomes, and organized the formation of numerous small businesses.

Sources: Myers 1988; Narayan 1988; Onyango 1988.

children initially, but the government had not considered how they would make up the revenues lost from fees, and local school systems began to levy other types of fees to cover costs. The overall cost to parents of schooling increased, quadrupling in certain districts, and the enrollment of poor children began to decline. The disparity between rich and poor communities grew even greater (Nkinyangi 1982).

China, Indonesia, Kenya, and Thailand have all initiated programs to reduce the direct costs to remote populations. The Thai government provides students in rural areas, where schooling ends at an early grade, with bicycles so that they can reach distant schools with higher grades. China, Indonesia, and Kenya have constructed boarding schools for the children of nomadic groups. Providing boarding facilities is very costly, however, and their success is unclear. In Kenya planners failed to consider how poverty would affect the ability of families to use boarding schools. Boarding students had to bring their own beds, blankets, pots, and other household items. The cost of these items was prohibitive for the populations targeted, such as nomadic children.

Instead, places were filled by the children of wealthy families who had been refused entry to the limited places in their own districts and who enrolled by giving false addresses (Nkinyangi 1982).

Several countries, including Bangladesh and China, have encouraged girls to attend school by providing cost incentives. In Bangladesh 50 percent of the primary scholarships given at the end of class five are reserved for girls. In rural, mountainous, and minority areas, China operates a program that uses incentives such as boarding schools, books, stationery, medical allowances, and educational guidance. The participation rates of girls have increased to over 90 percent (Unesco 1984).

REDUCING INDIRECT COSTS. Reducing indirect costs is often as important as reducing direct costs. Strategies to reduce them include changing the school calendar, providing child care for younger siblings, and instituting labor-saving technologies.

The timing of the school year in many countries does not take into account agricultural cycles, which limits attendance in rural areas. Revising the school year to accommodate the seasonal demands for child labor on the farms and in the fields is a relatively painless and inexpensive solution. The school day may also be changed to accommodate daily work schedules by instituting multiple shifts or providing classes early in the morning or in the evening. Flexible scheduling is a key strategy for improving the schooling of girls and rural children in Bangladesh (see box 6-4).

Girls are the caretakers of their younger siblings in many countries, which severely discourages them from attending school. Schools must recognize this fact and provide facilities for day care or make other arrangements for their female students. In China, where enrolling girls has been a problem, offering child care at places of employment has improved the enrollment of girls in urban areas. In rural areas some schools provide day care for younger siblings (Colletta and Sutton 1989). Another strategy is to establish preschools close to primary schools, which not only increases the attendance of girls in primary schools, but also benefits the education of younger siblings.

Labor-saving technologies are important for alleviating the significant time constraints that keep poor working children, especially rural girls, from attending school. In 1967 three such technologies were introduced in an experimental program in Burkina Faso (Upper Volta at that time): mechanical grain mills, easily accessible water wells, and carts. Although the project did increase the participation of girls in nonformal education, it did not seem to increase the attendance of girls in formal primary school. One tentative explanation for the failure to increase enrollment was that the schools were inconveniently located, which suggests that labor-saving technologies should be combined with other strategies (McSweeney and Freedman 1980).

Box 6-4. Improving Schooling for Girls and Rural Children in Bangladesh

Educational opportunities for girls and rural children have been particularly limited in Bangladesh. To improve the enrollment and retention rates of these groups, the government has instituted a number of measures, some of which incorporate features of the Nonformal Primary Education Program sponsored by the Bangladesh Rural Advancement Committee.

The Nonformal Primary Education Program (NFPE) began in 1983 as a pilot project to develop and test a low-cost alternative to primary education. The result was a learning environment that did not alienate rural children and a schedule that could be adapted to the needs of the local community. The school schedule is arranged so as not to interfere with the work of rural children. School is open two and a half hours a day, 280 days a year. During harvest, classes are held in the early morning or late afternoon, according to the preference of the parents. The NFPE offers a three-year curriculum that includes language, mathematics, basic science, social studies, and health and hygiene. The teachers are paraprofessionals (preferably village residents and mothers or heads of household) who are trained and supervised by professional staff.

The NFPE has achieved considerable success. The dropout rate of less than 2 percent indicates that rural communities find the program appropriate to their needs. In fact, the demand from poor communities has been so great that by 1988 the program was operating 730 centers that enrolled over 20,000 students. More than 90 percent of the first year's graduates were admitted to the fourth grade in government primary schools. With 60 percent of enrollees female, the program has succeeded in attracting and retaining girls.

Drawing from the NFPE, the government is launching a satellite school project that will build about 200 lower primary schools. These pilot schools will hire a predominantly female teaching force, offer the flexible scheduling appropriate for rural areas, and involve the local community in education.

The government is also involved in the following efforts: (a) to recruit, train, and deploy more women teachers by implementing aggressive recruitment strategies, assigning teachers to schools within their own localities, and improving accommodations for female teachers in both teacher training institutions and rural areas, (b) to provide physical facilities that meet the privacy requirements of girls (such as closed latrines), (c) to collect gender-sensitive data that allow the government to monitor progress toward achieving equal access for girls, (d) to extend support for the Female Education Scholarship Program, whose scholarships have, where offered, decreased the primary school dropout rate from 15 percent in 1979 to 4 percent in 1987.

Source: Mallon 1989.

MOBILIZING COMMUNITY SUPPORT. Increasing the demand for education depends largely on persuading parents that education is valuable. Thus one of the most significant ways to increase demand is to improve the quality of education and therefore to increase the opportunity costs of not attending school. Parents should also be involved in schooling. When parents are active in the educational process, their children are more likely to attend school. Efforts to mobilize community support for education can take many forms: establishing parent-teacher associations, holding school open houses, involving the community in building schools, and broadcasting radio and television programs that highlight the positive aspects of schooling. To be effective, such programs must use the indigenous language.

Many countries have instituted a variety of these measures to garner community support. China has conducted mass "consciousness raising" campaigns to introduce the concept of gender equality and to make a frontal attack on prejudices and feudal cultural norms. In Ethiopia village peasant associations were established to encourage parents to send their children to school. In Saudi Arabia, when the government sought to lower the 99 percent illiteracy rate among women, it faced stiff opposition from Islamic scholars. The education of girls began only after the government convinced the scholars that there was no conflict with the country's religious or Arab traditions and agreed that the schools would be run by a religious functionary (Al-Hariri 1987).

Equalizing the Learning Process

Redressing discriminatory treatment requires a genuine commitment to understanding the sources of unequal treatment and taking corrective measures. In some cases discrimination is easy to identify and remedy. For example, when boys' schools receive more educational resources than girls' schools or urban schools receive more than rural schools, the obvious solution is to reallocate resources. Texts that portray certain population groups (such as females or minorities) in negative ways should be rewritten. Teachers who differentiate among students based on gender, religion, or family income and who therefore inhibit the learning achievement of some students should be retrained to use new teaching methods. As self-evident as appropriate solutions may seem, they must be designed and implemented in ways that take into account cultural and religious factors. Failure to do so could undermine solutions that would otherwise succeed.

One of the most prevalent forms of discrimination in the classroom is teaching in a language that children do not understand or with which they are uncomfortable. In many developing countries children speak a language other than the national or official language at home. In such countries, the government must choose an appropriate language of instruction. The debate over language policy is long-standing. On one side, the advocates of instructing pupils in their first language argue that the literacy skills acquired in one language can be transferred to other languages and that developing these skills is easiest in the child's home language. On the other side some argue that teaching children in a local language places them at a disadvantage for further educational opportunities and that developing literacy skills in their first language reduces the time available for teaching them the official or national language.

Determining a sound language policy is of necessity heavily influenced by the unique economic, cultural, and linguistic factors of each country. In some situations, early primary education should be conducted in the first language. This might be the case if the children's first language has developed so that they have the conceptual and linguistic prerequisites for acquiring literacy skills (Dutcher 1982). In other situations, early immersion (that is, instruction in the second language) is more appropriate—for example, when students do not share a common language and more than half of all instruction is carried out in the second language (Eisemon, Prouty, and Schwille 1989).

On balance, the circumstances prevailing in many developing countries suggest that the most effective approach is to begin with the home language as the medium of instruction and to add or switch to a second language later. With this approach, children are able to acquire basic literacy, learn the fundamentals in various subjects, and adjust to the school and its demands before they confront the task of learning a new language. However, some bilingual programs teach a second language from the outset. Box 6-5 describes two successful bilingual programs.

Governments must eventually address the educational needs of mentally and physically handicapped children if they are to meet their goals of universal access. Realistically, low-income countries will probably have to give the special needs of these children low priority since the demand for education is overwhelming and they are struggling just to create enough places for the out-of-school children without special needs. In middle-income countries, building a special education infrastructure, training teachers in special education, and providing special education classes would be an important step toward equalizing educational opportunities for all children. Box 6-6 describes one country's efforts to provide education for the handicapped.

Summary

The vast majority of the 114 to 145 million children who do not attend school in developing countries come from one or more of the traditionally disadvantaged groups in society: rural dwellers, females, the poor, or minori-

Box 6-5. Improving Equity through Bilingual Education in Guatemala and Nigeria

Bilingual programs in Guatemala and Nigeria have helped children whose home language is different from the official or national language to participate in school.

Guatemala. Historically, language was a major reason that the indigenous Mayan population had limited access to educational opportunities in Guatemala. Although most Mayan children spoke a Mayan language—more than 200 exist in Guatemala—the language of instruction was Spanish. In 1980 PRONEBI, a pilot bilingual education program, was introduced to forty rural schools. Children were taught in one of four major Mayan languages for the first three primary grades and in Spanish thereafter. As is often the case at the outset of bilingual education programs, some parents feared that their children would receive an inferior education and would not learn Spanish. These concerns were overcome when parents saw their children improve in their mastery of Spanish *and* the indigenous language. In 1986 the bilingual program expanded to 400 schools, and by 1990 the number was expected to have reached 800. In fact, national law now requires bilingual education in rural communities that speak Mayan languages.

Nigeria. A project in Nigeria shows that teaching in the native tongue throughout primary school can improve learning without keeping children from acquiring the second, national, language. Although English is the official language of Nigeria, 15 million people in western Nigeria speak the native language of Yoruba. Beginning in 1970, the Institute of Education at the University of Ife launched the Six-Year Primary School Project, which used Yoruba as the sole medium of instruction throughout primary school and taught English as a second language. Nonproject classes had three years of instruction in Yoruba and three in English. When their primary education was complete, project children were significantly ahead of nonproject children in all subjects, including English. Project and nonproject children performed equally well on the public examinations. In secondary schools, project students retained an academic advantage over nonproject students in English, mathematics, and Yoruba.

Sources: Chesterfield and Seelye 1986; Fafunwa 1987.

ties. They do not participate in school for three principal reasons: inadequate supply of school places, lack of parental demand, and discriminatory treatment in school. Overcoming these obstacles, which overlap and interact with one another in complex ways, requires a combination of policies.

Measures to increase the supply of school places usually include building new schools, expanding and renovating existing facilities, recruiting and deploying teachers more effectively, instituting multiple-shift schooling and multigrade classes, providing single-sex schooling, and adopting nontraditional education schemes. Strategies to increase the demand for education

Box 6-6. Providing Education to the Handicapped in Zambia

The Zambian government officially recognized the educational needs of the handicapped for the first time in its Second National Development Plan (1972–76). The Third Plan (1977–80) gave priority to offering preservice and in-service teacher training and establishing new schools and units for handicapped children. In 1980 the government received assistance from Sweden for creating a special education system and launched a national campaign for educating disabled children.

The immediate objectives of that campaign were to educate the public about the special needs of disabled children, to establish provincial registers of disabled children, to lay the foundations of national health and educational services for them, and to supply technical aids and prosthetic devices to as many disabled children as possible.

District ascertainment teams, comprised of a local primary school teacher, a medical assistant or nurse, and a community development worker, were formed and dispatched throughout the nation. These teams identified disabled children and designed home-based intervention programs. The national campaign used 3,000 reporting centers in 57 districts. Ascertainment officers examined 11,000 children and identified 7,247 as severely disabled. Of these, 3,209 were physically impaired, 1,549 were visually impaired, 1,390 were hearing-impaired, 626 were mentally retarded, and 473 were multihandicapped.

As the Ministry of Education was assuming responsibility for educating the handicapped, the University of Zambia included courses in special education in its curriculum for the associate certificate in education. In 1984 the National College for Teaching of the Handicapped was established in Lusaka, enrolling sixty-nine students in its two-year program. Since primary education for the handicapped is expanding slowly, the college produces more teachers than the system can absorb. At the central level, a special education inspectorate has been established, and three special education inspectors are devoted to staff development. As of 1985, thirty-five institutions were serving 2,095 handicapped students at the primary level, more than double the number served in 1980.

Source: Csapo 1987.

usually focus on reducing the direct and indirect costs of schooling and mobilizing community support. Equalizing the treatment of students in school usually involves carefully examining discriminatory practices and making genuine efforts to eliminate them.

The slow growth of enrollment, the continued, rapid growth of population, and limited resources combine to make extending access to disadvantaged groups increasingly difficult and expensive. Nonetheless, if governments want to achieve their stated goals of universal and equitable access to education, they must commit the necessary resources and design policies that will open school doors to the disadvantaged.

Notes

1. The prevailing situation in Morocco is complex. Moroccan Arabic is beyond question the most commonly used language in everyday life and the only means of communication available to the 75 percent of the adult population who are illiterate. Yet it is systematically excluded from all official aspects of public life, as is Berber, another common first language. French is the de facto language of the modern sector. Classical Arabic is the official language for political and religious reasons, yet the majority of Moroccans neither understand nor use it. The opening words of a speech given by the king after an attempt on his life recognize this linguistic isolation: "We are talking to you in dialectal Arabic so as to be understood perfectly by all elements of our Royal Army and, generally speaking, by all our people" (Palazolli 1974).

2. Koranic schools generally provide instruction in the Koran, whereas mosque schools (supervised by an imam) generally offer the same curriculum taught in government schools. The distinction between them, however, is not always clear.

3. Among eighth-grade students in Nigeria and Thailand, however, girls in single-sex schools outperformed girls in coeducational schools; see Jimenez and Lockheed (1989) and Lee and Lockheed (1990).

7

Strengthening the Resource Base for Education

CHAPTER 1 ARGUED that primary education is critical to social and economic development and thus deserves high priority in the distribution of resources. Despite education's social profitability, however, many countries underinvest in primary education (World Bank 1986). Moreover, funding for primary education is especially vulnerable to macroeconomic adjustments and austerity measures. As chapter 2 shows, real recurrent spending per student on primary education among low-income developing countries declined significantly from the 1970s to the 1980s, largely in response to growing enrollments (Schultz 1985). This decline made it more difficult for primary school systems in many developing countries to improve student achievement, increase the number of graduates, and meet the enrollment demands of a rapidly growing population. In addition, serious equity and efficiency problems continue to affect the way that many developing countries raise and allocate education resources (Birdsall 1989; Jimenez 1987; World Bank 1986). This chapter asks how countries can efficiently and equitably ensure that primary schools obtain adequate resources.

Problems and Issues

To expand enrollment, increase the number of graduates, and improve the effectiveness of primary education requires a supportive financial environment. Some low-income countries have difficulty raising sufficient domestic resources. For many countries, however, the crucial issue is the absence of broad political consensus for committing the necessary resources to primary education. Mobilizing additional resources to improve primary schools in ways that are not only economically sound but politically feasible is therefore the principal challenge to achieving sustained educational development.

Existing Shortcomings

Educational financing in developing countries frequently suffers from three interrelated shortcomings. First, the financial base of the education system is often narrow and highly dependent on the general revenues of the central government. These revenues often come from a small number of taxpayers, who are taxed at a high rate and would not support additional taxes (World Bank 1988c). Given such weak tax systems, it can be dangerous for the education sector to depend too heavily on revenue from the central government. Without alternative sources of funding and mechanisms for tapping the willingness of individuals to pay for specific kinds of education, schools are severely limited in their ability to respond to unmet and changing needs.

Second, the incentive structure underlying the funding of public education is often weak, as is the link between funding and school performance. The fund-raising initiatives of schools and local communities are reactive rather than anticipatory, responding largely on an ad hoc basis to serious shortfalls in the resources promised by the national government. Local communities are often severely limited in their power and ability to mobilize resources and are unable to use their natural advantages fully or to turn parental commitment to education into financial support for schools.

Third, in many countries a considerable proportion of educational subsidies goes not to the neediest but to middle- and upper-income families, especially at the postprimary level. The education minister of one developing country questioned the wisdom of allowing this to continue:

> There is no sense in subsidizing those who are capable of paying for the education of their offspring. If 60 to 70 percent of our population can afford to pay something then why should we subsidize that 60 to 70 percent? . . . There must be a formula which would allow us to ensure that the children of the poor 30 to 40 percent are educated whilst not subsidizing the very wealthy or middle class. (Chung 1989b, pp. 25–26)

Overall Objectives

Given these shortcomings, efforts to strengthen the financing of primary education usually pursue one or more of the following policy objectives: use existing primary education resources cost-effectively; expand and improve funding for primary education; and promote more equitable financing.

Improving cost-effectiveness, the first objective, requires that educational policymakers select the most cost-effective inputs, devise a funding arrangement that encourages good school performance, and increase the availability of flexible local funds for schools. The need to establish strong and appropriate incentive systems that promote improved quality, cost-effectiveness, and local funding cannot be overemphasized.

Expanding and improving funding, the second objective, involves increasing the central government's general budget for education, and giving primary education higher priority within that budget. Of course, competing claims must also be considered—especially the need to expand and improve postprimary education. In many developing countries, however, particularly those whose education expenditures are high relative to the gross domestic product, reducing the overdependence of postprimary education on general tax revenues and using public subsidies more selectively would permit more resources to be earmarked for primary schools. Diversifying the sources of funding is also vital. A broader financial base would make primary education less vulnerable to economic shocks and more flexible in its ability to allocate resources. Stronger local financing, in particular, could expand the funding of primary education.

The third objective, promoting more equitable cost-sharing, is necessary to ensure that all segments of the population have access to education and to the economic and social mobility it affords (see Bray and Lillis 1988; Lewin and Berstecher 1989). Specific policies for generating funds for education should not impose undue hardship on the poor (Birdsall 1989). Equitable policies are especially important in countries where children from urban and well-to-do families are disproportionately enrolled in publicly supported postprimary education programs. Under those circumstances, relying on community finance or user charges at the primary school level would be tantamount to asking the poor to subsidize the educational privileges of the wealthy.

To achieve the objectives of cost-effective use of resources, expanded funding, and equitable financing, each country must develop policy measures carefully tailored to fit its unique conditions. Since country-specific policy analysis is difficult and outside the scope of this chapter, the following discussion is limited to broad policy measures that will be relevant in varying degrees to particular countries. The intent is to describe the possibilities and limitations of key strategies for reforming educational finance, and to make national authorities more aware of their options. Whatever specific measures they adopt, developing countries should have a coherent financial policy framework for coping with the changing educational and financial environment. The following section presents the key principles of educational finance.

Guiding Principles of Educational Finance

Education is the responsibility of parents and government. On the one hand, since reproductive decisions are generally left to parents, the primary responsibility for child rearing and education must remain with them. On the other hand, society has a legitimate interest in children's education and socialization. Adequate financial help from the government is imperative for two reasons. First, the investment in human capital will not be sufficient if parents must bear the full financial burden of educating their children. This is because parents rarely consider the societal benefits that education generates when they calculate the costs and benefits of sending their children to school. The divergence between societal benefits and private costs is further exacerbated by severe imperfections in the capital markets. These two kinds of market failure provide the traditional efficiency rationale for subsidizing education publicly.

Second, honest concern for the welfare of the poor demands public education. Because education is the single most powerful tool for advancing economic and social mobility, it is imperative that poor children have opportunities for schooling. School improvement programs, therefore, should not place excessive financial burdens on the poor, who have already been hard hit by the general economic decline of recent years.

Although the arguments supporting government subsidies apply to all levels of education, primary education has the strongest claim on public monies. The divergence of private and societal benefits is the most pervasive in primary education, and the social rate of return to primary education is generally much higher than that of postprimary education in many developing countries, although this could change over time (World Bank 1986). Moreover, the unit cost of primary education is a small fraction of the expenditure per student at the tertiary and secondary levels. Therefore, public resources invested in primary education would benefit more children and more poor families than resources invested in higher levels of education.

In light of the importance of primary education for development, public primary schools should avoid enrollment fees.[1] Fees discourage parents from sending their children, particularly daughters, to school. They obstruct social equity and prevent equalization of the marginal social costs and benefits of primary school enrollment, thus creating inefficiencies. In lieu of charging school fees, education systems should concentrate on the financing strategies already mentioned: (a) using existing resources more cost-effectively; (b) expanding and diversifying their sources of funding; and (c) redistributing the financial burden more equitably among the population. Each of these measures is discussed more fully in a later section. First, however, let us examine the cost of improving and expanding primary education.

The Cost of Improving Primary Education

Although increasing both enrollment and school effectiveness is desirable, tight budgetary constraints often force poor countries to choose between providing quantity or quality in education (Beeby 1979; Mingat and Tan 1985, 1988). In his study of Indonesia, Beeby (1979, p. 274) writes that "the most obvious conflict in the deciding of educational objectives appeared to be between the expansion of the schools and improving the quality of the work done in them." Governments may choose to expand education before improving its quality, but reaching the desired level of enrollment without improving student achievement may not be meaningful if vast numbers of schoolchildren do not learn the primary education curriculum. Moreover, if governments do not invest in improving school effectiveness, expansion may also be severely constrained. Consider, for example, cases in which repetition is common in the first or second grade and completion rates are very low. Repeaters then occupy many of the school places created, and the ability to expand the system further is limited.

Improving student learning in primary school, while also increasing enrollment, is often regarded as too costly for poor countries. The recurrent per-student expenditures associated with greater effectiveness are considered to be expensive: in upper-middle-income countries they are, on average, three times as high as per-student expenditures in lower-middle-income countries and ten times as high as those in low-income countries (see table 2-6). Learning can be increased and the cost of producing primary graduates can be lowered by investing in cost-effective inputs. In northeastern Brazil a program investing in quality-enhancing inputs, such as learning materials, realized significant savings by reducing the time students took to progress from one grade to the next (see box 7-1).

There is little doubt that the up-front costs associated with providing quality-enhancing inputs are significant. The size of these costs depends on the particular country and its level of provision. Although the cost of specific inputs varies among countries, differences in the average expenditure per student for nonsalary recurrent costs reflect differences in the availability of inputs, at least in the order of magnitude. The average central government expenditure per student for educational materials in upper-middle-income countries was, in 1980 and 1985, about \$5–\$6 more than that in low-income countries and \$4 more than that in lower-middle-income countries.

It would be simplistic to claim that this spending increase alone could significantly improve school effectiveness. Consider, however, the nonsalary school inputs that this amount could buy. Evidence from northeastern Brazil on various educational inputs found to enhance learning suggests that spending \$5 more per student—\$1.50 on textbooks, \$1.50 on writing materials, and \$2.00 on teacher training—could produce substantial gains in achievement (see table 7-1). The achievement gain depends on the subject matter and year, but the overall impact is positive. Increasing the availability and use of nonsalary inputs in primary schools to the level described in table 7-1 could dramatically improve student achievement in low-income countries, especially if adjusted to the needs of each country. The question is whether providing all children with these or equivalent inputs is financially feasible.

Consider how a budgetary increment of \$4–5 would affect the aggregate cost of education.[2] The effect of increasing expenditures \$5 per student in low-income countries (excluding China and India) and \$4 per student in lower-middle-income countries is indicated in table 7-2. These figures assume that dropout and repetition rates, as well as the entry rate (defined as the pro-

Box 7-1. Efficiency Gains in Northeastern Brazil

With financial assistance from the World Bank, the Brazilian government undertook in 1980 a program of massive educational investment in rural schools in the northeast, the poorest region in the country. A six-year evaluation showed increased student learning, which in turn reduced student repetition. The result was a significant cost saving, as fewer resources were consumed by repeaters. For example, investing $1 per student in writing materials and textbooks shortened the average length of time for a student to progress from second to fourth grade by 0.134 years. Since the annual cost per student was $30, this $1 investment reduced the total cost of producing a fourth-grade student by $4.02 (table 7B-1).

7B-1. Cost and Savings for Selected Educational Investments in Brazil, 1980s
(dollars)

		Savings per dollar invested[a]			
		Rural northeast		*All Brazil*	
Input	*Cost per student*	*Low-income regions*	*All regions*	*Low-income regions*	*All regions*
Textbooks and writing materials	3.41	4.02	3.12	2.86	0.81
Buildings, furniture, and equipment	5.45	2.39	1.84	1.69	0.47
In-service distance teacher training	1.84	1.88	1.45	1.33	0.37

a. The estimated savings are based on years saved, valued at $30 per student year, in producing a fourth-grade student.

Source: Harbison and Hanushek (forthcoming).

portion of children at the age of entry who are enrolled), do not change over time. The number of children at the age of entry is assumed, however, to grow at the rate of 2.6 percent and 1.4 percent for low- and lower-middle-income countries, respectively. Under this scenario, enrollment would increase 42 percent between 1985 and 2000. Given this projection, the total cost for a $5 increase in annual expenditures per student would amount to approximately $845 million in 1995, increasing to $947 million by the year 2000. The effects would be different for low- and lower-middle-income countries.

For low-income countries, increasing nonsalary recurrent expenditures per student $5 would increase the budget for primary education by 20 percent and the total education budget by 10 percent. These increases are substantial and unlikely to be funded from local sources. The effect of this budget increase is much less pronounced for lower-middle-income countries. The $4 increment would represent a 4 percent increase in the primary education budget and a 2 percent increase in the total education budget. For lower-

Table 7-1. The Costs and Achievement Gains Associated with Selected Educational Inputs in Second Grade in Northeastern Brazil, 1985

Educational input	Cost (1985 dollars)	Achievement gain per dollar[a] In reading 1983	1985	In mathematics 1983	1985
Writing materials available in the classroom	1.76	2.67	(1.24)	1.86	2.70
Textbooks available at school and at home	1.65	3.88	3.96	2.56	1.52
In-service teacher education					
Curso de Qualificação	2.50	?	(0.27)	?	(0.55)
LOGOS II (distance secondary education)	1.84	1.95	(0.79)	1.42	(0.69)

a. Numbers represent coefficients in the underlying regression models and can be interpreted as the marginal increment in test score (in points) associated with the respective input, divided by the cost of that input (in U.S. dollars). Numbers in parentheses represent statistically insignificant coefficients; question marks represent negative coefficients.

Source: Harbison and Hanushek forthcoming.

middle-income countries, therefore, financing the quality-enhancement package would not constitute a major burden.

A similar calculation could be made for increasing budgets 10 percent (or any other figure); in fact, the $5 for low-income countries amounts to about 10 percent of the total education budget (but about 20 percent of the primary education budget) for recurrent expenditures. The total annual cost under either scenario is less than $1 billion, or about 25 percent of the total annual assistance given to education by bilateral and multilateral donors during the 1980s (see chapter 8). These calculations represent a worst-case scenario. They do not take into account the potential savings associated with the gains in learning nor the improvement in the flow of students through the system

Table 7-2. Incremental Cost Effects of Increasing Expenditure per Student in Low- and Lower-Middle-Income Countries

Unit of measure	Low-income countries ($5 increase per student) 1995	2000	Lower-middle-income countries ($4 increase per student) 1995	2000
Millions of 1985 dollars	505	552	340	395
Percentage of primary education budget	20.6	19.9	4.2	3.9
Percentage of education budget	9.8	9.5	2.0	1.9

Note: Calculations are based on the recurrent education expenditure per primary student in 1985. Population at the age of entry is projected to grow at an annual rate of 2.6 percent in low-income countries and 1.4 percent in lower-middle-income countries.

Sources: For expenditure data, Unesco; for population growth rates, World Bank data.

produced by investments in more cost-effective inputs. The next sections examine such investments in detail.

Using Existing Primary Education Resources Efficiently

Understanding the issues of cost and effectiveness involved in pursuing a quality-oriented primary education policy is important in a constrained financial environment. These issues include the costs of high repetition and dropout in primary schools, the benefits of improving school effectiveness, and the measures and resource expenditures required to improve effectiveness.

In a constrained financial environment the primary education system must maximize its output for a given budget. This means that schools must be managed efficiently, must add only inputs that contribute significantly to learning, and must choose inputs that are the least expensive relative to their educational contributions.[3] Chapter 3 showed large differences in the effectiveness of various educational inputs, some of which are totally ineffective under the conditions prevailing in most developing countries. Chapter 5 showed management and organizational factors that inhibit the efficient use of resources. School authorities should carefully choose the inputs on which they spend additional funds, and the education system should be organized better to ensure greater efficiency in the use of those inputs. Failure to do so will mean that the marginal returns to primary education will be low and society may become reluctant to increase the funding of primary education significantly over time.[4] Before requesting increased funding, national education authorities should seriously consider the suggestions put forward in chapters 3 and 5.

Using Cost-Effective Inputs

One source of funds is the primary education subsector itself, which can reallocate existing funds. Four major opportunities exist for lowering costs without affecting student learning negatively: building schools with low-cost (often local) materials, increasing the student-teacher ratio (to 40–50:1), shortening the length of preservice teacher training, and using school facilities more intensively. Resources saved in these areas could then be reallocated to the inputs identified in chapters 3 and 4. The following discussion highlights policy decisions on resource use that can make a significant difference in cost-effectiveness.

CHOOSING SUITABLE STANDARDS FOR SCHOOL FACILITIES. An elaborate school facility is one example of an ineffective input. Learning can occur in a modest facility as easily as in an elaborate one. Building schools to meet international, rather than local, construction standards dramatically increases the cost of school construction, and equipping schools in low-income countries with the technologies common in industrial countries is prohibitively

Table 7-3. Cost per Student Place of Local and International-Grade Construction Materials in Six Sub-Saharan African Countries
(dollars)

Country	*International-grade brick and mortar*	*Local construction materials*
Burkina Faso	549	203
Central African Republic	478	176
Chad	460	156
Mali	417	226
Mauritius	355	183
Senegal	593	175

Source: World Bank data.

expensive. For example, the cost of constructing brick-and-mortar buildings in six Sub-Saharan African countries is more than double the cost of constructing buildings from local materials (see table 7-3).

ALTERING CLASS SIZE ONLY WHEN COST-EFFECTIVE. Reducing the number of students per class is a popular policy intended to improve learning. Some countries, such as Turkey, have targeted a class size of forty students. Reducing the class size marginally, while staying within a wide range of the average class size in most developing countries, has no significant effect on learning. Students do learn better in classes of fewer than twenty students, as noted in chapter 3. Reaching this level, however, requires considerable expenditure and is not economically feasible for most developing countries.

Indeed, *increasing* class size might be a better strategy. In most developing countries, it could free resources for other inputs, since the average student-teacher ratio is less than or equal to 40:1. This was the case for 54 percent of low-income countries, 79 percent of lower-middle-income countries, and 96 percent of upper-middle-income countries in 1985; only nine low-income countries (23 percent) reported average ratios greater than 50:1 (see table A-9 in the appendix).

The value of raising student-teacher ratios depends on the country's existing ratios and the dispersion of the school-age population. Enlarging classes is most feasible in countries such as China with student-teacher ratios below 40:1. In the few countries that have student-teacher ratios greater than 50:1 (for example, Burkina Faso), increasing the ratio is unlikely to be feasible or educationally desirable.

ADOPTING SUITABLE TRAINING AND HIRING POLICIES. Chapter 4 argues that the costs of training teachers can be significantly reduced by shifting the secondary education component of preservice teacher training to general secondary schools, reducing the length of preservice teacher training accordingly, and providing in-service training. Roughly 60 percent of the content of teacher training is taught in general secondary schools. As table 4-3 shows, the ratio of the recurrent unit cost of teacher training institutions to that of general secondary education is high. For a sample of twenty-seven countries, the ratio averaged 7.9, with only five countries having a ratio of less than 2. Since effective in-service teacher training can be done inexpensively, this approach could reduce training costs.

In-service training could yield savings in teacher salary costs as well. Some countries try to upgrade the quality of their schools by requiring teachers to have postsecondary academic degrees. As teacher pay scales typically reward higher levels of formal education, hiring teachers with college or university degrees instead of secondary or junior secondary degrees drives up recurrent expenditures significantly, doubling them in some countries. Using general secondary education as a hiring standard and giving graduates in-service pedagogical training would minimize the growth in teacher salaries that is unrelated to productivity.

USING SCHOOL FACILITIES ECONOMICALLY. Double shifts have often been used to increase the number of school places in countries whose school-age population is growing rapidly and whose budget is severely constrained. Using school facilities in two shifts is a sensible strategy in densely populated school districts with a rapidly growing population of school-age children, especially if the classes are extremely large and enrollment rates are low to begin with. Double shifts minimize the capital cost of providing more primary school places, allowing more students to enroll and more money to be used to purchase instructional materials and other inputs that improve student learning. For a fuller discussion of multiple shifts, see chapter 6.

Table 7-4 summarizes the opportunities for reallocating funds within the primary subsector. These opportunities vary across countries, but most countries can make some improvement in internal efficiency by using resources better.

Reducing Repetition and Dropout Rates

Repeating grades and dropping out of school can result from two broad sets of factors: family and student characteristics that affect the demand for education, and schools and educational policies that are ineffective. If parents are not interested in educating their children beyond the lower primary grades, perhaps because they perceive few benefits of education, and if they do not support academic learning at home, children will perform poorly in school and eventually drop out. Poor health and nutrition in impoverished communities also take their toll on school attendance and performance. The following discussion focuses on repetition and dropout caused by school ineffectiveness. The problem of low demand for education is addressed in chapter 6. It is important to recognize, however, that the attitudes that parents hold toward schooling may depend on school effectiveness; hence, demand factors are not entirely independent of school-related factors.

Table 7-4. Cost Effects of Specific Inputs

Input	*Reduce costs*	*No change*	*Increase costs* *Fixed*	*Increase costs* *Recurrent*
Inputs that raise learning				
Improved curriculum[a]	—	—	✓	—
Textbooks and materials	—	—	✓	✓
Sufficient instructional time	—	✓	—	—
On-site in-service teacher training	—	—	—	✓
Interactive radio instruction	—	—	✓	—
Programmed materials	—	—	✓	—
Micronutrients and health interventions	—	—	—	✓
School snacks	—	—	—	✓
Teacher training reform[b]	✓[c]	—	—	✓[c]
Inputs that have no negative effect on learning				
Multigrade classes	✓	—	—	—
Multiple shifts	✓	—	—	—
Low-cost school construction	✓	—	—	—
Class size of 40–50 students	✓[d]	—	—	✓[d]
Parental outreach program	—	—	—	✓
Reduction of discriminatory practices	—	—	—	✓[e]
Private provision of schooling	✓	—	—	—
Inputs that improve management				
Career ladders for teachers and managers	—	—	—	✓
Salary and nonsalary incentives for teachers[f]	—	—	✓	✓
Rationalized management structures	—	✓	—	—
Information systems	—	—	✓	✓
Staff training	—	—	✓	—

Note: A check mark (✓) indicates where an effect is felt.

a. With appropriate scope and sequence.

b. Higher pretraining general education requirements, shorter preservice training, and better in-service training.

c. Teacher training costs will be reduced; salary costs will increase.

d. Costs will be reduced where classes are smaller than 40–50 students; costs will increase where classes are larger than 40–50 students.

e. Training and supervision costs will increase.

f. Nonsalary incentives include, for example, housing in rural areas.

FOCUSING ON COMPLETION AND COST PER GRADUATE. The goal of primary education systems is to produce graduates who have learned the skills prescribed by the curriculum. Thus the most relevant measure of a system's effectiveness is not the number of students enrolled, which is often used to evaluate educational progress in developing countries, but the number of graduates who have achieved the required level of learning. Although many developing countries have attained high gross enrollment rates (indicating that the system has a high level of capacity), their primary education completion rates and student learning remain low. Many children drop out during the early years of the primary cycle and never attain basic literacy or numeracy. Repeating a grade enhances the probability of dropping out, and a strong determinant of repetition is achievement. Improving learning can improve the probability of being promoted from one grade to the next on time; on-time promotion may in turn reduce the probability of dropping out.

Dropouts and repeaters raise the costs associated with producing a graduate of the primary education system. These costs have three components. First is the amount spent directly on schooling: both the cost to society for providing a place for each child in school and the cost to parents for items such as transportation and school supplies. Second is the opportunity cost of schoolchildren's time—the value of the labor foregone when children attend school instead of holding outside jobs or working at home.[5] Third is the future cost to dropouts and their parents: in the labor market, failure to complete a primary education translates into a lower rate of return for each year of schooling missed. A study of the Philippines, for example, reports a big difference in the social rate of return to complete and incomplete cycles of elementary education: 12 and 4 percent, respectively (Tan and Paqueo 1989). Preliminary findings in Indonesia show that the economic returns to primary education are reduced 34 to 52 percent when both repetition and dropout rates are taken into account in calculating the social rate of return (Behrman and Deolalikar 1988).

Even when grade repetition does not decrease the number of graduates, it delays completion of the primary education cycle and raises the cost associated with producing a graduate.[6] High repetition rates also hinder the school's ability to accommodate new students and its effectiveness. They tend to produce overcrowded classrooms and to reduce the number of educational materials per student. In addition, repetition increases the direct costs that parents pay for their child's education. Each year that a child spends repeating a grade becomes many more years of foregone income and work experience. Partly because they enter the labor market relatively late, repeaters earn less than other workers in their age cohort.[7]

A common indicator of the level of resources needed to have one child complete the primary cycle—student-years required to produce one graduate—shows the high cost of repetition and dropout rates combined (see table 7-5). To calculate the average time required to graduate a student, one must average the years spent in school by students who (a) graduate without re-

Table 7-5. Median Difference between the Length of the Primary School Cycle and the Time Needed to Produce a Graduate in Developing Countries, Selected Years, 1970–85
(years)

Country income level	*1970*	*1975*	*1980*	*1985*
Low	4.8	4.8	3.8	4.0
Lower middle	3.8	2.4	2.7	1.8
Upper middle	1.2	1.8	1.2	1.2

Source: See table A-12 in the appendix.

peating a grade, (b) graduate after repeating one or more grades (repeaters), and (c) begin the cycle but do not complete it (dropouts). In 1985, for example, the median extra cost of producing one primary school completer was 4.0 years in low-income countries, compared with 1.2 years in upper-middle-income countries.[8] This means that to produce a primary school graduate, the education system in low-income countries has to spend, on average, four more years of resources than would be the case if students did not repeat grades or drop out before finishing primary school. At present the median repetition rate is 16.3 percent in low-income countries, and it is much higher in some countries—for example, 30 and 35 percent in Mali and the Central African Republic, respectively. The dropout rate prior to grade five is 8 percent in low-income countries, with countries such as Bangladesh and Haiti reporting about triple that rate.

INCREASING STUDENT FLOW AND COMPLETION. How much can be saved by reducing the repetition and dropout rates? The savings depend both on the extent of existing repetition and dropout and on the strength of three linkages: between teaching inputs, such as textbooks, and learning; between learning and school attendance; and between learning and promotion. To substantially lower the repetition and dropout rates, schools must develop an effective intervention program that considers the causes of repetition and dropout. These causes may include (a) family-related factors, such as low income and low educational attainment of the parents, (b) student characteristics, such as low regard for the future utility of education, poor motivation, and low academic ability, and (c) school-related factors such as physical remoteness, poor classroom environment, or school ineffectiveness (Haddad 1979; Shepard and Smith 1989; Unesco 1980). School-related factors appear to be an important determinant of achievement, and poor achievement is one determinant of dropout and repetition. Not enough is known, however, to understand which specific interventions would reduce repetition and dropout rates and by how much. Until such knowledge is available, educational policymakers should introduce interventions gradually.

Just as poor achievement leads to repetition and dropout, high cognitive achievement appears to be positively correlated with promotion and completion of the primary education cycle. Cases in northeastern Brazil, Nicaragua, the Philippines, and several Sub-Saharan African countries suggest that adopting educational measures to improve school effectiveness would improve the cognitive achievement of children, reduce repetition and dropout, and thereby increase the number of graduates (see chapter 3). It would also increase the economic productivity of these graduates (Behrman and Birdsall 1983; Boissiere, Knight, and Sabot 1985; Knight and Sabot 1990).

Reducing the repetition rate, the dropout rate, or both could substantially lower the cost of producing each primary school graduate in low- and lower-middle-income countries. Table 7-6 presents estimates of the savings realized

Table 7-6. The Cost of Producing a Primary School Graduate and the Recurrent Savings from Reducing Repetition and Dropout Rates
(dollars)

		Cost savings[b] per graduate for each percentage point reduction		
Country or country group	*Average cost of producing a graduate[a]*	*In repetition rates*	*In dropout rates*	*In repetition rates excluding years spent in school by dropouts*
Low-income countries[c]	279	7.9	11.5	6.0
Lower-middle-income countries[c]	734	22.3	98.7	10.4
Bangladesh, 1980–81	232	4.5	8.0	0.6
Central African Republic, 1986–87	929	14.1	55.0	8.0
Haiti, 1984–85	143	2.7	3.8	1.2
Mali, 1986–87	870	16.8	48.3	6.6

a. The cost includes the years that both graduates and dropouts spend in school, but does not take into account the benefits—if any—of partially completed primary school. The data on recurrent expenditure per student per year are based on the latest available Unesco information.

b. Cost savings include the cost of intervention.

c. The figures for low-income and lower-middle-income countries are unweighted averages based on samples of sixteen and twenty countries, respectively. Estimates of repetition and dropout rates for these countries were provided by Unesco.

Source: See tables A-12 and A-20 in the appendix.

for each percentage point that the repetition and dropout rates are reduced in low- and lower-middle-income countries. The total number of years that noncompleters spend in school are included in the cost of producing a graduate; this overstates the cost per graduate by assuming that dropouts are unlikely to acquire sustainable literacy or numeracy and that the years they spend in school are therefore wasted. It also considers only the direct costs of schooling, in this case, the annual recurrent costs of the central government.

If the repetition rates of completers were reduced by 1 percentage point, the number of student-years needed to produce a graduate would also decrease, thereby saving, on average, $8 in low-income countries and more than $22 in lower-middle-income countries. These figures reflect the difference in annual recurrent costs for low- and lower-middle-income countries.

Decreasing the probability that students will leave school before graduating would save even more. The savings per graduate associated with a 1 percentage point decline in the dropout rate could, on average, equal $12 in low-income countries and $99 in lower-middle-income countries. Since the potential savings associated with reducing the repetition rate as well as the number of noncompleters are large, the contribution that school ineffectiveness makes to dropping out should be separated from that of weak demand. Such a distinction would help policymakers shape appropriate school policies.

If the benefits of completing some years of primary school are taken into account (that is, if the years of schooling for dropouts are not included in computing the cost in student-years per completer), the savings realized by lowering the repetition rate are less. In this case, low-income countries would save $6, and lower-middle-income countries would save more than $10. The true savings would probably lie between the two sets of estimates.

The savings gained by achieving greater efficiency are not without cost. For example, Brazil, the case described in box 7-1, made a significant initial investment of more than $92 million to improve schools in the northeast. A consequence was that the number of repeaters and dropouts fell, requiring the commitment of additional resources to expand the capacity of the educational system and accommodate more students. Even though increasing inputs may produce considerable savings, these savings are not likely to be realized for two to five years. The challenge, therefore, is to find the resources needed to begin the process.

Mobilizing Domestic Resources for Primary Education

The savings realized by improving school effectiveness may or may not provide sufficient additional funding for primary education. That should be examined on a country by country basis. In general, a menu of other policies for expanding resources for primary education will be needed as well. Experience suggests that the following key policy measures should be considered seriously: (a) shifting public resources to education from activities in other sectors that have a low social priority, (b) increasing the priority of primary education within the education budget, and (c) diversifying the sources of funding for primary education. Of course, the usefulness of each measure will vary from country to country.

Reallocating Funds from Other Sectors

National education authorities should actively seek national financial support corresponding to the high priority that primary education deserves. This involves studying the overall allocation of government resources and developing an effective strategy for strengthening political and bureaucratic support for the education sector. Public resources in developing countries frequently finance activities of low social priority, for which the government does not have a comparative advantage. Prime examples are public subsidies to unprofitable state-owned enterprises.[9] Net transfers to these enterprises can be quite large, as indicated in figure 7-1. The flow of funds to primary education may be improved by, for example, reducing subsidies to nonfinancial

Figure 7-1. Average Annual Net Transfer from Government to Nonfinancial State-Owned Enterprises in Selected Countries, 1978–85

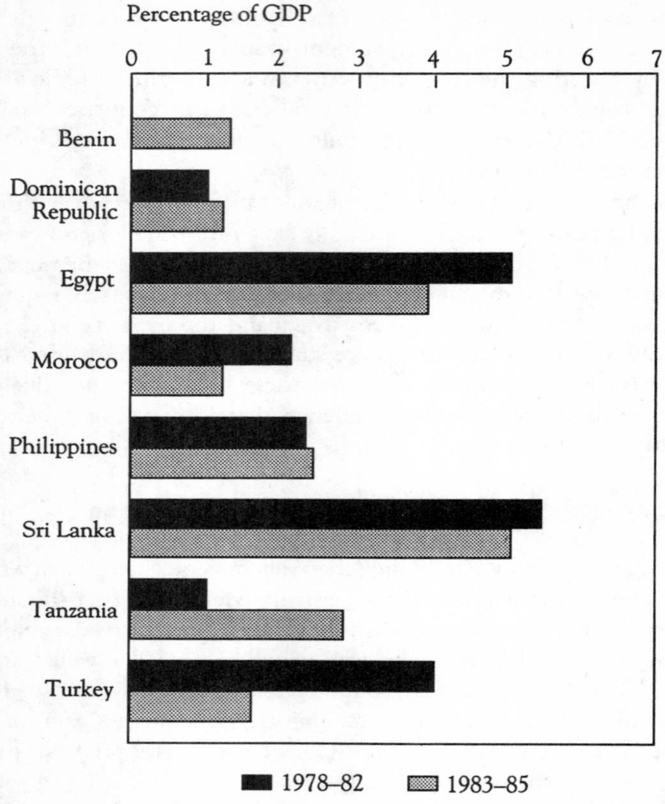

Source: World Bank 1988c.

state-owned enterprises. Governments can do this by planning their public expenditure programs while considering both the overall priorities for the economy and the appropriate division of responsibility between private and public activities.

Determining intersectoral priorities and reallocating resources depend on a host of factors, including each country's economic structure, culture, politics, and development strategy. No clear, practical rules guide intersectoral choices; spending decisions across sectors are based largely on intuitive judgments about what the overall balance should be.[10] Yet identifying bottlenecks, estimating rates of return, and assessing the value of different activities as a public good can provide useful clues to what the decisions should be (World Bank 1988c).

Improved primary education should be very competitive, given these criteria. Its rate of return is relatively high and is expected to remain so, as the pace of technological progress quickens. Moreover, the rate of return can be even greater if internal efficiency and learning achievement increase simultaneously—a strategy that is a real possibility. Moreover, minimizing bottlenecks in developing a trainable and mobile work force is imperative in a world driven increasingly by technology. Primary education is also a public good with widespread benefits for society.

The Philippines provides a good example of how a common goal that unites the government and its citizens can produce the necessary resources despite economic difficulties. One of the early, dramatic actions of President Corazon Aquino's administration was to double the budgetary allocation to the education sector. The public agreed that education had been shortchanged in the past and that the government had subsidized unproductive public enterprises instead. By privatizing state-owned enterprises and reducing subsidies to activities of low social priority, the government generated resources to finance huge increases in the education budget.

Making Primary Education a Priority within the Education Budget

Since the potential contribution of primary education to society is high and the risk of private underinvestment is substantial, governments should not make access to good primary schools contingent on the ability to pay. Public subsidies for primary education might have to rise faster than subsidies for education at higher levels. The case for such a strategy is strongest when one or more of the following conditions prevail: primary enrollment rates are low, the social rate of return to primary education is substantially higher than the return to postprimary education, the government does not subsidize education selectively, the beneficiaries of postprimary education are predominantly children of upper-income families, and tertiary education has a relatively large share of the total government budget.

A sound policy framework for strengthening the financing of primary education recognizes that social demand for more and better postprimary education will rise in the future. This is an inevitable result of both socioeconomic development and a more effective primary education system. Unless countries develop alternative sources of funding to meet this demand and rationalize their entitlement programs for secondary and tertiary education, they will find it increasingly difficult to expand investment in primary education.

Three major elements are essential to reducing the financial dependence of postprimary education on the national government and thus freeing more resources for primary education: (a) more efficient use of the public subsidies allocated to secondary and tertiary education, (b) greater participation in education by the private sector and local communities, and (c) establishment of an efficient and fair system of cost-sharing in government-funded schools.

USING SUBSIDIES MORE EFFICIENTLY. Public secondary schools in some countries have ample room to increase student learning achievement, to lower costs, or both. This is strongly suggested by comparisons of private and public secondary schools in Colombia, the Dominican Republic, the Philippines, Tanzania, and Thailand (Jimenez, Lockheed, and Paqueo 1991). On the whole, children in private schools have higher learning achievement than children in public schools, even though private schools spend less per student than public schools do. The lesson is not that public schools should be privatized, but that the secondary public school budget could be stretched much further.

Government subsidies could be used more efficiently if the public resources allocated to the postprimary education system were redeployed from low- to high-priority items. Many low-income countries spend too much on nonpedagogical expenditures and too little on teaching and learning materials. Student living allowances receive a substantial share of the budget in many countries, and teacher salaries take up more than 90 percent of the total budget in a few others. Subsidies should be allocated, over time, to developing and purchasing teaching materials; this would benefit school effectiveness (World Bank 1986).

ENCOURAGING POSTPRIMARY PRIVATE AND COMMUNITY SCHOOLS. Excess (or unmet) demand for education can be absorbed by private and community schools, as the experience of countries in various regions of the world shows. The Harambee schools in Kenya are one example of community response to the need for more secondary education. In the Republic of Korea, the Philippines, and some Latin American countries, the private sector has made a substantial contribution to providing secondary education. The history of the Community Junior Secondary Schools in Botswana demonstrates the fruitfulness of a joint initiative between the community and primary schools (Schiefelbein 1987; Swartland and Taylor 1988).

The policy environment for developing private and community schools must, however, be reformed in some countries. This largely means easing undue restrictions on private and locally run schools and facilitating their access to capital for financing expansion and purchasing teaching materials (James 1991).

IMPROVING COST SHARING. Cost recovery at postprimary levels of education, particularly at the tertiary level, is useful not only for raising resources but also for improving efficiency. It gives students the incentive to learn and to finish their schooling faster. It also improves equity, if an appropriate means-tested scholarship program or fee discount system is instituted. In many countries current subsidies for student living allowances direct a disproportionate share of the government education budget to high-income students (World Bank 1986).

Diversifying the Sources of Funding for Primary Education

Education budgets are often overly dependent on a single source of funding: the central government. Increasing the ability of the education budget to withstand fluctuations in central revenues requires diverse sources of funding. Funding can be diversified in two ways: by targeting new taxes for education and by strengthening local finance.

RAISING NEW NATIONAL TAXES. For developing countries with low ratios of tax to gross domestic product, the potential for increasing the flow of funds to primary education (even by shifting budgetary allocations) may be seriously limited by low national tax revenues. In many countries, a comprehensive fiscal reform program is desirable in order to rationalize the tax system and mobilize local resources. In addition to such reforms, which take time to implement if they are politically feasible at all, the education sector can seek alternative interim solutions that are consistent with the broad directions of the tax reform.[11]

One such solution would be to levy new taxes that are targeted to fund improvements in primary education. Identifying specific sources of revenue for specific programs that are perceived to have a high social value can overcome taxpayers' resistance to new taxes. In some cases such taxes have helped improve and protect the funding of target programs. Perhaps because education is so highly valued in many developing countries, taxpayers who would undoubtedly oppose untargeted tax increases may support taxes used to improve schools.

Dedicated, or more specifically, earmarked taxes that finance education and training are not uncommon (Deran 1965; McCleary 1988). They can be found in countries at different levels of development and at various times in their history. Furthermore, they take different forms, including taxes on property, business, selected commodities, payroll, imports, and interest or dividend income. Some are local taxes, while others are national (see box 7-2).

Earmarking does not in itself guarantee that targeted programs will be adequately funded nor that efficiency will be increased. In fact, earmarking can have the opposite effect. Governments can divert earmarked funds to other purposes or simply use them to replace past allocations from the general budget without increasing the total resources devoted to primary education. Because it is rarely the best approach, earmarking should be used selectively and under strict conditions. (See box 7-3 for more on the pluses and minuses of the practice.)

Earmarking is most successful when taxes are closely linked with the beneficiaries of the program—the projects for which the revenues are dedicated.[12] In other words, earmarking measures are more successful when the targeted program is driven by demand and good projects have been identified, often within the scope of a prepared investment program. In contrast, failure is high when the link between revenue and benefits is weak or absent and the

Box 7-2. Taxes for Education

Earmarked taxes supplement allocations in the general budget. When used to finance good education initiatives, they can be politically rewarding despite the public's usual aversion to taxation.

In the United States, several state governors found it politically expedient to institute taxes for education reform, even though raising federal taxes to reduce the deficit was politically unpopular. In 1987 Indiana passed a $750 million increase in corporate and personal income taxes to fund the governor's A+ Program of Education Excellence. In 1984 Tennessee adopted a $0.01 sales tax, which raised more than $1 billion for the Better Schools Program over three years. In 1983 Arkansas passed a comprehensive education reform package financed by a $155 million increase in the sales tax. That same year, South Carolina passed an education initiative that collected $217 million by raising the sales tax $0.01.

In 1982 the Republic of Korea discovered that the general budget appropriations would not meet the total costs of its national education system. The government responded by instituting a five-year education tax on liquor, tobacco, interest and dividend income, and the banking and insurance industry. By 1987 the education tax accounted for 15 percent of the Ministry of Education's budget. The government extended the education tax for another five years.

Brazil levies a 2.5 percent salary tax on the wages of employees in the private sector, and the funds are earmarked for primary education. The federal government collects the tax, two-thirds of which go to the states. In 1986 Pakistan introduced the Iqra surcharge on some imports and earmarked the proceeds for education. However, due to the fungibility of money, education did not benefit much from the money collected. In 1989 Turkey also introduced a

desire for project development is driven by the availability of earmarked funds—a case of good money chasing after bad projects. A national textbook fund and a community fund for primary education are examples of programs that have good potential for earmarking. The mandate of a national textbook fund would be to purchase the minimum number of textbooks needed by all primary schoolchildren,[13] while that of a community fund would be to provide locally mobilized resources (and matching grants, if any are given by the national government) to schools of the communities generating them.

STRENGTHENING LOCAL FINANCE. Developing countries vary in the degree of their reliance on local finance and in their methods of mobilizing local resources. Some countries rely heavily on financing primary education locally; others raise meager amounts of local funds. Some use equitable and efficient methods; many do not. This section discusses local finance from two perspectives: (a) expanding primary school resources by mobilizing local retax earmarked for education. Whether this tax is advisable is doubtful in view of the excessive proliferation of dedicated taxes in Turkey.

Nepal, the Philippines, Guinea, China, and Botswana also have local earmarked taxes. In an attempt to decentralize the decisionmaking system, Nepal passed the Village and Town Panchayat Amendment Acts of 1964 and 1965, which empowered local authorities to raise an education tax. The Philippines has a Special Education Fund based on the real property tax. In Guinea the subprefecture collects a $4 poll tax from all persons fifteen years of age or older. The effective coverage of this tax, which finances education and other social expenditures, is very high: 96 percent. In China a 1985 reform of the education system empowered local governments to levy an extra tax for education in cities and towns. The proceeds fund compulsory education. These taxes include an education tax on local enterprises and individual peddlers as well as a 3–5 percent tax on the annual income of collectively run enterprises.

Finally, the experience of Botswana provides an interesting example of an indigenous response to the educational demands of the community: levying an earmarked tax on housing. At the turn of the century, the demand for schools that taught academic subjects (especially English) rather than religion (taught by the missionaries) rose as the country increasingly came into contact with travelers and traders. As a consequence, the chiefs and their people established a number of independent tribal schools. In 1900 one tribal chief imposed an extra levy, collected with the tax on huts, that would finance a community school. This approach was so successful and popular that other chiefs soon followed suit, and a unique tax earmarked for education began.

Source: Internal World Bank documents.

sources and (b) improving the efficiency and equity of methods popularly used to generate local funds for primary schools.

Countries where the central government can no longer adequately fund social programs and where local resources are underutilized are becoming increasingly interested in the possibilities of local finance (Bray and Lillis 1988; Cornia, Jolly, and Stewart 1987; World Bank 1986). Some countries have not explored this avenue fully because they assume that no effective mechanisms exist to tap parental and community support for educating local children. However, getting parents more involved in their children's schools can lead them to provide additional financial support. This is suggested by a recent study in Thailand, which found a strong, positive correlation between parental participation in school activities and household contribution to schools and their teachers (Tsang and Kidchanapanish forthcoming). Therefore, schools and their communities should be given the encouragement and the means to mobilize household and other local resources.

Box 7-3. Arguments for and against Earmarking Funds

Many economists and specialists in public administration are opposed to earmarking. They argue that earmarking imparts inflexibility to the process of allocating resources, hampers effective control of the budget, and allows no remedy for misallocated funds. Resource misallocation arises when activities that have been financed with earmarked taxes are left with excess funds that cannot be transferred to underfunded programs. In Brazil, for example, the inflexibility of earmarked funds appears to be impeding the development of a more effective social program.

It would be unwise to rule out earmarked taxes under all circumstances. Criticism of earmarking is based on a particular outlook on public finance: that resources are allocated by a benevolent social planner attempting to exercise a well-defined social welfare function. Such a model of social choice is often inappropriate.

Public choice theory, which looks at social choice as a process involving economic agents with differing preferences and private interests, emphasizes that earmarking often provides a way out of a suboptimal situation. It does not always constrain expenditure unnecessarily. Suboptimal conditions can occur under general fund financing, which forces consumers to buy a bundle of complex and diverse products with their tax payments. As Buchanan (1963, p. 459) points out, "any requirement that one stick of butter be purchased with each loaf of bread would surely produce inefficiency in choice." By linking a tax increase to a particular public expenditure, earmarking allows economic agents to express and resolve their preferences. Earmarking can often build a consensus or convince voters to support an increase in the funding of a highly valued public good (such as high-quality primary education) more easily than increasing the general government revenue can. Many consumers resist a general increase that

There are, however, limitations to relying on local financing of education, and even policymakers who have called for the development of community resources recognize the need to fund primary education nationally and publicly. The fear is that greater reliance on local finance can worsen inequality in education and that primary schools in poor communities will be of substandard quality (Winkler 1989). The issue is not only whether raising local funds is efficient, but whether the mechanisms for addressing equity concerns effectively redistribute national education subsidies. This distributional issue will be discussed later. Here, though, it is interesting to note that even some countries with strong egalitarian values, such as Sri Lanka, use community financing to supplement tight national budgets (Lewin and Berstecher 1989). Moreover, countries like Pakistan, which had adopted policies abolishing or discouraging local funding of schools, are again attempting to generate local and private sources of funds for education (Jimenez and Tan 1987).

forces them to pay for public goods and services they do not find useful (such as expanding a bloated bureaucracy or inflating an excessively large military budget). Thus general tax increases dilute the net benefit that consumers expect to receive from an increase in spending for primary education. Earmarking should be permitted when the alternative would be to forgo an activity that is in the national interest.

Earmarking can also force the bureaucracy to undertake activities that it has few incentives to pursue. Funding some of these activities (such as purchasing learning materials at the expense of giving salary increases) may even run counter to the private interests of members of the bureaucracy. When discretionary public policy does not improve social welfare, a country may have to rely on rules and institutions. Teja (1988, p. 20) points out that often "the only way a legislature can agree upon supporting a vital social activity is by decentralizing and removing it from routine legislative consideration." Socially important expenditures, such as nonsalary expenditures, that lack strong, well-organized political backers are usually at risk during times of budget cuts. At such times, earmarking may help focus funding on primary education.

Three fundamental questions must be addressed by policymakers seeking to decide whether and how taxes should be earmarked for education. First, what is the logical stopping point (if any exists) that would, in principle, prevent an excessive amount of taxes from being earmarked? Second, what conditions would have to be imposed to ensure the success of earmarking? Third, what specific kind of taxes should be earmarked?

Sources: Buchanan 1963; Jimenez 1990; McCleary 1988; Teja 1988.

National governments usually depend on standard financial instruments, such as national income taxation, that tend to be rigid and unresponsive to the specific requirements and conditions of particular communities. Empowering communities and relying on them for financing can create funding sources that are not currently tapped efficiently by national governments.

These resources may be limited. Depending on the local economy, culture, and politics, some of these resources may only be available in kind, while others may be too irregular for the central government to depend on. More important, contributions often can only be raised on a voluntary basis, so compliance requires social pressure as well as local initiative and management.

However, there are benefits that come from greater reliance on the community. Dedicating local funds to schools in the community motivates individuals to increase their own contributions and to comply with educational tax measures. School authorities have more access to and more control over community resources, which they can spend to meet the changing needs of local schools (Bray and Lillis 1988). The need for educational materials and furniture, for example, can be met more expeditiously. Perhaps even more important, the level of spending can reflect the community's demand for education. Further, parental and community involvement enhances the accountability of schools and probably improves their cost-effectiveness. Local contributions can promote savings by enabling the education system to adjust educational inputs to local and regional differences in prices.[14]

Does greater local financing of primary education provide stronger incentives for schools to be more efficient? The effect of local financing on efficiency has not been adequately tested. A case study from the Philippines suggests, however, that when school quality and other socioeconomic characteristics are held constant, primary schools whose local funding was high compared with their total expenditure had lower recurrent unit costs than other schools (Jimenez, Paqueo, and de Vera 1988). Critics of locally funded schools have cited cases linking poor cognitive achievement to dependence on local fi-

Box 7-4. Financing Education in Zimbabwe: Partnerships between the Government and the Community

When Zimbabwe attained independence in 1980, the government declared that every child had the right to a primary (and secondary) education. To meet this objective, between 1979 and 1989 Zimbabwe nearly doubled the number of primary schools, from 2,401 to 4,504, and increased enrollment 178 percent, from 819,586 to 2,274,178 students. Given Zimbabwe's depressed economic base, this phenomenal increase would not have been possible without mobilizing the community to help shoulder both the capital and the recurrent costs of primary education. Annual parental contributions average Z$50 million (about Z$10 per family) for capital expenditures and Z$310 million (about Z$100 per student) for recurrent expenditures.

Partnerships between the government and the community are observed in both government and government-aided schools. Government schools, which comprise 6 percent of all primary schools, are constructed by local authorities and parents with their own resources. The government is responsible for the regular maintenance and repair of the school buildings and grounds; for teacher and staff salaries; for furniture, equipment, textbooks, and other instructional materials; and for operational expenses, such as electricity, water, and postal services. Extracurricular activities and additional facilities are supported by annual fees, which range from Z$1.50 to a maximum of Z$27.00, depending on local conditions and parents' ability to pay.

Voluntary parents' associations, active in nearly all government schools, provide additional resources for covering capital and recurrent costs. The parents decide which projects they wish to undertake for the school and then levy a tax

nancing. Although studies have found that students in community-funded schools have poorer achievement than those in nationally funded schools, those studies did not account for differences in the quality of the students entering, nor for differences in the cost per student of the education offered.

Direct parental expenditure on educational materials and household contributions to schools can finance some of the necessary improvements in primary education. Zimbabwe, for example, has made great progress in improving its primary education by mobilizing community participation and finance, as described in box 7-4. In the Republic of Korea, more than 20 percent of primary school expenditures in the early 1960s were financed by local sources and a large share was invested in learning materials (McGinn and others 1980). More dramatically, in 1949, parent-teacher associations in Korea contributed 75 percent of the funds for local schools (McGinn and others 1980). In Thailand a 1988 survey revealed that direct in-cash and in-kind contributions from parents to schools and teachers paid for about one-

on themselves to raise the necessary funds. Capital projects supported by parents' associations include construction of additional classrooms, libraries, school halls, swimming pools, and security fences. All facilities must be approved by the government and must conform to government standards; they are eventually handed over to the government for maintenance. Parents' associations may also supplement government funds for recurrent expenditures, such as paying the salaries of additional government-qualified teachers employed to lower the student-teacher ratio.

Government-aided schools are established and maintained by authorities other than the central government, but the government makes a grant for each student enrolled. The grant, which ranges from Z$11.50 for a first-grade student to Z$25.00 for a seventh grader, is commensurate with the assessed expenditure per student in government primary schools. In addition to these tuition grants, the government provides building grants that cover up to 25 percent of the total costs of construction. The residual costs are borne by local communities through contributions of labor, local building materials, and expertise.

The major disadvantage of relying on community contributions is that wealthier parents are able to pay substantially more than poorer parents. The government has attempted to overcome severe inequities by establishing a Disadvantaged Schools Program that provides capital development for extremely poor communities. This is, however, a small program with only Z$5 million per year.

Source: Chung 1989b.

third of school expenditures per primary school student (Tsang and Kidchanapanish forthcoming). In Rwanda, according to an internal World Bank report, about one-quarter of recurrent expenditures on primary education are funded by fees and other local contributions. And in Zaire, parents contribute at least 60 percent of the recurrent cost per unit in primary schools.

Other examples in Africa, although often associated with poor school quality, nevertheless indicate the willingness of communities to help finance primary education. In Lesotho 98 percent of primary schools are under the auspices of church missions; their student fees and the contributions of church or donor agencies accounted for 11 percent of costs in 1979–80. In Malawi 26 percent of all primary schools are not assisted by government subsidies, being wholly owned and administered by local communities. In Mali in 1986, parent associations financed 21 percent of the cost of primary education; school cooperatives and other community organizations funded another 4 percent (Fouilland 1987).

Parents and communities are obviously willing to pay for educating their children; the challenge is to blend local and central resources and target them to the most cost-effective inputs. The three broad, institutionally based approaches to local fund-raising include community taxation, compulsory or quasi-obligatory user fees, and school-based fund-raising organizations.[15]

Local taxation has been a good source of funding for primary and secondary education in some countries. China, Guinea Bissau, Mali, Nepal, Pakistan, the Philippines, and Yugoslavia have granted formal statutory powers to local governments for collecting local taxes to be used in whole or in part for community schools.

More common in developing countries are levies imposed and collected in accordance with traditional and less formal procedures (Bray and Lillis 1988). In a village in eastern Nigeria, the Council of Elders raises school funds by taxing members of the community. Mwethya work groups in Kenya require individuals to contribute labor to schools.

Perhaps the most common fund-raising operations in modern times are school-based organizations such as parent-teacher associations, boards of governors, and school committees. Their composition, responsibilities, and powers vary, but they generate resources by collecting school fees, soliciting voluntary contributions, and organizing festivals, dances, and other social events. Table 7-7 presents the kinds of contributions that local communities in Zambia have made to primary school projects.

Despite its virtues, mobilizing local resources is often problematic. Collective local contributions to neighborhood schools are often hampered by the existing system of taxation and the structure of incentives, and some countries have adopted measures that are arguably inefficient and inequitable. Furthermore, national governments are increasingly shifting the financial burden of providing other public services, like primary health care, to the community, which makes raising local funds for education more difficult.

Table 7-7. Community Support for Self-Help Primary School Projects in Zambia, 1979–84

Type of support	*Percentage of projects receiving support*
School committee work	58.7
Building materials	52.2
Skilled and unskilled labor	50.0
Voluntary cash contributions	43.5
Compulsory cash levies	41.2
Building plans	21.7
Land	17.4
Lodging facilities for boarders	17.4
Cooked meals for students	2.2
Teaching and supervision	0.0

Source: Kaluba 1988.

Community and parental involvement cannot work effectively unless the national government improves the capacity and willingness of the community to meet its increased financial responsibilities. In particular, countries must help local governments identify and use broader, more productive, and buoyant sources of tax and nontax revenues. In many cases, providing incentives for mobilizing local resources is also important.

SEEKING BETTER WAYS TO RAISE LOCAL RESOURCES. To improve the local financing of education, governments must carefully choose the means they use to mobilize local resources. Fund-raising strategies should be both fair and economically neutral, which implies that they should not discourage primary school enrollment nor the use of educational materials in schools. Moreover, governments should avoid taxes such as high local sales taxes that are potentially damaging to the local economy.

In an effort to raise additional resources for education, some countries have resorted to measures that are neither efficient nor equitable. In general, these measures include obligatory fees or contributions that depend on primary school enrollment—for example, direct fees for enrollment and textbooks as well as cash or in-kind contributions that parents of primary school students must pay directly to the school or indirectly to parent-teacher associations.

These measures are inefficient because they raise the effective price of primary education and discourage parents from enrolling their children in school. They can have a particularly negative impact on the enrollment of poor children and, hence, run counter to the equity objective of primary education. Countries with these kinds of obligatory contributions should search for alternative ways to mobilize local resources that do not depend on a child's enrollment or use of textbooks. The policy implication is that efforts to generate funds locally should rely on voluntary contributions (often in-kind) and, if feasible, on local taxation. Although fees may be the only way to marshal local contributions in some cases, they are obviously a second-best approach. Reforming local taxes and providing incentives for voluntary community contributions are, in general, more equitable and efficient alternatives.

In many developing countries, improving the ability of communities to garner additional school resources by raising taxes requires local tax reforms in two areas. One area for improvement is the capacity of local governments to administer taxes. In many countries, local governments are unable to realize the full potential of a given tax base because of administrative weaknesses (Dillinger 1988). These administrative shortcomings include, among others, the inability to determine, enforce, and collect taxes.

The other area for improvement is the assignment of tax sources. National governments generally reserve the broader, more elastic, and higher-yielding revenue sources for themselves, leaving local governments with a very narrow and unproductive tax base (Schroeder 1988). Furthermore, the yields from tax instruments available to local governments are often tightly constrained by the central government. Severe constraints on property taxation, one of the common sources of local taxes, take the form of strictly limited assessment ratios and maximum tax rates, generous exemption policies, and delayed—or canceled—general revaluations (Dillinger 1988).

Benefit taxation can play an important role in financing education. The benefit principle, which has a long history in public finance, can be modified to accommodate the special circumstances of the poor. It has been applied successfully to funding local activities such as irrigation and municipal public improvement projects (McCleary 1988). Municipalities in Colombia, for example, have successfully undertaken local public infrastructure projects by asking the potential beneficiaries to pay their share of the cost in accordance with their estimated benefits.

Governments should seriously consider developing an analogous approach to community financing of schools. Although substantial benefits can be gained from primary education by persons and institutions outside the community, the benefits accrue mostly to the parents, their children, and other members of the community.

Box 7-5 presents two examples of benefit-related tax measures that national governments might want to consider in assigning additional fiscal powers to local authorities. These potentially efficient and equitable ways of raising local funds for improving neighborhood schools include a poll tax on households (other than poor households) and an education surcharge on the real property tax on houses and land (to be determined by local governments within limits).

Empowering communities fiscally and assigning them greater responsibility for education may be stalled by poor attitudes and a lack of incentives. First, local leaders incur political and other costs when they support local taxation. Second, the financial mechanisms that reward school authorities for using funds cost-effectively are often absent. Past policies encouraged communities and their local governments to adopt an attitude of dependency on

Box 7-5. Using Poll and Property Taxes for Education

Property and poll taxes may be useful, when combined, for funding school improvements wanted by the community. In 1987 Guinea Bissau adopted a poll tax to raise additional funds for education and other social programs. All persons fifteen years of age and older had to pay $4; in at least one province, 96 percent paid the tax.

Poll taxes are efficient and economically neutral because they do not discourage individuals from enrolling in school or from using textbooks. Nonetheless, poll taxes are often rejected by policymakers who consider them to be regressive, and they should not be used to fund general budgetary expenditures.

Poll taxes can be made more equitable by exempting the poor and using the revenue to help finance programs that broadly benefit the poor (like improving primary education). Poll taxes of this sort are usually preferable to the regressive indirect taxes that support public education in most developing countries and favor the nonpoor. Using poll taxes to improve local primary schools is also probably more efficient and equitable than charging fees or requiring family contributions that are contingent on school enrollment—common methods of local fund-raising in the developing world.

Surcharges on the real property tax on houses and land (determined by local governments within limits) are another method of paying for school improvements. In many developing countries, local property taxes are a major source of educational financing. Land taxes, which are an equitable but often underused policy instrument of taxation, are economically neutral because they are only used for unimproved land. Property taxes require, however, a strong political will and administrative mechanisms to be adopted and implemented properly. Although taxes on houses are not neutral in themselves, their distortionary effects are largely attenuated when the benefits of having an improved primary school in the community are taken into account.

Source: World Bank 1988c.

the central government. One consequence of this attitude is that communities tend to wait for the national government to act, except in periods of serious and prolonged crisis. The political and social costs of marketing a tax increase may be so high that leaders are reluctant to propose new local taxes or user fees to finance educational and other public service improvements.[16]

One way of dealing with the problem is for the central government to send a strong financial signal that local funding is important and, indeed, essential. This signal can come from a matching grant system in which higher levels of government reward communities by contributing a certain amount for every dollar raised locally for the improvement of primary education. Governments may also establish a centrally financed school improvement fund and give grants to schools with meritorious proposals to improve their school quality. To show the government's commitment to mobilizing local resources, countries could require a contribution (in cash or in kind) from the community as a condition of the grant.

The use of incentives, which are particularly important for raising voluntary contributions, is critical for countries that do not allow schools to retain the funds raised by parents and their community. Such a prohibition is clearly inefficient. People are more willing to make financial contributions if the money supports improvements in their own schools. Mali, for example, probably weakened incentives for local giving when it adopted a new tax system that no longer permits schools to keep the funds raised by parent associations. To encourage continued contributions, the government recently agreed to develop a matching grant system.

Ensuring Equitable Financing

Two questions arise at this point. How can a policy of mobilizing local funds for primary education help schools in poor communities that have few resources? More generally, how can educational resources be distributed more fairly, and how can local financing be strengthened without creating greater disparity between the education systems of poor and well-to-do communities?

As earlier chapters have made clear, opportunities for schooling, and schools themselves, are far from equal in developing countries. Addressing this problem calls for greater sensitivity to the inequities in educational financing. It is particulary important to remember that making the postprimary education sector less dependent on the central government can result in more equitable distribution of the overall education budget. Consequently, strategies to reform the financing of postprimary education must be carefully considered and included, whenever relevant, in efforts to broaden the resource base for primary education.

The Equity Implications of Local Financing

Policymakers are wary about mobilizing local resources for three reasons. First, poor communities in countries like Haiti and Mali are already contributing substantially to schools and other services (such as water and health care). Second, some policymakers fear that inequality in educational opportunities will worsen as the reliance on local funding increases. Third, some think that relying on the community to finance primary schools will produce substandard schools in poor villages.

Some of the fears about increased local funding of education are exaggerated and focus on marginal cases. Clearly, financially overburdened communities can rarely spend more on education. Wealthier communities, however, are in a position to bear a larger share of local education costs. The reality is that different countries face different situations and will elect different policies. In countries with a centralized system of financing education, the central government will continue to shoulder the major part of funding primary schools. Many other countries can implement policies that enable and encourage community fund-raising to supplement various resources supplied by the central government. Unless increasing the amount of local financing automatically induces a drastic cut in the national budget for schools in poor communities, poor schools are unlikely to become substandard as a result of better local funding of education.[17] Improving the mobilization of local resources can, however, worsen educational inequality because of differences in the endowments and spending priorities of communities. This remains a valid cause for concern and should be addressed by policymakers seeking to design financial policies for education.

Directing National Resources to Poor Communities

Governments in many countries, especially those that rely on local financing, can reduce the inequities in education by improving their distribution of national funds for education. The challenge is to design a system of allocating central government resources that would favor disadvantaged communities and would complement locally generated resources. The policy of mobilizing more local resources could, in fact, significantly improve the funding of disadvantaged areas. For example, encouraging and enabling higher-income communities to raise additional revenue for their primary schools would give the central government more scope to target its resources to disadvantaged schools.

One approach to reforming distribution is a system of targeting and indexing in which the national government carefully selects a group of poor communities or school districts for special attention. The government's per capita allocation to primary schools in these areas could then be linked and equated to the average expenditure (from all sources) in a sample of schools in nonpoor communities. The amount of resources mobilized by any one community in this system would not significantly affect the amount of national subsidy that a community receives. Hence, this system would not be a disincentive for mobilizing local resources.

The literature on fiscal federalism and decentralized education suggests other ways in which inequality can be avoided or at least minimized (Guthrie, Garms, and Pierce 1988; Winkler 1989). The appropriate method will depend on each country's objectives and institutional constraints. The various approaches, which involve the use of grants-in-aid, differ in their design, complexity, and efficiency. Moreover, their ability to reduce inequities depends on how each country defines educational equity. A foundation plan that is appropriate for much of the United States may be adequate for a country whose overriding equity objective is to guarantee a desired minimum (foundation) level of educational spending for all school districts; it may not be sufficient, however, for a country that strongly believes in completely equal provision of education.[18]

Building National Commitment to Equity

Many developing countries must strengthen their national commitment to distributing central government funds through a system that explicitly favors schools in disadvantaged communities. Many countries allocate national funds for primary schools equally on a per capita basis, without taking into account differences in community needs and resources (Winkler 1989). The systems in other countries demonstrate perverse biases. For example, in some countries the allocations per student are lower in rural areas, particularly in poor regions, than in urban areas. Even within cities some districts (such as slums) are grossly disadvantaged.

In a decentralized education system, which countries may prefer for historical, cultural, and economic reasons, inequity can rarely be avoided completely. Much of the inequity can be remedied, however, if the country's political determination is strong enough. Yugoslavia provides an interesting example. Under its 1974 constitution Yugoslavia created the self-governing interested communities of education (SGICEs), which fund about 85 percent of the cost of elementary education for the country as whole. These associations of citizens, employed persons, and teaching personnel receive tax funds to administer the schools. In many respects, they resemble the school boards of North America. The republic defines the minimum program standards, which the SGICEs can exceed. Each association submits to the taxpayers a proposal on spending level per student, which may be funded through local taxation, voluntary contributions, or both. Taxpayers then vote on the proposal. SGICEs that are unable to meet the minimum standards receive supplementary resources through a system of general grants. Thus some of the poorer communities pay only 65 percent of the expenses for elementary education, compared with the average contribution of 85 percent.

The national government obviously plays a central role in reducing educational inequality. In countries that are taking greater advantage of local financing, it should help keep inequality from worsening, and in countries with highly decentralized educational systems, it should encourage equity (Winkler 1989). Solving the equity problem requires reducing or eliminating the advantage associated with a higher tax base. In effect, this means redistributing tax money in one form or another from high- to low-income communities. That task, which is not an easy one, requires consensus building and can only be undertaken by the national government. Consequently, the national government must commit itself to developing and implementing an adequate system of resource allocation that is consistent with the national definition of equity. A vigorous campaign to mobilize local resources means that the central government must play a stronger role in redistributing the resources that finance education.

Summary

Although many low-income countries need external assistance for improving primary education, a reasonable program of expansion and quality enhancement can be undertaken by and is within the means of many developing countries. Such a program requires an integrated, multipronged, and flexible approach to financing education. National authorities should pay close attention to the following issues. First, they should examine the factors that promote high repetition and dropout and render primary schools inefficient, and they should seek solutions. Second, they should reorder intersectoral and intrasectoral priorities, making primary education a priority in the budget for education. Access to primary education should not be rationed according to the ability to pay. Moreover, countries that spend less than 4–5 percent of their gross national product for education should seriously consider increasing the share of education in the national income. Third, countries should seriously consider broadening their educational resource base and instituting financial mechanisms that strengthen the commitment of parents and communities to primary education and fully exploit their willingness to pay. Finally, in reforming education finance, national authorities should address the issue of equity.

A concrete example of how a multipronged approach might be implemented is presented in box 7-6, which describes the Republic of Korea's success in financing universal primary education. A key lesson from Korea's experience is that balancing the responsibilities of parents, local institutions, the private sector, and the national government is essential. By doing so, Korea's education system is internally efficient, its finance is equitable and broadly based, and the central government's budget for education is concentrated on primary education.

Box 7-6. Financing Good Primary Education for All: The Korean Experience

The experience of the Republic of Korea reveals that successful educational development depends on many factors. In 1960 Korea was a poor country, with a per capita GNP equivalent to that of Indonesia, Morocco, or the Philippines in 1985. When Korea began rebuilding its war-ravaged economy, national leaders made education a priority. First, they took a broad, flexible approach to educational finance and invested a large proportion of the country's gross national product in education. The policy relied on parents' sense of responsibility, took advantage of their willingness to pay for their children's education, and allowed the private sector to respond to educational demands as they evolved.

Including all sources of finance, 9 percent of the gross domestic product went to education in 1966, and this rose to 10 percent in 1970. Students and parents paid for about 71 percent of educational expenses. Their contributions were used to construct and operate schools as well as to cover out-of-school expenses such as school supplies, transportation, extracurricular activities, and room and board. In the mid-1960s, out-of-school expenses accounted for 80 percent of all household educational expenditures, with close to half the total going toward primary education. A large share must have been used to purchase textbooks, since free books were provided to only one-quarter of the students.

Second, Korea concentrated central government funds on primary education, which is compulsory for children. By allocating three-fourths of its national education budget to primary schooling and relying on parents' willingness to pay for private secondary and tertiary education, Korea achieved primary education for all while satisfying the strong and growing demand for postprimary education. In 1965, public schools accounted for 99 percent of primary school enrollment but served only 43 and 27 percent of all students enrolled at the academic secondary and tertiary levels of education, respectively.

Third, Korea used local institutions to finance and provide primary education. As early as 1949, parent-teacher associations played an important role in financing primary education. Despite an ambitious compulsory education law,

Notes

1. Educated adults have higher individual earnings, more frequent employment in the urban labor markets, greater agricultural productivity, lower fertility, better health and nutritional status, and more "modern" attitudes—all considered dimensions of development.

2. The $5 amount is purely arbitrary. In 1985, upper-middle-income countries spent approximately that amount on nonsalary recurrent expenditures for primary education. In 1990 dollars, providing a minimum set of material and managerial inputs for schools could cost as much as $30 per student.

3. Research on cost-effective inputs for developing countries is limited. For a recent review, see Lockheed and Hanushek (1988).

the central government could provide only 15 percent of the revenues needed to finance primary education. Thus parent-teacher associations, which were organized to supplement teacher salaries and to increase parental involvement in school decisionmaking, provided 75 percent of the funds for local schools. Local governments contributed another 10 percent.

In the 1960s local sources provided between 20 and 25 percent of the total expenditures for local primary education. In 1970 parent-teacher associations were reorganized as the Yuksonghoe (voluntary parent-teacher association), with the same objectives as before. In 1974, after the reorganization, the Yuksonghoe fees amounted to 28 percent of the public budget for compulsory education.

Fourth, the central government provided grants to local schools for compulsory education. These grants amounted to 77 percent of the expenditures of the local government in 1966 and 73 percent in 1970. Conscious of the inequality existing among communities, the national government began in 1962 to equalize public expenditures among primary school districts across the country. It did so by means of formulas that distribute national funds on the basis of local need and ability to pay.

As in industry, Korea's educational development depended on the use of inexpensive but highly productive labor. Teachers were paid relatively low wages, and classes were very large. They averaged sixty or more students, with 37 percent of the primary school classrooms holding more than eighty students in 1965. Salaries remained low because (a) teachers were highly respected, and the teaching profession was thus attractive in and of itself, (b) teachers had many opportunities to supplement their official salary, chiefly working as examination tutors after school hours, and (c) the growing abundance of high school and college graduates, and the development of short, low-cost teacher training programs, created a large pool of low-cost labor for primary education.

Source: McGinn and others 1980.

4. Increased internal efficiency can lead to perverse consequences. The extent to which it frees up resources for funding high-priority inputs needed for improving and expanding the primary education system depends on the political environment of public finance. Weak representation of the education sector in the budgetary process and inadequate interministerial coordination can lead to lower allocations. The planning and finance ministries could, for example, assume that if primary education were more efficient, it would require less money. The education officials of some Asian countries called attention to this situation and its disincentive effect in a World Bank–sponsored seminar on Planning and Mobilization of Financial Resources for Education in the East Asia Region, Philippines, which was held in October 1987.

5. The opportunity cost of schooling may also be the reason why some students drop out of school early. Past studies (see, for example, King and Lillard 1983;

Rosenzweig and Evenson 1977) have shown that opportunities for farm or wage work decrease the educational level of rural children.

6. However, past studies found that children who repeat one or more grades in their early years of schooling are likely to withdraw from the educational system later. Furthermore, repeating a grade does not necessarily help learning. For a discussion of these findings and other consequences of repetition, see Behrman and Deolalikar (1988), Haddad (1979), Shepard and Smith (1989), and Unesco (1980).

7. According to data from the Côte d'Ivoire, the social internal rate of return to primary education falls 15 percent when the cost to the school and the student for one year of repetition is factored in.

8. Although some countries have more than five grades in the primary cycle, the lowest common denominator is used here to illustrate the differences in the severity of the repetition and dropout problem across countries.

9. Defense spending is another area in which government spending is unnecessarily high.

10. Conceptual problems are involved in developing empirically based criteria for allocating resources among sectors, and these problems remain high in policy-oriented research. See Jimenez (1990).

11. For an analysis of the weaknesses of the tax system in developing countries and a discussion of the general directions that tax reforms should take, see World Bank (1988b).

12. Teja (1988, p. 13) writes that "earmarking may be loosely thought of as an application of the benefits principle of taxation since, as in private markets, taxes are paid according to perceived benefits."

13. To minimize the inflexibility associated with the usual form of earmarking funds, taxes could be raised ostensibly to finance laudable educational programs with two provisos: (a) that earmarked funds in excess of the amount needed to finance minimum levels of educational materials and services can be allocated to other programs and (b) that shortfalls are to be covered automatically by allocations from the government's general revenue.

14. Frequently, adjusting educational inputs to local prices is hampered by applying the ministry of education's national standards on construction, teacher quality, salaries, and so forth.

15. Quasi-obligatory fees are voluntary contributions that, because of very strong social pressures (from teachers, for example), are often compulsory in practice.

16. Social marketing is "the design, implementation, and control of programs calculated to influence the acceptability of social ideas or practice" in a target group (Fine 1981, p. 25).

17. Local participation in school financing has been linked to poor school quality in Indonesia, where the organizational structure is being decentralized and it is not clear who is responsible for quality control.

18. In foundation plans, communities may tax and spend as they wish beyond a minimum level of spending guaranteed by the government on an equal basis.

8

International Aid to Education

ALTHOUGH FEW DISPUTE THE NEED to improve primary education systems in developing countries, the costs involved are formidable. As chapter 7 pointed out, domestic resources must—and can—be expanded and used more efficiently. But for real progress to be made in many countries, external assistance is critical.

Patterns of Giving to Primary Education

During the 1980s, 9.2 percent of all bilateral and multilateral aid for development was targeted to education, producing an average annual allocation of $4.2 billion (Naumann 1984).[1] Although its effect on development has never been analyzed systematically, the maximum benefit may not have been realized for two reasons. First, more than 95 percent of the assistance was directed toward secondary and tertiary education, while primary education was generally neglected. Second, support for the most cost-effective educational inputs was meager.

The Volume of Aid

Despite providing significant support for education overall, multilateral and bilateral donors have supported primary education only minimally. Between 1981 and 1986, the international aid disbursed to primary education amounted to a mere $181.3 million annually and represented 4.3 percent of the annual total aid to all levels of education for the same period (see table 8-1).[2] Regional differences do exist. In Sub-Saharan Africa between 1981 and 1983, 16 percent of aid went to secondary education, 17 percent to secondary technical education, and 34 percent to tertiary education. The remaining 33 percent went to primary education. However, this still amounted

Table 8-1. Aid Disbursed to 143 Countries for Primary Education, 1981–86
(billions of dollars)

Country group	*1981*	*1982*	*1983*	*1984*	*1985*	*1986*	*Annual average 1981–86*
Low-income	58,510	60,874	54,262	55,422	60,182	58,185	57,995
Lower-middle-income	102,527	77,654	68,406	99,325	91,908	95,324	103,826
Upper-middle-income	14,599	11,842	2,426	10,845	5,574	42,187	14,568
Miscellaneous[a]	4,377	3,405	7,722	4,775	4,953	4,702	4,939
Total	180,013	253,775	132,816	170,367	162,617	200,398	181,328

a. Includes high-income oil exporters and countries with populations under 1 million.

Sources: IREDU 1989; World Bank 1988c.

to investing over 500 times more per student in higher education than in primary education.

Most aid to primary education went to low- and lower-middle-income countries. Low-income countries received about 32 percent of all aid, and lower-middle-income countries received 57 percent (see figure 8-1). This represents $57.9 million or $0.87 per student for low-income countries—excluding China and India[3]—and $103.8 million or $0.99 per student in lower-middle-income countries. Upper-middle-income countries received $14.6 million, the equivalent of $0.16 per student. The region receiving the most aid per primary student—$1.34—was Europe, the Middle East, and North Africa ($41.7 million total). Africa and the Latin America and Caribbean region received $0.86 and $0.80 per student, respectively ($58.4 million and $54.3 million total). Aid to Asia amounted to less than 10 cents per student ($27 million total). Even if China and India are excluded from this regional grouping, a mere $0.29 was allocated to each child.

Figure 8-1. Average Annual Aid Disbursed to 143 Countries for Primary Education, 1981–86

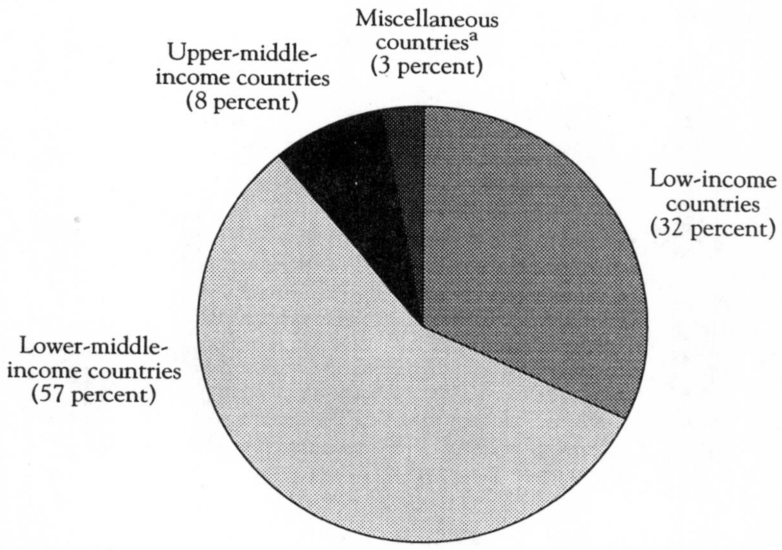

Total: $181.3 million

a. Includes high-income oil exporters and countries with populations under 1 million.
Sources: IREDU 1989; World Bank 1988c.

The flow of aid to primary education was not only small during the 1980s, but declining. To adjust for the sharp annual changes in the level of support provided, as shown in table 8-1, three-year running averages of aid flows were calculated (figure 8-2). These data show that aid to the poorest countries declined by 19 percent between 1981 and 1986, despite growing populations. Lower-middle-income countries received 38 percent less in aid, despite a slight increase since 1984. Only in upper-middle-income countries was there an increase in aid flows, of 25 percent since 1981, although aid had declined significantly until 1984. Since the data in figure 8-2 represent constant dollars, they do not take into account local purchasing power, which declined because of domestic inflation.

The Sources of Aid

Multilateral agencies provided about two-thirds—$120 million—of the annual aid disbursed to primary education from 1981 to 1986. The World Bank

Figure 8-2. Aid Dispersed for Primary Education by Country Income Level, 1981–86

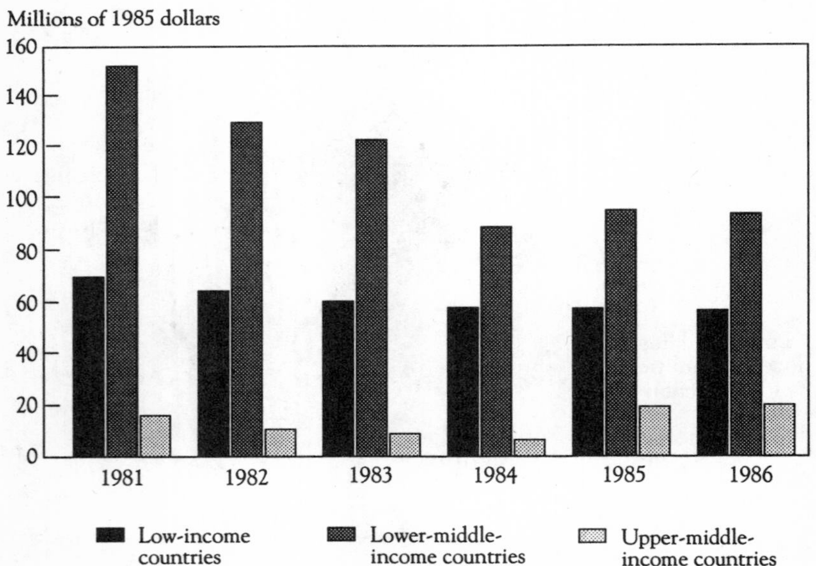

Note: Figures for 1982–85 represent three-year running averages, centered on the base year specified; figures for 1981 and 1986 are two-year averages.

Sources: IREDU 1989; World Bank 1988c.

Table 8-2. Average Annual Contribution of Major Bilateral and Multilateral Donors to Primary Education, 1981–86

Donors	*Millions of dollars*	*Percentage of total lending to primary education*
Bilateral donors	60.7	33.5
United States	25.8	14.3
France	7.4	4.1
Saudi Arabia	6.2	3.4
Sweden	5.1	2.8
United Kingdom	3.1	1.7
Germany[a]	2.9	1.6
Nongovernmental organizations	3.8	2.1
Other	6.3	3.5
Multilateral donors	120.6	66.5
World Bank	48.5	27.3
World Food Program	25.4	14.0
Interamerican Development Bank	6.3	9.1
UNICEF[b]	12.7	7.0
UNDP[c]	2.5	1.4
European Development Fund	2.2	1.2
Other	11.8	6.5
Total	181.3	100.0

a. Data refer to the Federal Republic of Germany before unification.

b. United Nations Children's Fund.

c. United Nations Development Programme.

Source: IREDU 1989.

provided 40 percent of multilateral aid and more than 27 percent ($48.5 million) of all donor aid during this time (see table 8-2). The World Bank began lending to education in 1963 and to primary education in 1970. Since then, the Bank has steadily increased its commitments to primary education, and that aid now averages more than 23 percent of total Bank lending for education (see table 8-3). Since 1970 the World Bank has committed over $2 billion to primary education; however, disbursement figures for the same period are significantly lower, since disbursements represent earlier commitments spread over time.

The remaining one-third of international aid to education—$60 million annually—was provided by bilateral donors and nongovernmental organizations. The largest donor was the United States, which allocated $25.8 million—more than France, Germany, Saudi Arabia, Sweden, and the United Kingdom combined. Nongovernmental organizations contributed $3.8 million, or 2 percent of world aid to primary education. This may underestimate real flows from these organizations, since governments do not keep track of nongovernmental aid as carefully as official development assistance. A more accurate figure would probably be 5 percent.

Table 8-3. World Bank Lending Commitments for Primary Education, Fiscal Years 1963–90

Fiscal period	*Lending for primary education (millions of current dollars)*	*Total lending for education (millions of current dollars)*	*Lending for primary education as a percentage of total lending for education*
1963–69	0.0	243.9	0.0
1970–74	36.5	814.9	4.5
1975–79	236.7	1,681.9	14.1
1980–84	423.7	2,964.2	37.8
1985–90	1,278.2	5,530.7	23.1

Note: The World Bank's fiscal year begins on July 1 and ends on June 30.
Source: World Bank data.

The Use of Aid

The effectiveness of aid depends not only on the volume but also on the kind of assistance being given. From 1981 to 1986 little of the aid to primary education was allocated to the inputs that are the most cost-effective (see table 8-4). For example, only 19 percent of all aid was devoted to pedagogical inputs such as teachers and textbooks,[4] but 30 percent went to assist nonsalary recurrent expenditures (used at the discretion of the recipient), and 30 percent supported infrastructure (buildings, furniture, and equipment). Another 13 percent financed social expenditures such as school feeding programs; this is equivalent to the entire food aid provided annually by the World Food Programme.

The lack of support for pedagogical materials is particularly notable in low-income countries, which are expanding their educational systems. In these countries, buildings, furniture, equipment, and social expenditures accounted for 58 percent of all aid. In lower-middle-income countries, these expenditures amounted to 40 percent. For middle-income countries, however, analyzing the distribution of aid by purpose is confounded by the large share of budgetary assistance—financial support to ministries of education for nonsalary recurrent expenditures—that may include pedagogical materials. Budgetary assistance increases with the country's level of income, reaching 90 percent of the external assistance to upper-middle-income countries.[5] As their level of income rises, recipient countries are apparently allowed to exercise more discretion in their use of aid for primary education.

Breaking down aid by activity is also instructive (see table 8-5). From 1981 to 1986 teacher training received only 7 percent of total aid, the training and support of administrative staff received 2 percent (8 percent if combined with unspecified administrative support), and educational and curriculum reforms received 11 percent and 5 percent, respectively.

The messages to be drawn from these data are alarming. First and foremost, the flow of aid to primary education has been small in every way: in ab-

Table 8-4. Type of Aid for Primary Education by Country Income Group and Region, 1981–86
(percent)

	Country group				*Region*				
	Low-income	*Lower-middle-income*	*Upper-middle-income*	*High-income oil exporters*	*Europe, the Middle East, and North Africa*	*Latin America and the Caribbean*	*Asia*	*Africa*	*All countries*
Technical assistance									
Teachers	5.4	8.5	0.2	0.0	12.9	0.4	2.3	9.1	7.1
Other	12.1	4.2	0.1	98.4	1.2	3.5	8.7	12.6	6.6
Short-term consultants	0.1	0.4	0.6	1.6	0.1	0.4	0.8	0.3	0.3
Books	5.0	0.5	0.1	0.0	0.5	0.4	5.2	2.7	1.8
Other didactic materials	2.5	3.4	0.7	0.0	7.5	0.5	2.0	1.5	3.0
Furniture and small equipment	13.8	0.7	0.1	0.0	0.6	0.7	9.3	8.5	4.3
Social expenditure (such as food)	19.5	11.4	0.3	0.0	17.9	1.4	11.9	18.1	13.0
Scholarships	1.2	2.9	0.4	0.0	6.3	0.2	2.1	0.5	2.3
Budgetary assistance	5.7	34.8	90.3	0.0	29.5	58.1	30.9	7.0	29.9
Big equipment	2.4	1.9	0.9	0.0	2.0	1.8	2.7	1.9	2.0
Buildings	22.1	25.8	3.3	0.0	20.6	22.6	16.0	28.9	23.4
Other and undistributed aid	10.2	5.6	3.0	0.0	0.9	10.0	8.1	8.9	6.3
Total	100.0	100.0	100.0	100.0	100.0	100.0	100.0	100.0	100.0

Source: IREDU 1989.

Table 8-5. Aid for Selected Educational Support Activities, 1981–86
(percent)

Activity	*Low-income countries*	*Lower-middle-income countries*	*Upper-middle-income countries*	*All countries*
Teacher training	7.6	6.8	0.7	6.7
Preservice	5.8	2.4	0.3	3.3
In-service	1.8	4.4	0.4	3.4
Education reform	10.7	18.1	8.4	15.5
Curriculum reform	7.7	2.7	8.0	4.5
Unspecified	3.0	15.4	0.4	11.0
Administrative support	16.6	4.6	0.1	7.7
Staff training and support	4.2	1.0	0.1	1.8
Unspecified	12.4	3.6	..	5.9
Total	34.9	29.5	9.2	29.9

.. Less than 0.1 percent.
Source: IREDU 1989.

solute terms, in relation to total aid for all levels of education, and in expenditure per pupil. Second, the level of aid has stagnated since the early 1980s, with the aid per pupil decreasing as a result of the continuous growth of the primary school–age population. Finally, only a small proportion of aid has gone to the most critical inputs, such as educational materials and teacher training.

Why Donors Have Neglected Primary Education

The global average of 4–5 percent of external aid allocated to primary education indicates that donor countries have not considered assisting primary education to be a significant part of their mission. Their neglect of primary education can be explained by differences in how they generate and use resources for primary and higher levels of education and in what they perceive to be the most effective use of foreign aid. In general, external sources of finance have preferred to support investment projects that are capital-intensive and foreign exchange–intensive; that are limited in the number, scope, and geographic dispersion of their components so that implementation places a minimal burden on scarce managerial resources; that depend heavily on the expertise of professionals (including teachers) in the donor country; and that involve study abroad for nationals in the recipient country.

Primary schools meet few of these criteria. They are highly dispersed, the cost of administering an individual school is minuscule from the perspective of an international donor, and no single school enjoys high visibility or exerts a separate, identifiable impact. For tertiary and secondary education, the situation is the reverse. Fewer postprimary institutions exist, and they are therefore more visible. Each has a discernible influence on the educational landscape, and the foreign exchange requirements are higher for tertiary and secondary than for primary education (this is especially true for technical institutions). In addition, many external funding agencies can easily support administrative staff development by financing graduate study abroad and either providing graduate professors outright or supplementing the budgets of local institutions so that they can hire expatriates directly.

In fact, as early as the 1960s, bilateral donors (including the United States Agency for International Development) and agencies such as Unesco (United Nations Educational, Scientific, and Cultural Organization) indicated that supporting primary schools directly was less efficient than supporting them indirectly by funding teacher education programs and central agencies such as curriculum development centers and examination councils. Financing these central supports of primary education was expected to have multiplier effects throughout the system and to provide more dividends for the same amount of money. Donors let the developing countries themselves support primary education per se and tended to avoid financing local, and especially recurrent, costs. These trends have not changed much; the Canadian International Development Agency, for example, still limits itself to intervening in the foreign exchange components of projects. In conclusion, the distribution of direct aid to education by level may reflect a perceived comparative advantage of higher over lower levels. Similarly, project aid remains the preferred mode of lending because it is highly visible.

New Priorities for Donor Support

The current challenges facing educational development and the imperative to build strong, sustainable, and good-quality primary education suggest, however, that former patterns of aid may no longer be appropriate and that external donors should increase their support for broad programs to develop primary education. The change must be threefold. First, additional resources must be concentrated on primary education, both in absolute terms and relative to other levels of education. Second, donors must begin to emphasize subsectoral development programs instead of individual projects. And because of the considerable up-front investment necessary, aid must be carefully coordinated. Mobilizing the resources to launch programs that improve quality and increase access—especially in low-income countries—requires the joint effort of donors and recipient governments. Third, donor agencies need to respond with flexibility to conditions in different countries.

Increasing the Level of Aid

The low level of aid to primary education and the urgent need to stimulate educational progress make a compelling case for giving more aid to primary education. Sound programs to develop primary education typically require more resources, sustained over a longer period, than can be mobilized internally, especially in the short and medium term. Countries that have designed programs that lay the groundwork for sustained educational development during the 1990s need access to increased, longer-term, and more flexible international aid. Although countries must increase their own level of educational funding, start-up expenditures are often considerable and interim financing is often necessary to meet the budget. External support for primary education budgets to bridge the temporary gap between feasible and required outlays is amply justified.

New aid must, however, be targeted more effectively than aid was in the past. Although some low-income countries still require external support for their efforts to expand coverage, they must also focus on improving quality. In middle-income countries, donor support may be less critical; in any event, support should shift from buildings and furniture to pedagogical materials. Increasing aid and targeting it more effectively is crucial for revitalizing primary education systems.

Supporting Subsectoral Development Programs

For the proposed approach to be effective, external assistance to primary education must be put in the context of broad, subsectoral expenditure programs. Subsectoral development programs include both recurrent and investment outlays, often span a decade, define investment priorities according to the country's situation, and specify a policy framework. In the World Bank's experience, the most successful strategies for developing and implementing such programs include at least three elements: in-depth analysis of subsector issues, a small number of focused policy and institutional objectives, and a persistent commitment to these objectives over an extended period of time. At regular intervals, progress should be reviewed and necessary adjustments made.

The spending priorities and the policy framework of these subsectoral development programs must be designed by the recipient countries and implemented by national specialists. The staff of external agencies can only play a facilitating role and possibly provide analytical support. To pave the way for recipient countries to take more responsibility for designing and implementing primary education improvement programs, donors must encourage the development of institutions within those countries for planning and managing programs. Specific investment programs could be appraised by a national agency on the basis of agreed criteria. Depending on the country's conditions and on donor preferences, external financing could be targeted to specific subprograms or commingled with national resources.

Support for the development of the primary sector should be mobilized for a variety of purposes. Three priorities for external assistance are particularly urgent. First, seed money is needed to help defray both local and foreign costs of developing institutions and improving management, focusing specifically on strengthening the examination and achievement assessment systems and the training of education administrators. Second, the international community must facilitate communication among countries that are attempting to reform their systems of primary education. By discussing their efforts to formulate and implement education policies, and especially to establish new enterprises such as systems for providing textbooks in primary schools, countries may learn from each other's experience. Serving as a catalyst for this type of information exchange is a function well suited to donors. Third, the international community should establish and finance a source of high-quality technical expertise that is neither political nor beholden to any particular international donor or government. This expertise would facilitate the transfer of knowledge and the development and implementation of subsectoral development programs.

Responding with Flexibility

Priorities differ among countries. While providing textbooks is of the utmost urgency for some countries, upgrading the physical plant is more important for others. In some instances, the paramount concern is simply getting schools to function. Priorities often depend on the country's level of income. Donors will have to recognize the constraints on developing countries and be flexible in defining their own priorities. They must also be more flexible in the ways they allow recipients to use their aid. As table 8-4 indicated earlier, low-income countries receive only 6 percent of their aid in the form of budgetary assistance, while upper-middle-income countries receive 90 percent in that form. The move to support broad subsectoral development rather than specific projects implies more autonomy for aid recipients at all levels of income.

Summary

International donor assistance to primary education during the past decade was minimal and not targeted to the most cost-effective educational inputs. Donors have neglected primary education at least partly because it has been easier, and in their view more effective, to generate and use funds for projects at higher levels of education. The current challenges facing educational development and the need to build sustainable, good-quality systems of primary education demand that donors change their lending patterns. To support the development of primary education effectively in the 1990s, donor agencies must increase the level of aid, particularly for programs of expansion and quality improvement, support subsectoral development programs, and respond with flexibility to the unique conditions of each country.

Notes

1. According to Naumann (1984), the total amount is shared as follows: bilateral aid, \$2.8 billion; multilateral aid, \$1.0 billion; and aid from nongovernmental organizations, \$0.4 billion. IREDU estimates a total amount of \$3.76 billion.

2. These figures probably underestimate the actual flows because some countries did not provide any record of the aid received. According to IREDU (1989), \$220 million is a more reasonable estimate, although the extra \$40 million are not accounted for arithmetically. This would nevertheless remain a mere 5 percent of the total aid to education. Except when specified otherwise, all figures on aid refer to actual disbursements.

3. This figure falls to \$0.20 per pupil when China and India are included.

4. Including unspecified technical assistance, without which the figure would be 12 percent.

5. The number of high-income oil exporters is too small for a meaningful average.

9

Education Reform: Policies and Priorities for Educational Development in the 1990s

EVERY NATION FACES THE CHALLENGE of improving its education system, including its primary education system. The breadth of needed reform is, however, much greater for developing than for industrial countries. Developing countries must not only improve their schools and foster the development of higher-order thinking skills, they must also expand coverage to include the school-age children who are out of school. Since they begin with proportionately fewer effective schools than industrial countries, their reform programs must adopt the most effective strategies for change. The most successful reform efforts focus on the school as a whole rather than on students, teachers, curricula, or administrators in isolation (Berman and McLaughlin 1977; Crandall and Loucks 1983; Fullan 1982; Havelock and Huberman 1977; Verspoor 1989). One strategy that has been tried, without success, is radical reform. Indeed, the central lesson learned from three decades of research on school reform in both industrial and developing countries is that educational change is a complex, dynamic, lengthy, and idiosyncratic process that proceeds in incremental steps.

Challenges in Educational Development

Effective improvement strategies gradually build on existing strengths, which differ among and within countries. Some theorists hold that education systems evolve in stages (Beeby 1966). Initially, schools function with untrained staff and narrow subject content. Next comes a rigid and ordered stage of formalism, characterized by trained but poorly educated teachers, followed by a stage of transition with better trained teachers and more flexibility, and finally the most desirable stage, a stage of meaning, in which teaching methods foster problem solving and creativity while catering to the individual differences of students. Thus the discrepancy between extant school practices and those that teach higher-order thinking skills is greatest in schools at the first level; the discrepancy decreases at each subsequent level.

The "stages of development" metaphor does not imply uniformity for all schools. One type of school dominates each stage; as the number of schools with the features of the next level increases, the system moves to the next stage. School systems can speed up the evolutionary process, but they cannot skip a stage because of cumbersome linkages with teacher training. While newly trained teachers may be able to work at more developed stages, experienced teachers are not likely to change their practices easily. Training large numbers of new teachers and changing the instructional routine of practicing teachers takes considerable time and can only be done gradually.

All types of schools function within all types of countries. Even the most highly industrialized countries have schools that function at the first stages of development, and excellent schools can be found in even the poorest countries. In general, the stage of an education system corresponds closely to the level of national economic development: the more economically advanced a country is, the better a school system it generally has. As a result, countries at different levels of economic development face different challenges. Strategies for improving primary education depend on the initial condition of each country and its schools.

Most low-income countries face the overriding problem of establishing a mass education system that efficiently imparts core academic and other skills to a high proportion of the school-age population. These countries require more physical facilities and better instruction. Most countries understand the technology for building schools and delivering primary education. Many have piloted low-cost programs to build schools, and many have widely and successfully tested the use of structured materials and technologies, such as interactive radio, to help poorly educated teachers teach simple problem-solving skills. Yet most low-income countries have difficulty implementing on a large scale the technologies that are known to be effective. In nearly all of these countries, establishing a modicum of administrative effectiveness is a necessary first step toward further reform.

For lower-middle-income countries, many of which have the capacity to provide mass education, the key challenge is to increase the number of schools that teach children more advanced problem-solving skills in a broad range of curriculum areas. Lower-middle-income countries have a higher proportion of trained teachers and can therefore progress toward this goal and implement innovative programs more easily than low-income countries can. Progress is often uneven and varies significantly among schools, but there is good potential to apply low-cost educational technology successfully and on a large scale.

For upper-middle-income and high-income countries, the capacity to reach and teach most children is well developed, and teacher education and training are generally satisfactory. Mass education systems in these countries must, however, refocus attention on the routine teaching of problem-solving skills, rather than basic skills, to all students.

All developing countries also need to invest in complementary human resource areas such as health, nutrition, and population control. To be most beneficial, educational reforms must be part of an integrated strategy for developing human resources.

The Challenge of Large-Scale Reform

Education systems, regardless of type, all seek to attract students, retain them throughout the primary cycle, and teach them basic literacy, numeracy, and problem-solving skills. To attain these long-term objectives, the education systems of the poorest countries must change significantly and on a large scale. Large-scale change requires time: successful reforms have taken twenty years in Ethiopia, fifteen years in Thailand, and a decade or more in industrial countries. Four factors affect the likelihood of success: the complexity of the reform, differences in local and national commitment to the reform, assumptions about the uniformity of implementation, and the adequacy of resources.

The first challenge to instituting change is that reforms can be complex, combining innovations that have multiple, open-ended, and sometimes conflicting objectives. Complex reforms are often suggested for low-income countries—to increase access, improve teaching, and enhance resource availability simultaneously, for example. Although central ministries often envision the complete program, educators at the local level typically lack the capacity to implement the required changes; they need simple, well-defined reforms.

The second challenge arises when different parties have different degrees of commitment to the reform. A large-scale program that responds to issues that central policymakers and planners consider priorities may not respond to issues that are priorities for administrators and parents at the school level. Thus the local commitment to national programs is often limited.

The third factor is uniformity. Large-scale programs set forth intervention strategies that are meant to be applied generally. The specific conditions of each school or district may, however, mean that nationally defined strategies are not applicable locally.

Finally, lack of resources can create problems. Policymakers at the central level tend to emphasize the process of adopting the program (that is, agreeing in principle to the reform); they may pay less attention to implementing the program locally. Implementation generally requires significant funding for long-term support and local training, but resources—particularly resources for external assistance to schools—are often limited.

The Challenge of Managing Reforms

The most effective approaches to managing the challenges posed by large-scale change combine systematic and phased implementation strategies with decentralized decisionmaking (delegating authority to lower levels of the education hierarchy) and deliberate dissemination of information about the reform. A phased implementation strategy allows the lessons learned in the early phases to be incorporated before moving on to the next phase. Such a strategy gives administrators experience resolving problems that arise in large-scale implementation, particularly those stemming from the issue of complexity.

Weick (1976) has noted that the "loose coupling" of the education system—that is, the poor linkages between the system's various organizational units—tends to exacerbate the difficulty of implementing large-scale change in a uniform manner. Decentralizing the responsibility for implementing large-scale change while holding local levels accountable for their performance eliminates some of the conflicts that arise between central and local priorities. This type of accountability is essential to freeing schools to test and adapt a variety of procedures and treatments while making them responsible for meeting national standards. It also, however, produces significant variations among schools, especially in the early phases.

The implementation of large-scale change further requires that issues of dissemination and of the relationship between central authorities and implementors at the school and district level be explicitly addressed. Both direct and indirect strategies should be considered. Direct strategies use power or administrative action to bring about a desired goal. They often take the form of heavy-handed, official mandates that force schools to participate in the changes being introduced. Indirect strategies rely on incentives that persuade people that the program is valuable and should be adopted. The most successful implementation approach generally combines direct and indirect strategies by providing support to and exerting pressure on local authorities.

A final key to implementing large-scale change is long-term commitment. Such change will not occur overnight, and sustained interest is needed to ensure its eventual success.

The Challenge of Managing Uncertainty

Education reform poses three problems for management that differ from those encountered in managing an existing system; all three reflect the uncertainty inherent in educational reform. First, information on how the existing system operates is readily available or relatively easy to generate, whereas information on a proposed reform in a particular country (its ease of implementation and its effectiveness) is nonexistent. Second, reforms require many people in the education system—students, parents, teachers, administrators, and technical specialists—to follow new procedures, whereas the existing system does not. Many find the changes threatening for personal, political, or technical reasons, while others welcome change. Third, many innovations cannot be transferred easily from one situation to another. Reform guidelines cannot direct the behavior of each teacher, because teaching is a process of problem solving, experimenting, evaluating, and adapting to new contexts and evolving goals (Murnane and Cohen 1986). Variation is a central condition of educational practice, and teachers must discover what works for them, for their students, and for particular subjects.

Because the education sector is so complex and unstructured, straightforward solutions to educational problems are frequently impossible. Achieving educational change is thus difficult under any circumstance; it is especially challenging in the unstable environments that often prevail in developing countries. In these countries, the generic problems are aggravated by policymakers who have only a limited knowledge of how teaching and learning actually take place in the classroom and how schools are actually managed. Teaching is an idiosyncratic activity, and no matter how well a reform program is designed, educational administrators and planners often cannot control its implementation or predict its effects.

Achieving educational change therefore requires a management approach that explicitly recognizes the need to manage uncertainty. Management strategies in developing countries must be flexible enough to deal with unexpected events and considerable variation in the implementation of school improvement programs. These strategies must join learning with action and be adaptable to the local context. The underlying assumption of such an approach reflects Hirschman's observation that "the term 'implementation' understates the complexity of the task of carrying out projects that are affected by a high degree of initial ignorance and uncertainty. Here, project implementation may mean in fact a long voyage of discovery in the most varied domains, from technology to politics" (Hirschman 1967, p. 35).

Successful development programs tend to have management structures that are a mixture of authority and bargaining (Johnston and Clark 1982; Paul 1982). Although much of the literature on these programs concerns rural development projects, the parallels with primary education development are many and the management lessons valuable. Moreover, studies of programs for educational change in the United States draw similar conclusions (Berman and McLaughlin 1977; Crandall and Loucks 1983; Huberman and Miles 1984). Successful operations feature broad, central guidelines that local authorities discuss, compromise on, and then flesh out or adapt both before and during implementation. It is the role of national ministries to design and promote the programs they want to see adopted. Local implementors then propose modifications based on specific local conditions, existing classroom routines, and early implementation experience. Program management is thus deliberately designed to minimize the risks associated with uncertain implementation by increasing the time and information available for decisionmaking.

Specifically, this calls for program management that combines an incremental, step-by-step approach to school improvement with methods for learning systematically from experience and assuring that schools are accountable for the goals set. An incremental approach provides the information and time necessary to better implement decisions; schools can be held accountable for achieving nationally defined objectives, but allowed and even encouraged to adopt local methods of attaining them. Commitment to change must be achieved at the school level.

Priorities for Reform

Educational and economic conditions are different in each country, as are educational priorities. Each country must, therefore, seek the educational and financial strategies that are appropriate to its goals and needs.

Table 9-1 summarizes the issues that are challenging countries at different levels of educational development. Low-income countries are typically moving into the stage of formalism, lower-middle-income countries into the transition stage, and upper-middle-income countries into the stage of meaning. Individual countries will, of course, fall outside these patterns.

Developing a Mass Education System in Low-Income Countries

The poorest nations of the world are primarily agricultural and have perilously low levels of educational attainment, widespread illiteracy, and rapidly increasing populations. Their development strategy is primarily to increase agricultural productivity, improve nutrition and health care, lower fertility, and expand the provision of education. Figure 9-1 ranks these countries by the capacity of their primary schools in 1985 (that is, their gross enrollment ratios) and indicates the system's actual coverage of the school-age population (net enrollment ratios). Twelve countries (31 percent) lacked the capacity to enroll half their primary school–age population. Only two countries—China and Sri Lanka—enrolled more than 90 percent of their primary school–age children. Nevertheless, the systems in all countries had the capacity to enroll more students than they actually did.

The critical challenge facing poor countries is to expand access while ensuring that the existing capacity is used to maximum advantage. Yet promoting access without improving the quality of instruction is meaningless and

Table 9-1. Challenges for Countries at Different Stages of Educational Development

Area of concern	*Challenges to reaching formalism stage*	*Challenges to reaching transition stage*	*Challenges to reaching meaning stage*
Effectiveness of schools (chapter 3)	Provide basic material and teaching inputs	Improve teacher knowledge, skills, and ability to teach problem solving	Concentrate on teaching advanced problem-solving skills
Teacher training (chapter 4)	Extend pretraining requirements for general education	Strengthen pedagogical skill development	Strengthen pedagogical skill development
Management (chapter 5)	Establish basic administrative infrastructure	Decentralize responsibility for school improvement; establish national management information system	Expand capability to generate information; conduct national assessments of achievement; establish school-based management information systems
Access to education (chapter 6)	Expand capacity to accommodate students	Target disadvantaged populations	Target disadvantaged populations
Resources (chapter 7)	Increase external aid; diversify sources of funding; reduce repetition rates	Support local initiatives with national funds	Allocate more resources to school-based innovation

Source: Authors' analysis.

wastes public and private resources. Reforms must therefore be implemented to ensure that children learn; in doing so, schools also speed the flow of students through the primary cycle and reduce the gap that exists between the system's capacity and its coverage.

For low-income countries in Sub-Saharan Africa, the challenge is especially daunting. The economic crisis of the past decade has left their education systems in disarray, and relatively few schools manage to teach the core skills in the national curriculum. To be able to enroll all children in primary school by the year 2000, these countries would have to increase the number of their physical facilities and teachers an average of more than 7 percent per year. Yet they face severe constraints on their human resources and finances, particularly since population growth remains high and the prospects for economic growth modest.

Furthermore, social demand and political pressure mean that most low-income countries must not only increase access to primary education but also

Figure 9-1. Gross and Net Enrollment Ratios of Children Age 6–11 in Low-Income Economies, 1985

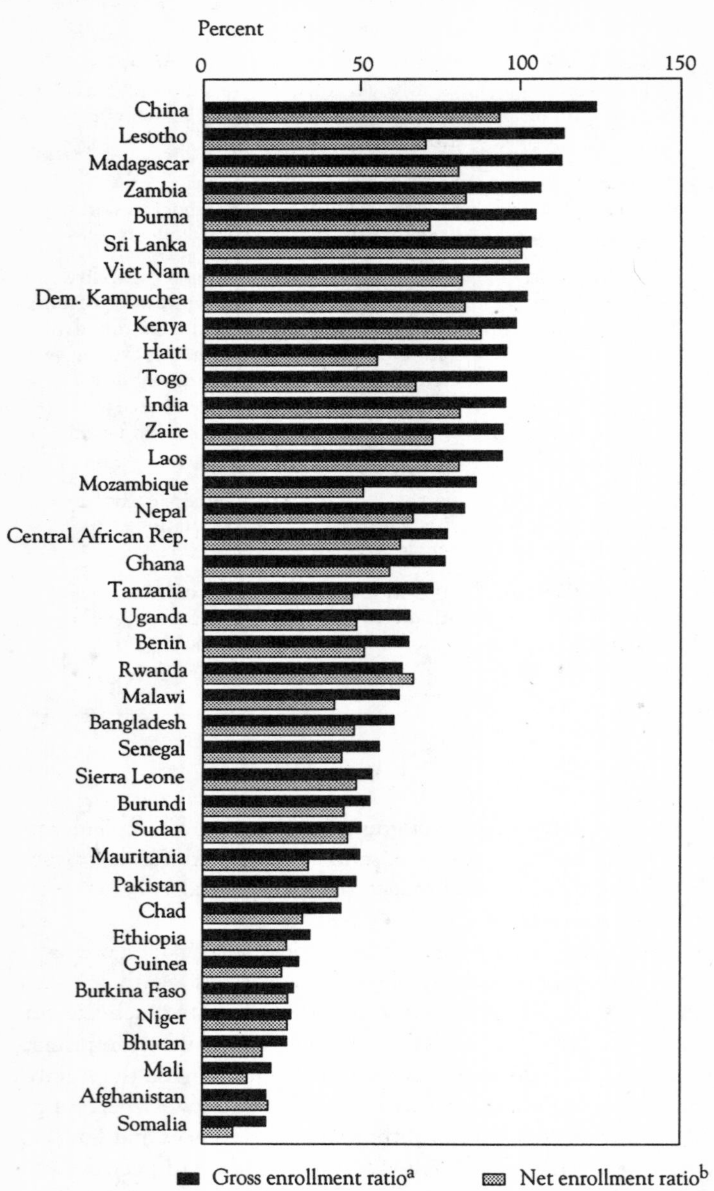

a. Student places available.

b. Coverage of the school-age population.

Source: See tables A-3 and A-4 in the appendix.

make fundamental improvements in its quality. This includes assuring that schools are operating and that they are supplied with at least the minimum instructional inputs. In addition, to ensure the fundamental teachability of students, these countries must enact nutrition and health measures both within and outside schools. The education systems in many low-income countries are barely entering the formalism stage of development; teaching problem-solving skills thus presents a major challenge requiring educational, administrative, and financial reforms. Such reforms are critical because many children do not receive more than a primary education. Consolidating and expanding the present system selectively is the only feasible short-term and possibly even medium-term strategy for many low-income countries.

The financial cost of expanding access to schooling and improving schools is considerable. Expanding the system would entail substantial investments in capital and recurrent costs for buildings and teachers. Increasing the availability of cost-effective inputs would have to be financed initially from the recurrent and capital budget because many of the poorest countries have difficulty mobilizing domestic public and private resources. In many countries the economic crisis of the 1980s, the burden of servicing debt, and the continuing high rate of population growth must be confronted by communities and the central government working together and by a significant infusion of external aid. If these are not forthcoming, many countries stand to lose the educational progress they realized in the 1960s and 1970s.

Although the problems to be addressed are formidable, the situation is not hopeless. Much was learned during the past decade to support informed decisions empirically. Policymakers and researchers understand the cost-effectiveness of alternative inputs in the education system much better than they did in the past. Primary education is now widely accepted as the joint responsibility of parents, communities, and governments. Governments increasingly recognize the importance of primary education for social, political, and economic development. The international community is reexamining the patterns of aid to education. This enhanced base of knowledge converges with the willingness to make a sustained commitment to developing primary education. The result is that poor developing countries have a unique opportunity to take effective action.

Striving for Quality and Equity in Lower-Middle-Income Countries

More lower-middle-income countries have excess primary school capacity (gross enrollment ratios greater than 100) than have insufficient capacity (see figure 9-2). Nearly half (including Botswana, the Dominican Republic, Indonesia, Jamaica, Paraguay, Peru, the Philippines, Thailand, and Zimbabwe) offer up to eight years of primary education to more than 90 percent of their school-age children. Some of these countries also have established education systems that not only provide near-universal access to

Figure 9-2. Gross and Net Enrollment Ratios of Children Age 6–11 in Lower-Middle-Income Economies, 1985

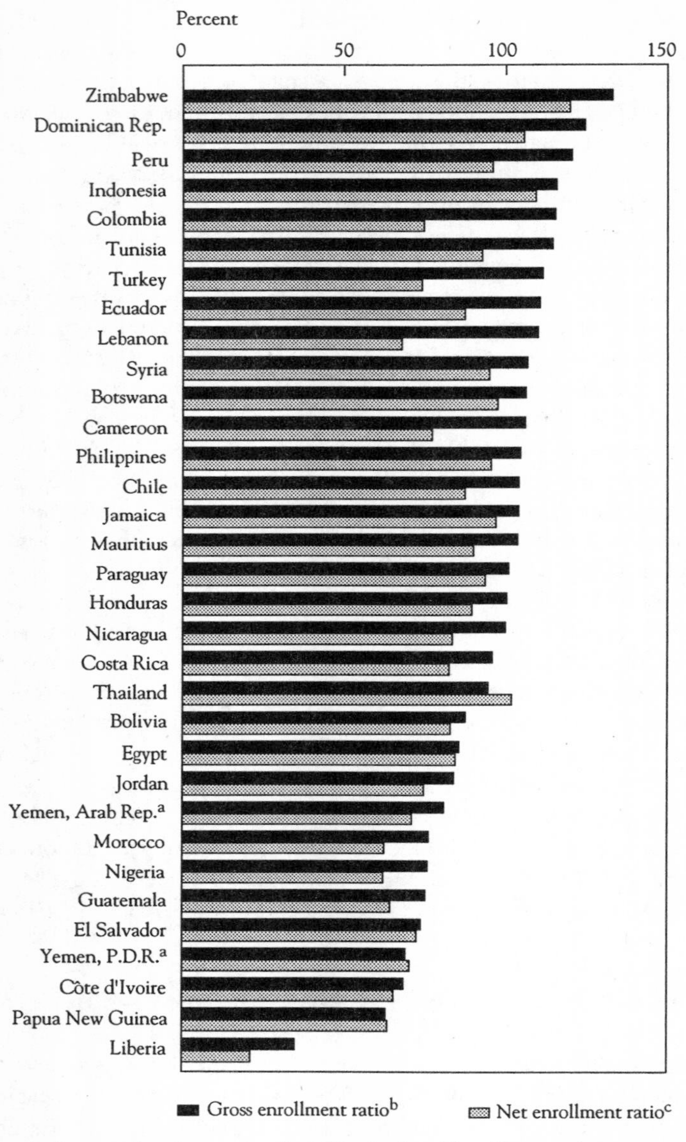

a. At the time that data were gathered for the Republic of Yemen, the country was divided into the Arab Republic of Yemen and the People's Democratic Republic of Yemen.

b. Student places available.

c. Coverage of the school-age population.

Source: See tables A-3 and A-4 in the appendix.

primary education, but also teach students the basic skills targeted in the curriculum.

In lower-middle-income countries, population growth rates are decreasing and the prospects for economic development are good for all but the highly indebted countries. Because of their interest in modernizing their economies, lower-middle-income countries give a high priority to mastering new technologies and building an up-to-date communications and computer infrastructure. In conjunction with reforming their trade and fiscal policies, they are transforming their industries and trying to occupy a more advanced niche in the emerging international markets.

The critical challenge for these countries in the years ahead is to increase the number of schools that emphasize the teaching of problem-solving skills and prepare students for further in-school and on-the-job learning. Many lower-middle-income countries are on the verge of the transition stage. They will have to make significant investments in developing and purchasing teaching materials; in upgrading the teaching force and, in some instances, the teachability of students; and in strengthening the national institutions that conduct education policy analysis and research.

Many of these countries still have large internal differences in access to and quality of education. In many countries the female enrollment rate is significantly behind the male enrollment rate; in others, considerable regional inequities persist. These disparities narrow the national human capital base and constrain efforts to increase the productivity of the labor force. They also inhibit social progress since the educational opportunities for girls in disadvantaged areas are even more limited than they are for boys. Addressing these internal differences requires not only the political will to support a sustained effort, but also the resources to invest in improving the access to and the quality of schools in disadvantaged areas. In many instances these countries must strengthen their national efforts to increase the cost-effectiveness of educational expenditures and to mobilize their own public and private resources as well as those of external funders.

Changing the Focus in Upper-Middle-Income Countries

Most upper-middle-income countries are considered to be industrialized or newly industrialized. That is, a larger share of their gross national product is contributed by industrial manufacturing and a smaller share by subsistence agriculture than in lower-middle-income or low-income countries. The development process in these countries is gaining momentum, the prospects for economic growth are good, and population growth is slowing. They are thus able to mobilize the resources necessary to finance primary education within the country. The school systems in most upper-middle-income countries have enough capacity to enroll more than 90 percent of the school-age population (see figure 9-3), and most primary school-age children are, in fact, enrolled.

Figure 9-3. Gross and Net Enrollment Ratios of Children Age 6–11 in Upper-Middle-Income Economies, 1985

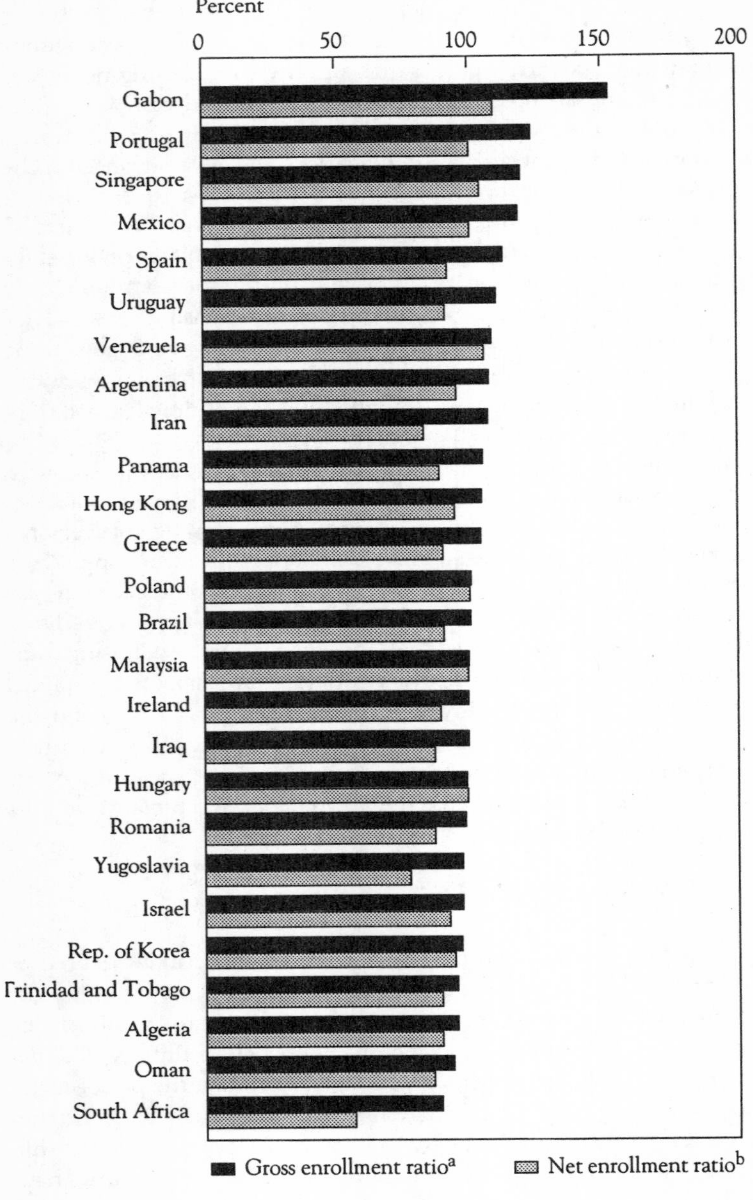

a. Student places available.

b. Coverage of the school-age population.

Source: See tables A-3 and A-4 in the appendix.

The number of students in upper-middle-income countries who receive no more than a primary education is small; most countries see primary education as the first step in the long process of building a broadly educated and trainable labor force. These countries—particularly newly industrialized economies such as Hong Kong, Malaysia, and Singapore—seek to develop a strong scientific and technical education program as a means to consolidate and improve their competitive position in a changing world economy. Most upper-middle-income countries have a strong and long-standing commitment to primary education as a fundamental building block of the development process, and most achieved high rates of enrollment and retention more than two decades ago.

At the primary level these countries should be concerned with (a) improving the instructional quality of primary education in order to reduce the rates at which students repeat a grade and (b) teaching advanced problem-solving and learning skills. To improve the learning achievement of children, these countries must develop new and often more challenging materials and they must upgrade the teaching force—the most difficult task. Many teachers were trained twenty and thirty years ago, and many lack both knowledge of the subject matter and skills to move students systematically toward higher levels of learning. A few upper-middle-income countries also need to increase the school participation of girls (Algeria and Oman) and the poor (Brazil, Iran, Oman, South Africa, and Yugoslavia).

In general, the upper-middle-income countries have the resources to carry out these improvements. They can, however, benefit from international cooperation, which would allow them to learn from the experience of developed countries that have already institutionalized a continuous process of changing and improving primary education. At the same time, upper-middle-income countries must share the lessons they learned during the past thirty years with educational planners and policymakers in low- and lower-middle-income countries.

Implementation Strategies: Steps toward Successful Reform

To implement the reform programs needed to increase significantly the proportion of effective primary schools in a given country requires substantial change in almost all settings. Education ministries are often the largest employer in the nation; counting students, the primary school system often engages more individuals in coordinated, purposeful activity than any other organization. Education is of great political and symbolic significance in all countries, and changing educational organizations is a slow, complex process that is difficult to predict and, therefore, difficult to plan in detail.

Recent research on educational change in developing countries has identified high-outcome reform strategies as those that "made implementation of the change program at the classroom level the primary concern and created conditions that facilitated application of the change in schools" (Verspoor

1989, p. 147). Adopting a phased, incremental approach to change is of central importance in this process (Rondinelli, Middleton, and Verspoor 1989; Verspoor 1989). Other recommendations to establish effective schools include (a) strengthening local administrative structures to provide effective delivery of training and support to schools, and (b) establishing strong testing, monitoring, and evaluation systems to inform efforts to adjust policies and programs to changing conditions.

Effective strategies also incorporate the key elements of implementing decentralization policies in developing countries: alter the role of the central ministry from one of control to one of support and train both central and local administrators (Rondinelli and Nellis 1986).

Changing systems as large and complex as primary education systems requires a long time and a significant amount of resources. High levels of national political commitment and sustained funding from national, and often international, sources are necessary. Commitment deepens with success, and success hinges on the program's ability to meet local needs and adapt to local conditions. To achieve this level of commitment, senior officials and political figures, teachers, principals, community leaders, and school support staff must help design and implement the reforms. Their efforts are vital for success.

Appendix

Contents

Introduction 237

Tables

A-1 Population Age 6–11, Selected Years, 1965–2000 240

A-2 Total Primary School Enrollment, Growth, and Female Students as a Percentage of Enrollment, Selected Years, 1965–85 247

A-3 Children Age 6–11 Enrolled in Primary School and Growth in Enrollment, Selected Years, 1965–85 253

A-4 Gross Enrollment Ratio in Primary School, Selected Years, 1965–85 259

A-5 Distribution of Total Primary School Enrollment by Grade, Latest Available Year, 1981–86 265

A-6 Distribution of Enrollment in Grades 1–5, Latest Available Year, 1982–86 271

A-7 Ratio of Female to Male Students in Primary School, by Grade, Latest Available Year, 1980–86 277

A-8 Number of Primary Schools and Student-School Ratio, Selected Years, 1965–85 283

A-9 Number of Primary School Teachers and Student-Teacher Ratio, Selected Years, 1965–85 289

A-10 Students Enrolled in Private Primary Schools, Selected Years, 1965–85 295

A-11 Repeaters in Primary Schools, Selected Years, 1965–85 301

A-12 Time Needed to Produce a Primary School Graduate, Selected Years, 1970–85 307

A-13 Percentage of Primary Cohort Reaching the Last Grade, Selected Years, 1970–86 313

A-14 Public Expenditure on Education, Selected Years, 1965–85 *319*

A-15 Expenditure on Education as a Percentage of GNP and of Government Expenditure, Selected Years, 1965–85 *325*

A-16 The Share of Recurrent Expenditure in Total Public Expenditure on Education, Selected Years, 1965–85 *331*

A-17 Public Recurrent Expenditure on Primary Education, Its Growth, and Its Share in Total Public Recurrent Expenditure on Education, Selected Years, 1965–85 *337*

A-18 Public Recurrent Expenditure on Primary Education as a Percentage of GNP, Selected Years, 1965–85 *343*

A-19 Recurrent Expenditure on Primary Education, by Purpose, Latest Available Year, 1969–86 *349*

A-20 Public Recurrent Expenditure per Student in Primary School, Selected Years, 1965–85 *355*

A-21 Total Recurrent Expenditure on Primary Education per Inhabitant, Selected Years, 1965–85 *361*

A-22 Expenditure on Primary School Teachers and Teaching Materials, Latest Available Year, 1979–86 *367*

A-23 Preschool Enrollment Rate, Selected Years, 1965–80 *373*

Technical Notes *378*

Introduction

In each appendix table, economies are listed within their income group in ascending order by 1986 GNP per capita. The exceptions are Guinea, the Islamic Republic of Iran, Iraq, Democratic Kampuchea, Lao People's Democratic Republic, Lebanon, Libya, Romania, and Viet Nam, for which GNP per capita data are unavailable. These countries are listed at the end of their corresponding income group according to the classification of *World Development Report 1988* (World Bank 1988c). Countries 121–29 are classified as "other economies" because they were nonreporting nonmembers of the World Bank in 1988. The reference numbers in the alphabetical list below reflect the order in the tables.

Education indicators have been computed only for countries with a population of more than 1 million in 1986. Data for the education indicators are from the Unesco data base, and data for the economic indicators are from the World Bank, unless otherwise specified. Notes about data for specific countries are from Unesco (1988b).

Totals reflect sums for constant cases. Constant cases are those countries for which data are available for all years surveyed (or years adjacent to those surveyed). For example, if there are no data for Viet Nam in 1975 (or 1974 or 1976), then the data for Viet Nam are omitted in calculating totals for all years surveyed.

Data for the previous or following year (when available) have been provided when data for the specified year are missing. Computations have been adjusted accordingly.

Although Germany and Yemen have both recently unified, separate data are presented for the former Federal Republic of Germany and the German Democratic Republic and for the Arab Republic of Yemen and the People's Democratic Republic of Yemen, in the interests of historical accuracy. Similarly, although the Soviet Union has recently broken apart, a single set of statistics is reported for the Soviet republics collectively.

For more information about the methodology, definitions, and data sources used in compiling the tables, see the technical notes at the end of this appendix.

Economies and Their Order in Tables

Economy	Order
Afghanistan	34
Albania	122
Algeria	86
Angola	121
Argentina	84
Australia	111
Austria	108
Bangladesh	5
Belgium	107
Benin	17
Bhutan	2
Bolivia	46
Botswana	57
Brazil	74
Bulgaria	123
Burkina Faso	3
Burundi	13
Cameroon	59
Canada	117
Central African Republic	19
Chad	35
Chile	69
China	22
Colombia	68
Congo, People's Republic of	61
Costa Rica	70
Côte d'Ivoire	51
Cuba	125
Czechoslovakia	124
Denmark	114
Dominican Republic	49
Ecuador	66
Egypt, Arab Republic of	53
El Salvador	56
Ethiopia	1
Finland	113
France	110
Gabon	88
German Democratic Republic	126
Germany, Federal Republic of	112
Ghana	30
Greece	89
Guatemala	60
Guinea	36
Haiti	27
Honduras	52
Hong Kong	95
Hungary	79
India	20
Indonesia	42
Iran, Islamic Republic of	97
Iraq	98
Ireland	92
Israel	94
Italy	105
Jamaica	58
Japan	115
Jordan	71
Kampuchea, Democratic	37
Kenya	23
Korea, Democratic People's Republic of	128
Korea, Republic of	85
Kuwait	101
Lao People's Democratic Repubic	38
Lebanon	73
Lesotho	29
Liberia	40
Libya	103
Madagascar	11
Malawi	6
Malaysia	75
Mali	8
Mauritania	32
Mauritius	67
Mexico	77
Mongolia	127
Morocco	45
Mozambique	10
Myanmar	9
Nepal	4
Netherlands	109
New Zealand	104
Nicaragua	54
Niger	16
Nigeria	48
Norway	118
Oman	91
Pakistan	28

Panama	83
Papua New Guinea	50
Paraguay	62
Peru	63
Philippines	44
Poland	80
Portugal	81
Romania	99
Rwanda	21
Saudi Arabia	100
Senegal	33
Sierra Leone	25
Singapore	96
Somalia	18
South Africa	76
Spain	90
Sri Lanka	31
Sudan	26
Sweden	116
Switzerland	120
Syrian Arab Republic	72
Tanzania	14
Thailand	55
Togo	15
Trinidad and Tobago	93
Tunisia	65
Turkey	64
Uganda	12
United Arab Emirates	102
United Kingdom	106
United States	119
Uruguay	78
Union of Soviet Socialist Republics	129
Venezuela	87
Viet Nam	39
Yemen, Arab Republic of	43
Yemen, People's Democratic Republic of	41
Yugoslavia	82
Zaire	7
Zambia	24
Zimbabwe	47

Key to Symbols Used in Tables

— Not available.
n.a. Not applicable.
m Median.
u Unweighted mean.
w Weighted mean.
w′ Weighted mean excluding China and India.

Table A-1. Population Age 6–11, Selected Years, 1965–2000

Economy	Total (thousands)					Annual growth (percent)			World Bank estimates: Total (thousands)		World Bank estimates: Annual growth, 1985–2000 (percent)
	1965	1970	1975	1980	1985	1965–75	1975–80	1980–85	1985	2000	
Low-income economies	**244,998**	**278,552**	**326,692**	**350,856**	**340,124**	**2.8 *m***	**2.6 *m***	**2.8 *m***	**342,404**	**422,762**	**3.1 *m***
						2.8 *u*	**2.4 *u***	**2.1 *u***			**3.0 *u***
						2.9 *w*	**1.4 *w***	**-0.6 *w***			**1.4 *w***
Excluding China and India	**61,958**	**71,997**	**82,432**	**92,864**	**104,710**	**2.9 *w′***	**2.4 *w′***	**2.4 *w′***	**104,080**	**154,448**	**2.7 *w′***
1 Ethiopia	4,399	4,971	5,572	6,273	7,608	2.4	2.4	3.9	7,220	11,124	2.9
2 Bhutan	143	157	172	191	208	1.9	2.1	1.7	185	279	2.8
3 Burkina Faso	769	881	993	1,139	1,260	2.6	2.8	2.0	1,265	2,109	3.5
4 Nepal	1,484	1,757	2,010	2,293	2,622	3.1	2.7	2.7	2,648	4,027	2.8
5 Bangladesh	9,359	11,191	12,894	15,195	17,546	3.3	3.3	2.9	17,091	20,761	1.3
6 Malawi	649	756	895	1,046	1,172	3.3	3.2	2.3	1,214	1,957	3.2
7 Zaire	2,843	3,153	3,636	4,445	5,279	2.5	4.1	3.5	5,118	7,991	3.0
8 Mali	789	909	1,040	1,174	1,351	2.8	2.5	2.8	1,257	1,960	3.0
9 Myanmar	3,773	4,218	4,708	5,301	5,703	2.2	2.4	1.5	5,434	6,389	1.1
10 Mozambique	1,259	1,475	1,708	1,922	2,188	3.1	2.4	2.6	2,270	3,585	3.1
11 Madagascar	938	1,089	1,248	1,429	1,675	2.9	2.7	3.2	1,719	2,833	3.4
12 Uganda	1,371	1,667	1,935	2,276	2,710	3.5	3.3	3.6	2,583	4,137	3.2
13 Burundi	469	523	584	638	772	2.2	1.8	3.9	727	1,266	3.8
14 Tanzania	1,893	2,304	2,756	3,311	4,009	3.8	3.7	3.9	3,722	6,721	4.0
15 Togo	254	322	367	416	485	3.7	2.6	3.1	509	861	3.6
16 Niger	573	648	752	881	1,033	2.8	3.2	3.2	1,057	1,791	3.6
17 Benin	372	427	495	574	686	2.9	3.0	3.6	688	1,144	3.4
18 Somalia	513	582	668	782	1,029	2.7	3.2	5.7	880	1,482	3.5
19 Central African Republic	254	276	305	344	404	1.8	2.5	3.2	417	666	3.2
20 India	74,879	84,640	93,993	103,418	108,828	2.3	1.9	1.0	115,938	131,272	0.8

Economy											
21 Rwanda	515	631	763	913	1,085	4.0	3.7	3.5	1,006	1,923	4.4
22 China	108,160	121,916	150,267	154,575	126,585	3.3	0.6	–3.9	122,386	137,042	0.8
23 Kenya	1,671	2,011	2,423	2,989	3,754	3.8	4.3	4.7	3,903	6,854	3.8
24 Zambia	598	703	820	1,053	1,221	3.2	5.1	3.0	1,189	1,991	3.5
25 Sierra Leone	361	400	449	511	584	2.2	2.6	2.7	568	866	2.9
26 Sudan	2,003	2,252	2,599	3,043	3,640	2.6	3.2	3.6	3,692	5,678	2.9
27 Haiti	613	706	805	872	914	2.8	1.6	0.9	913	1,103	1.3
28 Pakistan	9,832	11,285	13,058	14,243	15,562	2.9	1.8	1.8	15,911	24,611	3.0
29 Lesotho	144	163	183	208	241	2.4	2.6	3.0	239	382	3.2
30 Ghana	1,314	1,463	1,640	1,804	2,099	2.2	1.9	3.1	2,183	3,467	3.1
31 Sri Lanka	1,798	2,017	2,200	2,021	2,128	2.0	–1.7	1.0	2,103	2,428	1.0
32 Mauritania	169	190	216	248	285	2.5	2.7	2.9	275	446	3.3
33 Senegal	547	635	757	915	1,052	3.3	3.8	2.8	1,118	1,736	3.0
34 Afghanistan	1,735	2,090	2,458	2,622	2,215	3.5	1.3	–3.3	2,918	4,111	2.3
35 Chad	498	553	616	690	777	2.2	2.3	2.4	738	1,161	3.1
36 Guinea	605	674	763	844	950	2.4	2.0	2.4	650	1,355	5.0
37 Democratic Kampuchea	973	1,114	1,194	1,044	511	2.1	–2.6	–13.3	503	1,555	7.8
38 Lao People's Democratic Republic	385	429	473	500	560	2.1	1.1	2.3	607	876	2.5
39 Viet Nam	6,095	7,374	8,276	8,713	9,394	3.1	1.0	1.5	9,560	12,822	2.0
Lower-middle-income economies	**65,919**	**76,186**	**86,092**	**96,003**	**104,896**	**2.8** *m*	**2.7** *m*	**2.3** *m*	**105,769**	**136,570**	**1.8** *m*
						2.8 *u*	**2.0** *u*	**1.9** *u*			**1.8** *u*
						2.7 *w*	**2.2** *w*	**1.8** *w*			**1.7** *w*
40 Liberia	183	216	254	303	360	3.4	3.6	3.5	353	627	3.9

Note: *m*, median; *u*, unweighted mean; *w*, weighted mean; and *w*′, weighted mean excluding China and India.

a. The figures for the former Arab Republic of Yemen and the former People's Democratic Republic of Yemen, summed together, represent Yemen's projected population age 6–11 in 2000.

b. The figures for the former Federal Republic of Germany and the former German Democratic Republic, summed together, represent Germany's projected population age 6–11 in 2000.

Sources: United Nations 1988; World Bank data.

(*Table continues on the following page.*)

Table A-1 (*continued*)

Economy	Total (thousands)					Annual growth (percent)			World Bank estimates		
									Total (thousands)		Annual growth, 1985–2000
	1965	1970	1975	1980	1985	1965–75	1975–80	1980–85	1985	2000	(percent)
41 People's Democratic Republic of Yemen	220	248	290	330	350	2.8	2.6	1.2	334	452[a]	2.0
42 Indonesia	16,941	18,953	21,382	24,449	25,460	2.4	2.7	0.8	25,657	28,005	0.6
43 Arab Republic of Yemen	696	757	905	1,035	1,241	2.7	2.7	3.7	1,503	2,114[a]	2.3
44 Philippines	5,352	6,577	7,241	7,709	8,636	3.1	1.3	2.3	8,863	10,297	1.0
45 Morocco	2,306	2,791	3,028	3,184	3,573	2.8	1.0	2.3	3,384	4,557	2.0
46 Bolivia	608	687	781	897	1,037	2.5	2.8	3.0	1,031	1,395	2.0
47 Zimbabwe	803	952	1,120	1,313	1,446	3.4	3.2	2.0	1,368	2,248	3.4
48 Nigeria	8,039	9,599	11,727	14,119	16,726	3.8	3.8	3.4	17,380	29,109	3.5
49 Dominican Republic	666	806	908	946	975	3.1	0.8	0.6	975	1,194	1.4
50 Papua New Guinea	329	383	429	520	545	2.7	3.9	0.9	544	710	1.8
51 Côte d'Ivoire	757	933	1,162	1,447	1,810	4.4	4.5	4.6	1,666	2,704	3.3
52 Honduras	383	452	549	672	765	3.7	4.1	2.6	759	1,121	2.6
53 Egypt	4,882	5,262	5,578	5,995	7,142	1.3	1.5	3.6	7,346	9,566	1.8
54 Nicaragua	318	370	431	494	571	3.1	2.8	3.0	571	782	2.1
55 Thailand	5,388	6,195	7,205	7,486	7,457	2.9	0.8	–0.1	7,315	7,692	0.3
56 El Salvador	524	628	713	796	869	3.1	2.2	1.8	869	981	0.8
57 Botswana	98	117	142	163	183	3.7	2.8	2.4	192	298	3.0
58 Jamaica	281	352	381	357	327	3.1	–1.3	–1.7	316	324	0.2
59 Cameroon	950	1,035	1,152	1,332	1,583	2.0	2.9	3.5	1,727	2,996	3.7
60 Guatemala	784	899	1,024	1,176	1,379	2.7	2.8	3.2	1,390	1,806	1.8
61 Congo	163	185	211	244	279	2.6	2.9	2.8	304	590	4.5
62 Paraguay	387	422	463	509	570	1.8	1.9	2.3	570	817	2.4
63 Peru	1,866	2,197	2,502	2,774	3,032	3.0	2.1	1.8	2,799	3,391	1.3
64 Turkey	4,739	5,544	6,094	6,983	7,056	2.5	2.8	0.2	6,964	8,808	1.6

65 Tunisia	805	932	957	1,022	1,110	1.8	1.3	1.7	1,112	1,375	1.4
66 Ecuador	872	1,023	1,205	1,356	1,490	3.3	2.4	1.9	1,497	1,977	1.9
67 Mauritius	137	158	146	122	129	0.6	–3.6	1.2	133	121	–0.6
68 Colombia	3,175	3,694	3,995	3,909	4,087	2.3	–0.4	0.9	4,134	4,555	0.6
69 Chile	1,319	1,499	1,536	1,505	1,450	1.5	–0.4	–0.7	1,479	1,496	0.1
70 Costa Rica	269	318	338	333	373	2.3	–0.3	2.3	363	429	1.1
71 Jordan	314	385	446	557	622	3.6	4.5	2.2	629	877	2.2
72 Syrian Arab Republic	1,015	1,193	1,333	1,532	1,873	2.8	2.8	4.1	1,857	2,765	2.7
73 Lebanon	353	424	465	438	390	2.8	–1.2	–2.3	385	391	0.1
Upper-middle-income economies	**60,674**	**66,815**	**71,158**	**76,743**	**81,832**	**1.3 *m* 1.2 *u* 1.6 *w***	**1.5 *m* 1.4 *u* 1.5 *w***	**0.7 *m* 1.2 *u* 1.3 *w***	**82,870**	**97,642**	**0.5 *m* 0.3 *u* 1.1 *w***
74 Brazil	13,907	16,098	16,894	17,564	19,087	2.0	0.8	1.7	19,159	22,617	1.1
75 Malaysia	1,666	1,899	2,005	2,168	2,195	1.9	1.6	0.2	2,240	2,630	1.1
76 South Africa	3,103	3,578	3,885	4,195	4,579	2.3	1.5	1.8	4,934	6,559	1.9
77 Mexico	7,785	9,274	10,915	12,701	12,751	3.4	3.1	0.1	13,137	15,374	1.1
78 Uruguay	296	316	303	311	323	0.2	0.6	0.7	324	328	0.1
79 Hungary	1,016	818	813	924	1,006	–2.2	2.6	1.7	1,004	683	–2.5
80 Poland	4,223	3,601	3,069	3,277	3,746	–3.1	1.3	2.7	3,745	3,455	–0.5
81 Portugal	1,013	1,046	1,032	1,006	999	0.2	–0.5	–0.2	1,013	923	–0.6
82 Yugoslavia	2,330	2,238	2,187	2,159	2,239	–0.6	–0.3	0.7	2,255	1,989	–0.8
83 Panama	219	258	295	318	324	3.0	1.5	0.4	325	325	0.0
84 Argentina	2,663	2,763	2,897	3,203	3,722	0.8	2.0	3.1	3,713	3,994	0.5
85 Republic of Korea	4,905	5,560	5,239	5,148	5,036	0.7	–0.4	–0.4	5,167	4924	–0.3
86 Algeria	2,004	2,479	2,873	3,289	3,742	3.7	2.7	2.6	3,741	5,408	2.5

Note: *m*, median; *u*, unweighted mean; *w*, weighted mean; and *w′*, weighted mean excluding China and India.

a. The figures for the former Arab Republic of Yemen and the former People's Democratic Republic of Yemen, summed together, represent Yemen's projected population age 6–11 in 2000.

b. The figures for the former Federal Republic of Germany and the former German Democratic Republic, summed together, represent Germany's projected population age 6–11 in 2000.

Sources: United Nations 1988; World Bank data.

(*Table continues on the following page.*)

Table A-1 (continued)

Economy	Total (thousands) 1965	1970	1975	1980	1985	Annual growth (percent) 1965–75	1975–80	1980–85	World Bank estimates: Total (thousands) 1985	2000	Annual growth, 1985–2000 (percent)
87 Venezuela	1,551	1,863	2,159	2,360	2,622	3.4	1.8	2.1	2,628	3,289	1.5
88 Gabon	60	61	78	99	120	2.7	5.0	3.9	137	256	4.3
89 Greece	879	846	896	875	852	0.2	–0.5	–0.5	857	806	–0.4
90 Spain	3,462	3,813	3,938	3,974	3,737	1.3	0.2	–1.2	4,019	3,318	–1.3
91 Oman	91	105	124	154	201	3.1	4.4	5.4	202	297	2.6
92 Ireland	353	372	392	421	421	1.1	1.4	0.0	418	385	–0.5
93 Trinidad and Tobago	146	174	164	146	151	1.2	–2.2	0.7	149	172	1.0
94 Israel	351	377	424	509	549	1.9	3.8	1.5	560	556	–0.0
95 Hong Kong	600	635	537	507	514	–1.1	–1.1	0.3	495	557	0.8
96 Singapore	339	345	300	271	242	–1.2	–2.0	–2.2	237	254	0.5
97 Islamic Republic of Iran	4,186	4,880	5,671	6,515	7,506	3.1	2.8	2.9	7,249	11,666	3.2
98 Iraq	1,354	1,598	1,898	2,282	2,819	3.4	3.8	4.3	2,822	4,838	3.7
99 Romania	2,174	1,821	2,174	2,367	2,349	0.0	1.7	–0.2	2,340	2,039	–0.9
High-income oil exporters	**1,097**	**1,397**	**1,804**	**2,322**	**3,010**	**7.3 *m* 7.4 *u* 5.1 *w***	**5.2 *m* 7.2 *u* 5.2 *w***	**4.9 *m* 6.0 *u* 5.3 *w***	**3,010**	**5,268**	**3.7 *m* 3.4 *u* 3.8 *w***
100 Saudi Arabia	759	934	1,178	1,480	1,937	4.5	4.7	5.5	1,943	3,466	3.9
101 Kuwait	63	119	169	211	261	10.4	4.5	4.3	255	346	2.1
102 United Arab Emirates	20	27	51	100	160	9.9	14.1	10.0	162	272	3.5
103 Libya	255	317	405	532	652	4.7	5.6	4.2	650	1,184	4.1
High-income industrial market economies	**63,365**	**66,142**	**64,842**	**61,110**	**56,728**	**0.8 *m* 0.6 *u* 0.2 *w***	**–1.8 *m* –1.5 *u* –1.2 *w***	**–2.3 *m* –2.0 *u* –1.5 *w***	**56,515**	**56,365**	**–2.1 *m* –1.8 *u* –0.7 *w***

104	New Zealand	340	364	372	348	317	0.9	–1.3	–1.9	321	313	–0.2
105	Italy	4,817	5,305	5,519	5,344	4,727	1.4	–0.6	–2.4	4,630	3,390	–2.1
106	United Kingdom	4,730	5,523	5,523	4,919	4,207	1.6	–2.3	–3.1	4,203	4,632	–0.7
107	Belgium	898	946	901	810	748	0.0	–2.1	–1.6	741	691	–0.5
108	Austria	644	751	753	640	534	1.6	–3.2	–3.5	537	508	–0.4
109	Netherlands	1,355	1,439	1,467	1,331	1,118	0.8	–1.9	–3.4	1,108	1,151	0.3
110	France	4,939	5,056	5,067	4,981	4,625	0.3	–0.3	–1.5	4,623	4,571	–0.1
111	Australia	1,339	1,475	1,499	1,547	1,455	1.1	0.6	–1.2	1,454	1,629	0.8
112	Federal Republic of Germany	5,064	5,785	5,986	4,613	3,634	1.7	–5.1	–4.7	3,515	4,156[b]	1.1
113	Finland	496	467	435	376	377	–1.3	–2.9	0.1	378	364	–0.3
114	Denmark	438	462	471	449	397	0.7	–1.0	–2.4	396	356	–0.7
115	Japan	10,023	9,677	10,420	11,699	10,905	0.4	2.3	–1.4	10,910	8,270	–1.8
116	Sweden	628	663	694	683	608	1.0	–0.3	–2.3	608	711	1.0
117	Canada	2,647	2,784	2,469	2,122	2,138	–0.7	–3.0	0.1	2,147	2,266	0.4
118	Norway	366	372	392	386	344	0.7	–0.3	–2.3	343	339	–0.1
119	United States	24,106	24,479	22,272	20,343	20,147	–0.8	–1.8	–0.2	20,156	22,498	0.7
120	Switzerland	536	597	604	518	446	1.2	–3.0	–3.0	445	520	1.0
Other economies												
121	Angola	792	868	1,020	1,237	1,439	2.6	3.9	3.1	1,448	2,190	2.8
122	Albania	307	362	380	370	402	2.1	–0.5	1.7	417	482	1.0
123	Bulgaria	795	767	748	778	799	–0.6	0.8	0.5	806	712	–0.8
124	Czechoslovakia	1,458	1,326	1,284	1,454	1,622	–1.3	2.5	2.2	1,620	1,294	–1.5
125	Cuba	964	1,265	1,444	1,355	1,032	4.1	–1.3	–5.3	1,025	1,141	0.7

Note: *m*, median; *u*, unweighted mean; *w*, weighted mean; and *w′*, weighted mean excluding China and India.

a. The figures for the former Arab Republic of Yemen and the former People's Democratic Republic of Yemen, summed together, represent Yemen's projected population age 6–11 in 2000.

b. The figures for the former Federal Republic of Germany and the former German Democratic Republic, summed together, represent Germany's projected population age 6–11 in 2000.

Sources: United Nations 1988; World Bank data.

(*Table continues on the following page.*)

Table A-1 (continued)

Economy	Total (thousands) 1965	1970	1975	1980	1985	Annual growth (percent) 1965–75	1975–80	1980–85	World Bank estimates: Total (thousands) 1985	2000	Annual growth, 1985–2000 (percent)
126 German Democratic Republic	1,600	1,689	1,571	1,247	1,229	–0.2	–4.5	–0.3	1,234	1,239[b]	0.0
127 Mongolia	167	202	245	276	303	3.9	2.4	1.9	298	395	1.9
128 Democratic People's Republic of Korea	2,117	2,338	2,539	2,835	3,077	1.8	2.2	1.7	3,076	3,573	1.0
129 Union of Soviet Socialist Republics	28,926	29,501	26,078	25,268	27,098	–1.0	–0.6	1.4	26,909	28,446	0.4

Note: *m*, median; *u*, unweighted mean; *w*, weighted mean; and *w′*, weighted mean excluding China and India.

a. The figures for the former Arab Republic of Yemen and the former People's Democratic Republic of Yemen, summed together, represent Yemen's projected population age 6–11 in 2000.

b. The figures for the former Federal Republic of Germany and the former German Democratic Republic, summed together, represent Germany's projected population age 6–11 in 2000.

Sources: United Nations 1988; World Bank data.

Table A-2. Total Primary School Enrollment, Growth, and Female Students as a Percentage of Enrollment, Selected Years, 1965–85

Economy	Total (thousands)					Annual growth (percent)			Females as a percentage of total				
	1965	1970	1975	1980	1985	1965–75	1975–80	1980–85	1965	1970	1975	1980	1985
Low-income economies	**195,202**	**198,552**	**263,993**	**278,410**	**286,785**	**5.5 *m***	**6.0 *m***	**3.3 *m***	**35.5 *m***	**37.0 *m***	**39.0 *m***	**40.0 *m***	**41.0 *m***
						5.6 *u*	**6.2 *u***	**3.3 *u***	**33.4 *u***	**35.3 *u***	**38.0 *u***	**39.8 *u***	**41.1 *u***
						3.1 *w*	**1.1 *w***	**0.6 *w***	**35.0 *w***	**37.0 *w***	**40.0 *w***	**42.0 *w***	**43.0 *w***
Excluding China and India	**28,522**	**36,227**	**47,392**	**58,267**	**66,618**	**5.2 *w′***	**4.2 *w′***	**2.7 *w′***	**38.0 *w′***	**41.0 *w′***	**42.0 *w′***	**42.0 *w′***	**43.0 *w′***
1 Ethiopia	379	655	1,084	2,131	2,449	11.1	14.5	2.8	28.0	31.0	32.0	35.0	39.0
2 Bhutan	9	9	15[a]	30	45	5.2	14.9	8.4	11.0	0.0	20.0	30.0	36.0
3 Burkina Faso	90	105	141	202	352	4.6	7.5	11.7	32.0	36.0	37.0	37.0	37.0
4 Nepal	316	390	543	1,068	1,819	5.6	14.5	11.2	15.0	15.0	15.0	28.0	29.0
5 Bangladesh	4,284	5,284	8,350	8,240	8,920	6.9	–0.3	1.6	31.0	32.0	34.0	37.0	40.0
6 Malawi	338	363	642	810	943	6.6	4.8	3.1	37.0	37.0	40.0	41.0	43.0
7 Zaire	2,067	3,088	3,545	4,196	4,970[a]	5.5	3.4	3.4	32.0	37.0	40.0	42.0	43.0
8 Mali	140	204	252	291	292	6.1	2.9	0.1	33.0	36.0	36.0	36.0	37.0
9 Myanmar	2,250	3,178	3,476	4,148	5,000[a]	4.4	3.6	3.8	46.0	47.0	48.0	48.0	47.0
10 Mozambique	354	497	975[a]	1,307	1,248	10.7	6.0	–0.9	36.0	34.0	41.0	42.0	44.0
11 Madagascar	675	938	1,211	1,724	1,600[a]	6.0	7.3	–1.5	45.0	46.0	43.0	49.0	48.0
12 Uganda	569	720	974	1,292	2,015[a]	5.5	5.8	9.3	37.0	39.0	40.0	43.0	43.0
13 Burundi	147	182	130	176	386	–1.2	6.2	17.0	29.0	33.0	38.0	39.0	42.0
14 Tanzania	769	856	1,592	3,368	3,170	7.5	16.2	–1.2	38.0	39.0	42.0	47.0	50.0
15 Togo	156	229	363	506	463	8.8	6.9	–1.8	29.0	31.0	35.0	38.0	38.0
16 Niger	62	89	142	229	276	8.6	10.0	3.8	32.0	35.0	35.0	35.0	36.0
17 Benin	125	155	260	380	444	7.6	7.9	3.2	30.0	31.0	32.0	32.0	34.0
18 Somalia	29	33	198	272	194	21.2	6.6	–6.5	21.0	24.0	35.0	36.0	34.0
19 Central African Republic	128	176	221	246	310	5.6	2.2	4.7	26.0	33.0	36.0	37.0	39.0

— Not available.

Note: *m*, median; *u*, unweighted mean; *w*, weighted mean; and *w′*, weighted mean excluding China and India.

a. Estimate.

Source: Unesco data.

(*Table continues on the following page.*)

Table A-2 (continued)

Economy	Total (thousands) 1965	1970	1975	1980	1985	Annual growth (percent) 1965–75	1975–80	1980–85	Females as a percentage of total 1965	1970	1975	1980	1985
20 India	50,471	57,045	65,660	73,873	86,465	2.7	2.4	3.2	36.0	37.0	38.0	39.0	40.0
21 Rwanda	330	419	402	705	837	2.0	11.9	3.5	41.0	44.0	46.0	48.0	49.0
22 China	116,209	105,280	150,941	146,270	133,702	2.6	–0.6	–1.8	39.0	45.0	45.0	45.0	45.0
23 Kenya	1,042	1,428	2,881	3,927	4,702	10.7	6.4	3.7	36.0	41.0	46.0	47.0	48.0
24 Zambia	410	695	872	1,042	1,348	7.8	3.6	5.3	44.0	44.0	45.0	47.0	47.0
25 Sierra Leone	126	166	206	315	370[a]	5.0	8.9	3.3	36.0	40.0	39.0	42.0	41.0
26 Sudan	427	826	1,169	1,464	1,738	10.6	4.6	3.5	35.0	38.0	36.0	40.0	40.0
27 Haiti	284	367	487	642	873	5.5	5.7	6.3	43.0	44.0	44.0	46.0	47.0
28 Pakistan	2,823	3,548	4,668	4,795	6,613	5.2	0.5	6.6	23.0	26.0	29.0	32.0	33.0
29 Lesotho	168	183	222	245	314	2.8	2.0	5.1	61.0	60.0	59.0	58.0	56.0
30 Ghana	1,414	1,420	1,180	1,443	1,596	–1.8	4.1	2.0	42.0	43.0	43.0	44.0	43.0
31 Sri Lanka	2,216	1,671	1,436	2,081	2,243	–4.2	7.7	1.5	46.0	47.0	47.0	48.0	48.0
32 Mauritania	20	32	50	91	141	9.6	12.7	9.2	25.0	28.0	36.0	35.0	40.0
33 Senegal	219	263	312	420	584	3.6	6.1	6.8	36.0	39.0	42.0	40.0	40.0
34 Afghanistan	358	541	785	1,116	580	8.2	7.3	–12.3	15.0	14.0	15.0	18.0	31.0
35 Chad	164	192	213	246[a]	338	2.6	2.9	6.6	19.0	25.0	26.0	27.0	28.0
36 Guinea	172	191	199	258	276	1.5	5.3	1.4	31.0	32.0	34.0	33.0	32.0
37 Democratic Kampuchea	800	338	475[a]	495[a]	520[a]	–5.1	0.8	1.0	36.0	43.0	41.0	41.0	41.0
38 Lao People's Democratic Republic	161	245	317	479	523	7.0	8.6	1.8	37.0	37.0	42.0	45.0	45.0
39 Viet Nam	4,501[a]	6,551[a]	7,404	7,887	8,126	5.1	1.3	0.6	—	—	49.0	47.0	48.0
Lower-middle-income economies	**47,884**	**59,620**	**71,888**	**93,726**	**104,405**	**4.3 *m***	**3.7 *m***	**3.2 *m***	**43.5 *m***	**45.0 *m***	**46.0 *m***	**47.5 *m***	**48.0 *m***
						4.8 *u*	**4.1 *u***	**3.0 *u***	**41.5 *u***	**42.6 *u***	**44.1 *u***	**44.8 *u***	**45.6 *u***
						4.1 *w*	**5.4 *w***	**2.2 *w***	**44.0 *w***	**45.0 *w***	**45.0 *w***	**46.0 *w***	**47.0 *w***
40 Liberia	84	120	104	147	128[a]	2.2	7.2	–2.7	29.0	33.0	34.0	35.0	35.0
41 People's Democratic Republic of Yemen	50	135	229	267	320[a]	16.4	3.1	3.7	22.0	20.0	31.0	28.0	26.0

42	Indonesia	11,687	14,870	17,777	25,537	29,897	4.3	7.5	3.2	45.0	46.0	45.0	46.0	48.0
43	Arab Republic of Yemen	69	88	255	456	981	14.0	12.3	16.6	6.0	9.0	11.0	13.0	20.0
44	Philippines	5,816	6,969	7,597	8,506	8,926	2.7	2.3	1.0	48.0	49.0	51.0	49.0	49.0
45	Morocco	1,116	1,175	1,548	2,172	2,280	3.3	7.0	1.0	30.0	34.0	36.0	37.0	38.0
46	Bolivia	495	679	859	978	1,193[a]	5.7	2.6	4.1	41.0	41.0	45.0	47.0	47.0
47	Zimbabwe	676	736	863	1,235	2,215	2.5	7.4	12.4	42.0	45.0	46.0	46.0	48.0
48	Nigeria	2,912	3,516	6,166	13,760	12,915	7.8	17.4	–1.3	38.0	37.0	43.0	43.0	44.0
49	Dominican Republic	557	764	911	1,106	1,220	5.0	4.0	2.0	50.0	50.0	50.0	50.0	50.0
50	Papua New Guinea	140	191	238	300	345[a]	5.4	4.7	2.8	38.0	37.0	37.0	41.0	44.0
51	Côte d'Ivoire	354	503	673	1,025	1,200[a]	6.6	8.8	3.2	34.0	36.0	38.0	40.0	41.0
52	Honduras	284	382	461	601	766	5.0	5.4	5.0	49.0	50.0	49.0	50.0	50.0
53	Egypt	3,499	3,795	4,181	4,663	6,214	1.8	2.2	5.9	39.0	38.0	38.0	40.0	43.0
54	Nicaragua	206	285	342	472	562	5.2	6.7	3.6	50.0	50.0	51.0	51.0	52.0
55	Thailand	4,630	5,635	6,686	7,393	7,150	3.7	2.0	–0.7	47.0	47.0	47.0	48.0	48.0
56	El Salvador	382	510	738	834	915[a]	6.8	2.5	1.9	48.0	48.0	49.0	50.0	50.0
57	Botswana	66	83	116	172	224	5.8	8.2	5.4	56.0	53.0	55.0	55.0	52.0
58	Jamaica	263	354	372	359	345[a]	3.5	–0.7	–0.8	49.0	50.0	50.0	50.0	49.0
59	Cameroon	742	923	1,123	1,379	1,705	4.2	4.2	4.3	40.0	43.0	45.0	45.0	46.0
60	Guatemala	384	506	627	803	1,016	5.0	5.1	4.8	44.0	44.0	45.0	45.0	45.0
61	Congo	187	241	319	391	476	5.5	4.2	4.0	41.0	44.0	46.0	48.0	49.0
62	Paraguay	357	424	452	519	571	2.4	2.8	1.9	47.0	47.0	47.0	48.0	48.0
63	Peru	1,901	2,341	2,841	3,161	3,712	4.1	2.2	3.3	45.0	46.0	47.0	48.0	48.0
64	Turkey	3,972	5,012	5,464	5,656	6,636	3.2	0.7	3.2	40.0	42.0	45.0	45.0	47.0
65	Tunisia	734	936	933	1,054	1,291	2.4	2.5	4.1	34.0	39.0	39.0	42.0	45.0
66	Ecuador	775	984	1,216	1,534	1,675[a]	4.6	4.8	1.8	48.0	48.0	49.0	49.0	49.0
67	Mauritius	135	150	151	129	141	1.1	–3.1	1.8	47.0	49.0	49.0	49.0	50.0
68	Colombia	2,274	3,286	3,911	4,168	4,040	5.6	1.3	–0.6	50.0	50.0	50.0	50.0	50.0
69	Chile	1,525	2,040	2,299	2,185	2,062	4.2	–1.0	–1.2	49.0	50.0	49.0	49.0	49.0
70	Costa Rica	277	349	361	349	363	2.7	–0.7	0.8	48.0	49.0	49.0	48.0	48.0

— Not available.

Note: *m*, median; *u*, unweighted mean; *w*, weighted mean; and *w′*, weighted mean excluding China and India.

a. Estimate.

Source: Unesco data.

(*Table continues on the following page.*)

Table A-2 (*continued*)

Economy	Total (thousands)					Annual growth (percent)			Females as a percentage of total				
	1965	1970	1975	1980	1985	1965–75	1975–80	1980–85	1965	1970	1975	1980	1985
71 Jordan	295	278	386	454	531	2.7	3.3	3.2	42.0	44.0	46.0	48.0	48.0
72 Syrian Arab Republic	706	925	1,274	1,556	2,030	6.1	4.1	5.5	32.0	36.0	40.0	43.0	46.0
73 Lebanon	334	435	415[a]	405	360[a]	2.2	–0.5	–2.3	43.0	45.0	47.0	47.0	47.0
Upper-middle-income economies	**54,725**	**67,215**	**76,214**	**84,971**	**91,153**	**1.3 *m***	**1.0 *m***	**1.1 *m***	**48.0 *m***	**48.0 *m***	**48.0 *m***	**49.0 *m***	**48.5 *m***
						2.2 *u*	**1.5 *u***	**1.6 *u***	**46.4 *u***	**44.9 *u***	**46.4 *u***	**47.5 *u***	**48.0 *u***
						3.4 *w*	**2.2 *w***	**1.4 *w***	**47.0 *w***	**48.0 *w***	**47.0 *w***	**48.0 *w***	**48.0 *w***
74 Brazil	9,923	15,405	19,549	22,598	24,770	7.0	2.9	1.9	50.0	50.0	49.0	49.0	48.0
75 Malaysia	1,440	1,684	1,893	2,009	2,199	2.8	1.2	1.8	46.0	47.0	48.0	49.0	49.0
76 South Africa	3,043	3,989	4,635	4,353	4,400[a]	4.3	–1.2	0.2	50.0	50.0	50.0	50.0	50.0
77 Mexico	6,916	9,248	11,461	14,666	15,124	5.2	5.1	0.6	48.0	48.0	48.0	49.0	49.0
78 Uruguay	335	354	323	331	356	–0.4	0.5	1.5	49.0	48.0	49.0	49.0	49.0
79 Hungary	1,414	1,116	1,051	1,162	1,298	–2.9	2.0	2.2	48.0	48.0	49.0	49.0	49.0
80 Poland	5,178	5,257	4,310	4,167	4,801	–1.8	–0.7	2.9	48.0	48.0	48.0	49.0	48.0
81 Portugal	893	992	1,205	1,240	1,235	3.0	0.6	–0.1	49.0	49.0	48.0	48.0	48.0
82 Yugoslavia	1,643	1,579	1,495	1,432	1,449	–0.9	–0.9	0.2	48.0	48.0	48.0	48.0	48.0
83 Panama	203	255	335	338	340	5.1	0.2	0.1	48.0	48.0	48.0	48.0	48.0
84 Argentina	3,125	3,386	3,571	3,917	4,589	1.3	1.9	3.2	49.0	49.0	49.0	49.0	49.0
85 Republic of Korea	4,941	5,749	5,599	5,658	4,857	1.3	0.2	–3.0	48.0	48.0	48.0	49.0	49.0
86 Algeria	1,358	1,887	2,663	3,119	3,481	7.0	3.2	2.2	38.0	38.0	40.0	42.0	44.0
87 Venezuela	1,453	1,770	2,108	2,530	2,771	3.8	3.7	1.8	50.0	50.0	49.0	49.0	49.0
88 Gabon	79	101	129	155	184	5.0	3.7	3.5	46.0	48.0	49.0	49.0	49.0
89 Greece	964	907	936	901	888	–0.3	–0.8	–0.3	48.0	48.0	48.0	48.0	48.0
90 Spain	3,358	3,930	3,653	3,610	3,484	0.8	–0.2	–0.7	48.0	50.0	49.0	49.0	48.0
91 Oman	0	3	55	92	178	—	10.8	14.1	—	0.0	27.0	34.0	44.0
92 Ireland	378	400	405	420	420	0.7	0.7	0.0	49.0	49.0	49.0	49.0	49.0
93 Trinidad and Tobago	212	226	199	167	168	–0.6	–3.4	0.1	49.0	49.0	50.0	50.0	50.0
94 Israel	445	479	535	622	699	1.9	3.1	2.4	49.0	48.0	49.0	49.0	49.0

95 Hong Kong	597	740	643	540	537[a]	0.7	−3.4	−0.1	46.0	47.0	48.0	48.0	48.0
96 Singapore	357	364	328	292	290[a]	−0.8	−2.3	−0.1	46.0	47.0	47.0	48.0	47.0
97 Islamic Republic of Iran	2,574	3,416	4,468	4,799	6,788	5.7	1.4	7.2	32.0	35.0	38.0	40.0	44.0
98 Iraq	964	1,099	1,776	2,616	2,816	6.3	8.1	1.5	30.0	29.0	33.0	46.0	45.0
99 Romania	2,932	2,879	2,889	3,237	3,031	−0.1	2.3	−1.3	48.0	49.0	49.0	49.0	49.0
High-income oil exporters	**507**	**875**	**1,398**	**1,828**	**2,458**	**10.2 *m***	**6.2 *m***	**5.6 *m***	**28.0 *m***	**36.0 *m***	**46.0 *m***	**47.5 *m***	**47.5 *m***
						9.6 *u*	**6.8 *u***	**6.4 *u***	**31.0 *u***	**36.3 *u***	**43.5 *u***	**45.5 *u***	**46.8 *u***
						10.7 *w*	**5.5 *w***	**6.1 *w***	**27.0 *w***	**35.0 *w***	**41.0 *w***	**43.0 *w***	**45.0 *w***
100 Saudi Arabia	256	423	678	927	1,344	10.2	6.5	7.7	22.0	31.0	36.0	39.0	43.0
101 Kuwait	54	76	112	149	173	7.6	5.9	3.0	43.0	42.0	46.0	48.0	49.0
102 United Arab Emirates	0[a]	26	52	89	152	—	11.3	11.3	—	35.0	46.0	47.0	48.0
103 Libya	197	350	556	663	789	10.9	3.6	3.5	28.0	37.0	46.0	48.0	47.0
High-income industrial market economies	**71,918**	**74,249**	**69,785**	**65,179**	**61,143**	**0.5 *m***	**−2.1 *m***	**−2.2 *m***	**49.0 *m***	**49.0 *m***	**49.0 *m***	**49.0 *m***	**—**
						0.1 *u*	**−1.8 *u***	**−1.8 *u***	**48.6 *u***	**48.9 *u***	**49.1 *u***	**49.2 *u***	**48.8 *u***
						−0.3 *w*	**−1.4 *w***	**−1.3 *w***	**48.0 *w***	**49.0 *w***	**49.0 *w***	**49.0 *w***	**49.0 *w***
104 New Zealand	368	400	391	381	329	0.6	−0.5	−2.9	48.0	49.0	49.0	49.0	49.0
105 Italy	4,520	4,857	4,833	4,423	3,716	0.7	−1.8	−3.4	48.0	48.0	49.0	49.0	49.0
106 United Kingdom	5,165	5,806	5,725	4,911	4,296	1.0	−3.0	−2.6	49.0	49.0	49.0	49.0	49.0
107 Belgium	962	978	942	842	730	−0.2	−2.2	−2.8	49.0	49.0	52.0	53.0	49.0
108 Austria	471	532	502	400	344	0.6	−4.4	−3.0	49.0	49.0	49.0	49.0	48.0
109 Netherlands	1,409	1,462	1,453	1,333	1,469	0.3	−1.7	2.0	49.0	49.0	49.0	49.0	49.0
110 France	5,524	4,940	4,602	4,610	4,116	−1.8	0.0	−2.2	49.0	49.0	51.0	49.0	48.0
111 Australia	1,557	1,691	1,633	1,718	1,542	0.5	1.0	−2.1	49.0	49.0	48.0	49.0	49.0
112 Federal Republic of Germany	3,516	3,973	3,903	2,784	2,272	1.0	−6.5	−4.0	49.0	49.0	49.0	49.0	49.0
113 Finland	460	386	454	373	381	−0.1	−3.9	0.4	47.0	47.0	48.0	49.0	49.0

— Not available.

Note: *m*, median; *u*, unweighted mean; *w*, weighted mean; and *w′*, weighted mean excluding China and India.

a. Estimate.

Source: Unesco data.

(*Table continues on the following page.*)

Table A-2 (*continued*)

Economy	Total (thousands) 1965	1970	1975	1980	1985	Annual growth (percent) 1965–75	1975–80	1980–85	Females as a percentage of total 1965	1970	1975	1980	1985
114 Denmark	440	443	491	435	403	1.1	−2.4	−1.5	49.0	49.0	48.0	49.0	49.0
115 Japan	9,927	9,631	10,365	11,827	11,095	0.4	2.7	−1.3	49.0	49.0	49.0	49.0	49.0
116 Sweden	607	615	699	667	613	1.4	−0.9	−1.7	49.0	49.0	49.0	49.0	49.0
117 Canada	3,641	3,736	2,440	2,185	2,255	−3.9	−2.2	0.6	48.0	49.0	49.0	49.0	48.0
118 Norway	412	386	390	390	335	−0.5	0.0	−3.0	49.0	51.0	49.0	49.0	49.0
119 United States	32,474	33,950	30,446	27,449	26,870	−0.6	−2.1	−0.4	48.0	49.0	49.0	49.0	49.0
120 Switzerland	465[a]	463[a]	516[a]	451	377	1.0	−2.7	−3.5	49.0	49.0	49.0	49.0	49.0
Other economies													
121 Angola	218	434	822[a]	1,301	750[a]	14.2	9.6	−10.4	33.0	36.0	38.0	47.0	45.0
122 Albania	348	497	579	553	544	5.2	−0.9	−0.3	47.0	47.0	47.0	47.0	48.0
123 Bulgaria	1,117	1,050	980	994	1,081	−1.3	0.3	1.7	49.0	49.0	48.0	48.0	48.0
124 Czechoslovakia	2,221	1,966	1,881	1,904	2,074	−1.6	0.2	1.7	48.0	49.0	49.0	49.0	49.0
125 Cuba	1,232	1,530	1,796	1,469	1,077	3.8	−3.9	−6.0	49.0	49.0	48.0	48.0	47.0
126 German Democratic Republic	1,413	1,236	1,060	852	860	−2.8	−4.3	0.2	48.0	48.0	48.0	48.0	48.0
127 Mongolia	104	146	130	145	155[a]	2.3	2.2	1.3	50.0	49.0	48.0	49.0	49.0
128 Democratic People's Republic of Korea	—	—	—	—	—	—	—	—	—	—	—	—	—
129 Union of Soviet Socialist Republics	25,152	25,798	21,366	21,714	23,585	−1.6	0.3	1.7	49.0	49.0	50.0	50.0	50.0

— Not available.

Note: *m*, median; *u*, unweighted mean; *w*, weighted mean; and *w′*, weighted mean excluding China and India.

a. Estimate.

Source: Unesco data.

Table A-3. Children Age 6–11 Enrolled in Primary School and Growth in Enrollment, Selected Years, 1965–85

Economy	Total (thousands)					Annual growth (percent)			Enrolled children age 6–11 as a percentage of population age 6–11				
	1965	1970	1975	1980	1985	1965–75	1975–80	1980–85	1965	1970	1975	1980	1985
Low-income economies	**146,576**	**159,879**	**194,738**	**213,603**	**228,299**	**5.4 *m***	**4.9 *m***	**3.5 *m***	**28.8 *m***	**29.6 *m***	**37.2 *m***	**45.9 *m***	**50.3 *m***
						5.7 *u*	**6.1 *u***	**3.5 *u***	**36.1 *u***	**38.8 *u***	**43.9 *u***	**50.5 *u***	**53.8 *u***
Excluding China and India	**22,790**	**29,197**	**38,365**	**46,852**	**53,991**	**2.9 *w***	**1.9 *w***	**1.3 *w***	**67.6 *w***	**66.2 *w***	**68.5 *w***	**69.4 *w***	**76.3 *w***
						5.3 *w′*	**4.1 *w′***	**2.9 *w′***	**37.6 *w′***	**42.2 *w′***	**48.2 *w′***	**52.6 *w′***	**54.1 *w′***
1 Ethiopia	206	378	693	1,452	1,641	12.9	15.9	2.5	4.7	9.3[a]	15.2[a]	28.3[a]	26.4[a]
2 Bhutan	6	6	10	21	32	5.2	16.0	8.8	4.3	4.0	6.0	11.0	18.8[a]
3 Burkina Faso	72	82	109	155	277	4.2	7.3	12.3	9.3	9.3	13.4[a]	16.6[a]	26.8[a]
4 Nepal	316	390	681	1,317	1,730	8.0	14.1	5.6	21.3	22.2	33.9	57.4	66.0
5 Bangladesh	3,576	4,407	6,955	7,581	8,353	6.9	1.7	2.0	38.2	39.4	53.9	49.9	47.6
6 Malawi	182	196	325	416	485	6.0	5.1	3.1	34.5[a]	25.9	36.3	39.8	41.4
7 Zaire	1,571	2,218	2,649	3,209	3,807	5.4	3.9	3.5	55.3	70.4	72.9	72.2	72.1
8 Mali	116	128	173	199	190	4.1	2.8	-0.9	18.0[a]	14.1	16.7	16.9	14.1
9 Myanmar	1,793	2,503	2,752	3,370	4,062[b]	4.4	4.1	3.8	47.5	59.4	58.5	63.6	71.2
10 Mozambique	299	421	550	784	899	6.3	7.3	2.8	23.8	28.6	32.2	40.8	50.3[a]
11 Madagascar	472	696	960	1,379	1,304[c]	7.4	7.5	-1.4	50.3	63.9	76.9	96.5	80.3
12 Uganda	378	494	623	791	1,310	5.1	4.9	10.6	27.6	29.6	32.2	34.8	48.3
13 Burundi	120	135	85	115	278	-3.4	6.2	19.3	25.6	25.7	14.6	22.2[a]	44.4[a]
14 Tanzania	432	455	834	1,578	1,526	6.8	13.6	-0.7	22.8	24.3[a]	37.2[a]	58.5[a]	46.8[a]
15 Togo	113	176	278	338	324	9.4	4.0	-0.8	44.6	54.7	75.9	81.3	66.8
16 Niger	49	71	115	177	225	8.9	9.0	4.9	8.6	13.4[a]	18.7[a]	24.7[a]	26.7[a]
17 Benin	93	120	200	293	348	8.0	7.9	3.5	25.0	28.0	40.5	51.0	50.7
18 Somalia	19	27	117	141	98	19.9	3.8	-7.0	3.7	4.6	17.6	18.1	9.6
19 Central African Republic	94	135	157	196	250	5.3	4.5	5.0	37.1	49.0	51.7	56.9	62.0

— Not available.

Note: *m*, median; *u*, unweighted mean; *w*, weighted mean; and *w′*, weighted mean excluding China and India.

a. Primary school starts at age seven; figures adjusted accordingly. b. Estimate. c. 1984 data.

Source: Unesco data.

(*Table continues on the following page.*)

Table A-3 (continued)

Economy	Total (thousands) 1965	1970	1975	1980	1985	Annual growth (percent) 1965–75	1975–80	1980–85	Enrolled children age 6–11 as a percentage of population age 6–11 1965	1970	1975	1980	1985
20 India	41,828	47,898	55,811	62,792	73,495	2.9	2.4	3.2	63.8	65.3	68.7	70.8	80.8
21 Rwanda	275	349	324	497	586	1.7	8.9	3.3	53.3	68.1[a]	52.2[a]	66.8[a]	66.2[a]
22 China	81,958	82,784	100,562	103,959	100,813	2.1	0.7	–0.6	90.5[a]	83.7[a]	81.4[a]	79.9[a]	93.1[a]
23 Kenya	672	987	2,171	2,690	3,278	12.4	4.4	4.0	40.2	49.1	89.6	90.0	87.3
24 Zambia	311	476	579	655	824	6.4	2.5	4.7	63.8[a]	83.1[a]	86.7[a]	76.1[a]	82.6[a]
25 Sierra Leone	83	114	145	229	282	5.7	9.6	4.3	23.0	28.6	32.3	44.8	48.3
26 Sudan	325	618	909	1,141	1,354	10.8	4.7	3.5	19.8[a]	33.6[a]	42.8[a]	45.9[a]	45.6[a]
27 Haiti	150	194	258	319	500	5.6	4.3	9.4	29.9[a]	33.6[a]	39.0[a]	36.5	54.7
28 Pakistan	2,823	3,548	4,668	4,795	6,613	5.2	0.5	6.6	28.7	31.4	35.7	33.7	42.5
29 Lesotho	87	98	126	132	169	3.8	0.9	5.1	60.4	60.3	69.1	63.3	69.9
30 Ghana	936	788	964	1,159	1,232	0.3	3.8	1.2	71.2	53.9	58.8	64.2	58.7
31 Sri Lanka	1,648	1,588	1,364	1,977	2,131	–1.9	7.7	1.5	91.7	78.7	62.0	97.9	100.1
32 Mauritania	13	20	33	60	95	9.8	12.7	9.6	7.4	10.6	15.3	24.3	33.2
33 Senegal	135	163	225	337	460	5.2	8.4	6.4	24.7	25.6	29.8	36.9	43.7
34 Afghanistan	223	366	493	727	379	8.3	8.1	–12.2	15.9[a]	21.5[a]	20.1	33.7[a]	20.7[a]
35 Chad	114	134	156	178	244	3.2	2.7	6.5	22.9	24.2	25.3	25.8	31.4
36 Guinea	142	159	163	214	193	1.4	5.6	–2.0	28.8[a]	28.8[a]	26.1[a]	30.9[a]	24.9[a]
37 Democratic Kampuchea	633	273	384[b]	400[b]	420[b]	–4.9	0.8	1.0	65.0	24.6	32.1	38.3	82.3
38 Lao People's Democratic Republic	140	211	273	414	451	6.9	8.7	1.7	36.3	49.2	57.7	82.7	80.5
39 Viet Nam	4,173	6,073	6,864	7,416	7,641	5.1	1.6	0.6	68.5	82.4	82.9	85.1	81.3
Lower-middle-income economies	**37,563**	**46,383**	**56,372**	**74,197**	**83,648**	**4.8 *m***	**3.6 *m***	**2.8 *m***	**67.1 *m***	**73.9 *m***	**78.1 *m***	**81.3 *m***	**85.4 *m***
						5.0 *u*	**4.2 *u***	**2.5 *u***	**64.5 *u***	**69.0 *u***	**74.4 *u***	**80.7 *u***	**84.7 *u***
						4.1 *w*	**5.6 *w***	**2.5 *w***	**63.8 *w***	**69.1 *w***	**73.2 *w***	**85.4 *w***	**90.3 *w***
40 Liberia	50	72	95	144	78	6.6	8.7	–11.5	27.2	33.5	37.4	47.6	21.5

41	People's Democratic Republic of Yemen	46	101	179	187	224[b]	14.6	0.9	3.7	25.4[a]	50.3[a]	75.4[a]	69.4[a]	71.5[a]
42	Indonesia	8,859	11,275	13,773	19,660	23,561	4.5	7.4	3.7	64.1[a]	72.2[a]	78.7[a]	97.7[a]	111.1[a]
43	Arab Republic of Yemen	52	69	204	366	732	14.6	12.4	14.9	7.5	11.3[a]	27.4[a]	43.4[a]	72.2[a]
44	Philippines	4,559	5,301	5,766	6,107	6,882	2.4	1.2	2.4	104.2[a]	97.8[a]	96.4[a]	96.1[a]	97.0[a]
45	Morocco	917	921	1,192	1,675	1,868	2.7	7.0	2.2	39.8	40.3[a]	47.7[a]	64.0[a]	63.5[a]
46	Bolivia	373	467	609	777	875	5.0	5.0	2.4	74.8[a]	67.9	78.0	86.6	84.3
47	Zimbabwe	555	595	659	906	1452	1.7	6.6	9.9	85.1[a]	76.3[a]	72.1[a]	84.2[a]	121.8[a]
48	Nigeria	2,217	2,676	4,800	11,225	10,575	8.0	18.5	−1.2	27.6	27.9	40.9	79.5	63.2
49	Dominican Republic	398	510	647	785	866	5.0	3.9	2.0	73.3[a]	77.3[a]	86.2[a]	99.9[a]	107.3[a]
50	Papua New Guinea	120	164	204	260	291	5.4	5.0	2.3	44.5[a]	52.3[a]	58.0[a]	60.7[a]	64.6[a]
51	Côte d'Ivoire	264	360	495	828	977	6.5	10.8	3.4	42.9[a]	47.6[a]	52.4[a]	70.4[a]	66.5[a]
52	Honduras	212	289	349	444	574	5.1	4.9	5.3	68.2[a]	78.4[a]	77.9[a]	80.8[a]	91.0[a]
53	Egypt	3,198	3,530	4,046	4,602	6,135	2.4	2.6	5.9	65.5	67.1	72.5	76.8	85.9
54	Nicaragua	140	194	234	312	397	5.3	5.9	4.9	53.9[a]	64.3[a]	66.5[a]	77.3[a]	84.9[a]
55	Thailand	4,024	4,823	5,610	6,642	6,423	3.4	3.4	−0.7	91.4[a]	94.8[a]	94.9[a]	106.5[a]	103.5[a]
56	El Salvador	260	351	460	517	519[c]	5.9	2.4	0.1	60.9[a]	68.3[a]	78.7[a]	79.0[a]	73.8[a]
57	Botswana	33	49	84	124	149	9.8	8.1	3.7	41.1[a]	51.5[a]	59.0	75.9	99.1[a]
58	Jamaica	248	335	342	335	332[b]	3.3	−0.4	−0.8	88.5	95.2	89.7	94.0	98.5
59	Cameroon	504	642	801	994	1243	4.7	4.4	4.6	53.1	62.1	69.5	74.6	78.5
60	Guatemala	281	371	457	581	737	5.0	4.9	4.9	43.7[a]	50.4[a]	54.4[a]	60.5[a]	65.4[a]
61	Congo	131	171	237	265	323	6.1	2.3	4.0	80.4	92.5	112.6	108.9	115.8
62	Paraguay	256	307	340	400	446	2.9	3.3	2.2	80.0[a]	88.4[a]	88.8[a]	95.2[a]	95.2[a]
63	Peru	1,315	1,707	2,105	2,399	2,955	4.8	2.6	4.3	70.5	77.7	84.1	86.5	97.5
64	Turkey	3,082	3,992	4,371	4,527	5,313	3.6	0.7	3.3	65.0	87.3[a]	71.7	64.8	75.3
65	Tunisia	553	704	757	848	1,045	3.2	2.3	4.3	68.7	75.5	79.1	82.9	94.2
66	Ecuador	640	804	942	1,191	1,299[c]	3.9	4.8	2.2	73.4	78.7	78.2	87.8	88.8
67	Mauritius	116	133	132	112	118	1.3	−3.2	1.0	84.3	84.2	90.6	92.0	91.5

— Not available.
Note: *m*, median; *u*, unweighted mean; *w*, weighted mean; and *w*′, weighted mean excluding China and India.
a. Primary school starts at age seven; figures adjusted accordingly. b. Estimate. c. 1984 data.
Source: Unesco data.

(*Table continues on the following page.*)

Table A-3 (continued)

Economy	Total (thousands)					Annual growth (percent)			Enrolled children age 6–11 as a percentage of population age 6–11				
	1965	1970	1975	1980	1985	1965–75	1975–80	1980–85	1965	1970	1975	1980	1985
68 Colombia	1,715	2,399	2,895	3,197	3,104	5.4	2.0	–0.6	65.9[a]	79.1[a]	87.2[a]	81.8	75.9
69 Chile	1,087	1,355	1,418	1,348	1,287	2.7	–1.0	–0.9	100.3[a]	90.4	92.3	89.6	88.8
70 Costa Rica	223	283	311	297	313	3.4	–0.9	1.1	101.4[a]	89.0	92.1	89.3	83.9
71 Jordan	254	241	351	406	472	3.3	3.0	3.1	81.0	62.5	78.6	73.0	76.0
72 Syrian Arab Republic	602	829	1,161	1,397	1,808	6.8	3.8	5.3	59.3	69.5	87.1	91.2	96.5
73 Lebanon	279	363	346	339	275[c]	2.2	–0.4	–5.1	79.0	85.6	74.5	77.2	69.0
Upper-middle-income economies	**42,120**	**49,577**	**57,000**	**64,009**	**69,554**	**1.5 *m***	**1.4 *m***	**1.4 *m***	**85.4 *m***	**88.1 *m***	**90.1 *m***	**92.5 *m***	**91.5 *m***
						2.3 *u*	**2.1 *u***	**1.7 *u***	**84.0 *u***	**82.5 *u***	**87.5 *u***	**89.8 *u***	**91.4 *u***
						3.1 *w*	**2.3 *w***	**1.7 *w***	**76.3 *w***	**80.2 *w***	**85.0 *w***	**88.5 *w***	**90.3 *w***
74 Brazil	7,520	9,717	11,320	12,858	14,187	4.2	2.6	2.0	65.9[a]	73.4[a]	80.5[a]	88.6[a]	90.1[a]
75 Malaysia	1,412	1,673	1,893	2,009	2,192	3.0	1.2	1.8	84.7	88.1	94.5	92.6	99.9
76 South Africa	1,765	2,308	2,702	2,537	2,564[b]	4.4	–1.3	0.2	56.9	64.5	69.6	60.5	56.0
77 Mexico	5,186	7,213	9,096	11,767	12,816	5.8	5.3	1.7	66.6	77.8	83.3	92.6	100.5
78 Uruguay	252	262	246	258	294	–0.2	1.0	2.6	85.4	82.8	81.2	83.0	91.0
79 Hungary	1016	771	777	868	995	–2.6	2.2	2.8	100.0	94.3	95.6	94.0	98.9
80 Poland	3,488	2,934	2,529	2,671	3,111	–3.2	1.1	3.1	98.4[a]	95.6[a]	98.6[a]	99.5[a]	100.7[a]
81 Portugal	868	899	931	991	1,004	0.7	1.3	0.3	103.3[a]	86.0	90.3	98.5	100.5
82 Yugoslavia	1,535	1,482	1,446	1,399	1,433	–0.6	–0.7	0.5	78.5[a]	79.3[a]	79.2[a]	77.9[a]	77.1[a]
83 Panama	147	191	258	281	287	5.8	1.7	0.4	82.6[a]	74.1	87.4	88.5	88.6
84 Argentina	2,466	2,674	2,838	3,046	3,545	1.4	1.4	3.1	92.6	96.8	97.9	95.1	95.2
85 Republic of Korea	4,496	5,251	5,212	5,371	4,729	1.5	0.6	–2.5	91.7	94.5	99.5	104.3	93.9
86 Algeria	1,060	1,491	2,201	2,676	3,203	7.6	4.0	3.7	52.9	60.1	76.6	81.4	85.6
87 Venezuela	1,048	1,356	1,649	2,014	2,277	4.6	4.1	2.5	82.6[a]	88.9[a]	92.6[a]	103.2[a]	105.6[a]
88 Gabon	55	70	90	110	132	5.0	4.1	3.7	92.5	115.7	116.4	111.0	109.5
89 Greece	866	824	868	844	766	0.0	–0.6	–1.9	98.5	97.3	96.9	96.4	89.9
90 Spain	2,714	3,364	3,536	3,521	3,433	2.7	–0.1	–0.5	78.4	88.2	89.8	88.6	91.9

91 Oman	0	3	40	77	155	—	14.0	15.0	3.0	32.2	50.0	77.0	—
92 Ireland	321	347	359	378	373	1.1	1.0	-0.3	91.0	93.2	91.5	89.8	88.7
93 Trinidad and Tobago	153	165	147	133	134	-0.4	-2.0	0.1	104.7	95.1	89.7	91.2	88.8
94 Israel	347	375	417	470	505	1.9	2.4	1.4	98.9	99.5	98.6	92.4	92.0
95 Hong Kong	468	550	495	483	484[c]	0.6	-0.5	0.1	78.0	86.7	92.1	95.3	94.4
96 Singapore	308	325	297	269	258[c]	-0.4	-2.0	-1.0	90.8	94.2	99.1	99.4	104.3
97 Islamic Republic of Iran	2,097	2,844	4,222	4,534	6,210	7.2	1.4	6.5	61.1[a]	71.2[a]	74.4	69.6	82.7
98 Iraq	707	885	1,495	2,275	2,436	7.8	8.8	1.4	63.9[a]	55.4	78.8	99.7	86.4
99 Romania	1,825	1,603	1,936	2,169	2,031	0.6	2.3	-1.3	100.1[a]	88.0	89.1	91.6	86.5
High-income oil exporters	**370**	**662**	**1,098**	**1,523**	**2,035**	**10.8 *m***	**7.0 *m***	**5.8 *m***	**56.7 *m***	**64.3 *m***	**72.2 *m***	**70.0 *m***	**67.7 *m***
						10.5 *u*	**7.6 *u***	**6.1 *u***	**51.7 *u***	**61.7 *u***	**74.7 *u***	**74.5 *u***	**74.3 *u***
						11.1 *w*	**6.8 *w***	**6.0 *w***	**34.3 *w***	**47.4 *w***	**60.9 *w***	**65.6 *w***	**67.6 *w***
100 Saudi Arabia	178	302	495	735	1,058	10.8	8.2	7.6	23.4	32.4	42.1	49.7	54.6
101 Kuwait	47	69	106	140	159	8.5	5.7	2.6	75.0	57.9	62.8	66.4	61.0
102 United Arab Emirates	0	19	42	73	119	—	11.7	10.3	—	—	70.7	81.6	73.6
103 Libya	145	272	455	575	699	12.1	4.8	4.0	56.7	85.7	112.2	108.2	107.1
High-income industrial market economies	**59,708**	**61,475**	**59,531**	**55,363**	**51,108**	**0.7 *m***	**-1.7 *m***	**-2.7 *m***	**96.2 *m***	**93.2 *m***	**93.2 *m***	**94.7 *m***	**91.4 *m***
						0.3 *u*	**-1.7 *u***	**-2.3 *u***	**93.4 *u***	**90.9 *u***	**90.8 *u***	**91.0 *u***	**89.9 *u***
						0.0 *w*	**-1.4 *w***	**-1.6 *w***	**94.7 *w***	**93.4 *w***	**92.3 *w***	**91.2 *w***	**90.6 *w***
104 New Zealand	302	322	325	325	278	0.7	0.0	-3.1	88.9	88.5	87.5	93.3	87.7
105 Italy	4,311	4,665	4,697	4,383	3,697	0.9	-1.4	-3.3	89.5	87.9	85.1	82.0	78.2
106 United Kingdom	4,294	4,852	4,834	4,015	3,327	1.2	-3.6	-3.7	90.8	87.9	87.5	81.6	79.1
107 Belgium	888	899	844	770	673	-0.5	-1.8	-2.7	98.8	95.1	93.7	95.0	90.0
108 Austria	466	527	498	399	343	0.7	-4.3	-3.0	72.4	70.2	66.2	62.4	64.2
109 Netherlands	1,269	1,338	1,345	1,232	1,022	0.6	-1.7	-3.7	93.7	93.0	91.7	92.5	91.4
110 France	4,535	4,582	4,491	4,449	3,967	-0.1	-0.2	-2.3	91.8	90.6	88.6	89.3	85.8

— Not available.
Note: *m*, median; *u*, unweighted mean; *w*, weighted mean; and *w′*, weighted mean excluding China and India.
a. Primary school starts at age seven; figures adjusted accordingly. b. Estimate. c. 1984 data.
Source: Unesco data.

(Table continues on the following page.)

Table A-3 (continued)

Economy	Total (thousands) 1965	1970	1975	1980	1985	Annual growth (percent) 1965–75	1975–80	1980–85	Enrolled children age 6–11 as a percentage of population age 6–11 1965	1970	1975	1980	1985
111 Australia	1,309	1,464	1,468	1,568	1,416	1.2	1.3	-2.0	97.7	99.3	98.0	101.4	97.4
112 Federal Republic of Germany	3,516	3,973	3,903	2,784	2,263	1.0	-6.5	-4.1	69.4	68.7	65.2	60.3	62.3
113 Finland	427	395	368	303	309	-1.5	-3.8	0.4	102.8[a]	101.0[a]	100.3[a]	95.6[a]	99.3[a]
114 Denmark	438	443	471	435	397	0.7	-1.6	-1.8	100.0	95.9	99.9	115.3[a]	118.1[a]
115 Japan	9,927	9,631	10,365	11,827	11,095	0.4	2.7	-1.3	99.0	99.5	99.5	101.1	101.7
116 Sweden	503	511	588	562	512	1.6	-0.9	-1.8	96.2[a]	93.2[a]	101.7[a]	98.0[a]	99.4[a]
117 Canada	2,690	2,696	2,301	2,010	2,066	-1.5	-2.7	0.6	101.6	96.9	93.2	94.7	96.6
118 Norway	304	306	325	321	274	0.7	-0.2	-3.1	100.3[a]	98.9[a]	100.0[a]	98.9[a]	93.8[a]
119 United States	24,106	24,479	22,272	19,588	19,138	-0.8	-2.5	-0.5	100.0	100.0	100.0	96.3	95.0
120 Switzerland	423	392	436	392	331	0.3	-2.1	-3.3	95.5[a]	79.4[a]	86.1[a]	89.1[a]	88.3[a]
Other economies													
121 Angola	187	374	549	871	502	11.4	9.7	-10.4	29.0[a]	52.7[a]	65.9[a]	86.3[a]	42.7[a]
122 Albania	—	—	—	—	—	—	—	—	—	—	—	—	—
123 Bulgaria	657	642	602	644	705	-0.9	1.4	1.8	98.7[a]	100.1[a]	97.0[a]	100.0[a]	105.7[a]
124 Czechoslovakia	1,458	1,322	1,283	1,454	1,622	-1.3	2.5	2.2	100.0	99.7	100.0	100.0	100.0
125 Cuba	872	1,197	1,436	1,323	967	5.1	-1.6	-6.1	90.5	94.6	99.5	97.6	93.7
126 German Democratic Republic	1,413	1,237	1,060	852	834	-2.8	-4.3	-0.4	105.5[a]	73.2	67.5	68.3	67.9
127 Mongolia	81	113	129	144	154	4.8	2.2	1.4	59.7[a]	68.2[a]	64.2[a]	63.2[a]	61.7[a]
128 Democratic People's Republic of Korea	—	—	—	—	—	—	—	—	—	—	—	—	—
129 Union of Soviet Socialist Republics	24,025	24,787	21,366	20,885	22,402	-1.2	-0.5	1.4	100.0[a]	100.0[a]	97.1[a]	100.0[a]	100.0[a]

— Not available.

Note: *m*, median; *u*, unweighted mean; *w*, weighted mean; and *w′*, weighted mean excluding China and India.

a. Primary school starts at age seven; figures adjusted accordingly. b. Estimate. c. 1984 data.

Source: Unesco data.

Table A-4. Gross Enrollment Ratio in Primary School, Selected Years, 1965–85

Economy	Total (percent)					Female (percent)				
	1965	*1970*	*1975*	*1980*	*1985*	*1965*	*1970*	*1975*	*1980*	*1985*
Low-income economies	**40.8 *m***	**37.8 *m***	**55.0 *m***	**62.9 *m***	**65.0 *m***	**27.0 *m***	**28.4 *m***	**43.0 *m***	**48.4 *m***	**55.4 *m***
	44.2 *u*	**47.7 *u***	**58.7 *u***	**67.3 *u***	**69.4 *u***	**32.8 *u***	**37.4 *u***	**48.0 *u***	**57.2 *u***	**59.8 *u***
	83.0 *w*	**76.5 *w***	**93.8 *w***	**91.1 *w***	**96.2 *w***	**64.6 *w***	**65.2 *w***	**82.0 *w***	**79.5 *w***	**84.9 *w***
Excluding China and India	**44.1 *w′***	**47.9 *w′***	**61.7 *w′***	**67.0 *w′***	**67.3 *w′***	**31.2 *w′***	**35.6 *w′***	**49.7 *w′***	**56.8 *w′***	**58.5 *w′***
1 Ethiopia	8.6	13.6	20.1	35.1	33.7	4.8	8.6	13.0	24.7	26.8
2 Bhutan	6.0	5.8	8.6	15.7	26.5	0.9	0.6	4.2	9.8	19.0
3 Burkina Faso	11.7	12.0	14.7	18.3	28.6	7.6	8.7	10.9	13.5	21.1
4 Nepal	25.0	26.2	51.5	88.3	82.2	7.9	8.0	16.2	51.5	50.0
5 Bangladesh	50.2	54.3	73.4	61.6	60.1	31.3	35.4	51.0	46.4	49.8
6 Malawi	42.1	35.8	55.9	60.0	61.7	30.1	25.8	43.0	48.4	53.4
7 Zaire	72.7	97.9	97.5	94.4	94.2	47.0	72.8	78.1	79.1	81.7
8 Mali	21.9	22.4	24.3	24.8	21.6	14.0	15.3	17.1	18.0	16.1
9 Myanmar	67.9	87.6	84.9	90.9	104.5	62.3	83.1	82.1	88.5	100.2
10 Mozambique	33.0	39.7	82.7	99.2	85.8	23.5	26.7	66.9	83.8	75.0
11 Madagascar	71.9	86.1	97.0	142.7	112.6	64.4	78.8	83.7	139.3	108.6
12 Uganda	41.5	37.7	44.0	49.5	65.0	30.3	29.5	35.1	42.7	55.4
13 Burundi	27.1	30.4	22.2	28.6	52.5	15.7	19.6	16.9	22.2	43.7
14 Tanzania	31.8	34.0	52.6	92.8	72.2	23.6	26.6	43.8	86.1	71.3
15 Togo	61.3	71.0	99.0	121.7	95.4	35.8	43.8	68.5	93.2	73.2
16 Niger	10.8	14.2	19.7	27.0	27.8	6.8	9.8	13.9	19.2	20.0
17 Benin	33.7	36.4	50.4	63.5	64.7	20.4	22.0	31.2	39.9	43.0
18 Somalia	8.2	8.1	56.0	34.0	20.0	3.4	4.0	40.0	24.0	13.0
19 Central African Republic	50.6	63.8	72.7	71.5	76.7	25.2	40.7	51.0	51.2	59.4
20 India	76.9	77.8	80.9	83.3	95.0	58.2	60.5	64.2	67.1	79.3
21 Rwanda	56.2	69.6	55.0	62.9	62.6	45.4	61.0	50.4	59.9	60.9

— Not available.

Note: *m*, median; *u*, unweighted mean; *w*, weighted mean; and *w′*, weighted mean excluding China and India.

Sources: United Nations 1988; Unesco data.

(*Table continues on the following page.*)

Table A-4 (continued)

Economy	Total (percent) 1965	1970	1975	1980	1985	Female (percent) 1965	1970	1975	1980	1985
22 China	105.8	90.9	122.2	112.4	123.5	85.2	84.3	114.2	103.3	114.1
23 Kenya	54.6	62.1	103.9	115.2	98.4	39.9	51.8	95.9	110.1	95.6
24 Zambia	55.5	89.7	96.6	98.0	106.0	48.6	80.0	88.3	92.0	95.0
25 Sierra Leone	29.4	34.8	38.5	51.7	53.2	20.6	27.7	30.2	42.8	43.3
26 Sudan	32.1	37.9	46.6	49.9	49.8	22.9	29.0	33.6	40.9	40.8
27 Haiti	47.9	53.7	62.3	73.6	95.4	41.3	47.8	55.6	67.9	89.4
28 Pakistan	33.0	35.8	41.0	39.3	48.2	15.5	19.3	24.6	26.7	33.8
29 Lesotho	90.7	87.0	105.6	102.5	113.3	107.1	101.6	123.4	120.2	125.2
30 Ghana	69.2	62.3	72.0	80.0	76.0	57.5	53.5	62.6	71.2	65.8
31 Sri Lanka	92.5	99.0	77.4	102.9	103.0	86.4	94.3	73.9	100.4	101.5
32 Mauritania	12.1	14.6	20.3	36.6	49.4	5.6	8.1	14.5	25.8	39.4
33 Senegal	40.0	41.4	41.2	45.9	55.5	28.8	31.8	34.5	36.4	44.6
34 Afghanistan	21.7	26.9	24.8	33.7	20.0	6.5	7.8	7.5	12.4	12.8
35 Chad	32.9	34.7	34.5	35.7	43.5	12.3	17.4	18.1	19.1	24.5
36 Guinea	29.3	29.4	27.0	31.4	30.2	17.8	18.5	18.0	20.8	19.0
37 Democratic Kampuchea	82.1	30.3	39.8	47.4	101.8	59.3	26.0	33.2	39.4	82.4
38 Lao People's Democratic Republic	41.9	57.0	67.0	94.0	94.0	31.2	42.6	56.3	86.0	85.0
39 Viet Nam	—	—	107.0	108.8	102.3	—	—	108.1	106.1	99.0
Lower-middle-income economies	**78.1 *m***	**81.4 *m***	**86.5 *m***	**98.9 *m***	**102.5 *m***	**67.8 *m***	**77.3 *m***	**79.9 *m***	**93.6 *m***	**101.1 *m***
	76.4 *u*	**81.6 *u***	**85.4 *u***	**92.9 *u***	**97.2 *u***	**67.6 *u***	**73.5 *u***	**78.3 *u***	**86.3 *u***	**91.3 *u***
	73.8 *w*	**79.7 *w***	**84.7 *w***	**99.7 *w***	**100.9 *w***	**65.1 *w***	**71.7 *w***	**77.8 *w***	**92.3 *w***	**95.4 *w***
40 Liberia	45.7	55.8	40.9	48.6	35.5	25.8	36.7	27.5	34.5	25.0
41 People's Democratic Republic of Yemen	20.6	57.1	81.0	65.0	70.2	8.9	22.8	51.2	36.3	37.4
42 Indonesia	71.9	80.0	86.0	107.2	117.6	64.0	72.7	78.4	99.7	114.3
43 Arab Republic of Yemen	9.9	12.4	28.8	45.9	82.3	1.0	2.3	6.5	11.6	33.0

44	Philippines	112.7	108.3	107.1	112.6	106.2	110.6	107.1	110.0	112.6	106.7
45	Morocco	56.7	51.5	62.0	83.0	77.5	34.7	36.0	45.2	62.9	60.5
46	Bolivia	84.0	76.2	84.9	84.3	89.0	67.8	61.8	75.6	78.5	83.4
47	Zimbabwe	77.8	69.8	69.9	84.9	135.1	64.8	62.5	63.8	78.7	130.6
48	Nigeria	31.7	36.6	52.6	97.5	77.2	24.2	27.0	44.7	84.5	68.0
49	Dominican Republic	87.3	98.4	102.2	117.6	126.5	87.6	99.2	103.2	118.8	129.4
50	Papua New Guinea	43.9	51.6	57.1	58.9	64.0	34.6	39.1	44.1	50.9	59.0
51	Côte d'Ivoire	48.9	56.5	60.7	74.2	69.6	33.0	41.0	45.7	59.4	57.4
52	Honduras	77.4	88.2	87.6	92.8	101.9	77.2	88.6	87.0	93.6	103.0
53	Egypt	71.7	72.1	75.0	77.8	87.0	57.5	56.4	59.7	64.7	77.5
54	Nicaragua	67.1	80.0	82.0	99.0	101.5	67.2	81.0	84.5	101.8	107.0
55	Thailand	78.4	81.4	83.6	98.9	96.1	74.3	77.3	79.9	97.4	93.6
56	El Salvador	75.8	84.0	74.4	74.9	75.1	74.0	81.7	73.2	75.0	75.9
57	Botswana	61.5	65.0	71.8	92.3	107.9	64.3	67.3	78.6	100.1	112.6
58	Jamaica	108.9	118.8	97.5	100.7	105.6	105.6	119.0	98.0	101.2	106.5
59	Cameroon	78.1	89.2	97.5	103.5	107.7	61.8	75.4	86.6	94.0	98.3
60	Guatemala	50.5	58.3	63.2	70.8	76.5	45.6	52.4	57.5	64.9	70.2
61	Congo	—	—	—	—	—	—	—	—	—	—
62	Paraguay	93.7	103.0	99.3	103.7	102.5	89.0	98.5	95.9	100.5	99.9
63	Peru	101.9	106.6	113.5	114.0	122.4	92.9	99.3	109.1	111.0	119.8
64	Turkey	100.8	109.6	107.6	96.4	113.3	82.5	94.5	96.6	90.4	109.9
65	Tunisia	91.2	100.4	97.4	103.1	116.4	65.0	79.5	78.0	87.6	106.0
66	Ecuador	88.9	96.2	101.0	113.1	112.4	86.1	94.5	99.5	111.5	112.0
67	Mauritius	95.0	93.4	107.3	107.9	105.3	92.3	92.6	106.2	107.4	105.8
68	Colombia	87.4	108.4	117.9	128.5	117.2	88.7	110.0	120.2	130.0	118.6
69	Chile	119.0	104.8	112.2	108.8	105.7	117.7	104.7	111.5	107.7	104.3
70	Costa Rica	107.6	109.7	107.0	104.8	97.4	106.1	109.1	106.5	103.8	96.2
71	Jordan	94.1	72.0	86.5	81.6	85.4	82.5	64.9	81.7	81.2	84.7
72	Syrian Arab Republic	69.6	77.5	95.6	101.6	108.4	46.4	58.9	78.3	88.6	101.1

— Not available.

Note: m, median; u, unweighted mean; w, weighted mean; and w', weighted mean excluding China and India.

Sources: United Nations 1988; Unesco data.

(Table continues on the following page.)

Table A-4 (continued)

Economy	Total (percent)					Female (percent)				
	1965	1970	1975	1980	1985	1965	1970	1975	1980	1985
73 Lebanon	110.6	121.4	106.4	111.4	111.8	97.7	111.8	102.4	106.8	105.6
Upper-middle-income economies	**99.5 *m***	**101.3 *m***	**103.1 *m***	**102.3 *m***	**101.0 *m***	**98.1 *m***	**99.8 *m***	**102.5 *m***	**102.1 *m***	**99.9 *m***
	92.6 *u*	**97.8 *u***	**100.9 *u***	**101.7 *u***	**104.7 *u***	**88.7 *u***	**94.2 *u***	**96.8 *u***	**99.4 *u***	**102.6 *u***
	95.4 *w*	**105.5 *w***	**98.3 *w***	**102.4 *w***	**103.3 *w***	**91.5 *w***	**102.2 *w***	**94.7 *w***	**99.8 *w***	**100.4 *w***
74 Brazil	107.2	143.1	87.8	98.9	100.9	106.9	143.7	87.0	96.6	96.5
75 Malaysia	86.4	88.7	94.5	92.6	100.2	79.9	83.4	92.0	92.0	99.7
76 South Africa	85.2	97.0	102.8	89.8	83.0	87.4	98.6	103.7	90.1	83.2
77 Mexico	88.8	99.7	105.0	115.5	118.6	86.4	97.4	101.8	114.6	117.5
78 Uruguay	113.4	112.1	106.6	106.4	110.2	113.3	109.1	105.8	105.5	109.0
79 Hungary	101.7	97.5	98.7	96.4	98.5	100.5	96.7	98.8	96.6	98.8
80 Poland	104.9	101.3	100.4	99.6	101.1	103.0	99.6	99.2	99.1	100.2
81 Portugal	88.7	94.9	116.8	123.2	123.7	88.1	94.7	115.6	122.6	120.5
82 Yugoslavia	105.9	105.7	102.6	99.7	96.9	103.4	103.5	101.4	99.0	96.6
83 Panama	96.8	99.1	113.6	106.2	105.0	94.9	96.9	110.9	104.1	102.5
84 Argentina	101.0	105.4	106.1	106.1	107.4	100.9	105.7	106.0	106.0	107.8
85 Republic of Korea	100.7	103.4	106.9	109.9	96.4	98.9	102.9	107.2	110.5	96.7
86 Algeria	67.7	76.1	92.7	94.8	93.0	53.5	58.2	75.5	81.0	82.3
87 Venezuela	97.2	98.6	99.7	108.7	108.5	98.1	99.8	100.4	109.3	108.1
88 Gabon	—	—	102.0	115.0	152.9	—	—	100.0	113.0	150.5
89 Greece	109.7	107.2	104.4	102.9	104.2	108.6	106.2	103.7	102.8	104.4
90 Spain	115.1	122.8	111.2	109.0	112.9	113.6	124.9	111.5	108.5	112.4
91 Oman	0.0	3.3	44.0	59.7	88.5	0.0	0.9	24.4	41.7	79.9
92 Ireland	107.3	107.4	103.3	99.9	99.9	107.8	107.4	103.2	100.2	100.1
93 Trinidad and Tobago	98.5	110.9	105.2	97.6	94.7	97.7	110.0	106.0	100.9	96.2
94 Israel	94.7	95.9	96.5	95.0	96.8	94.8	94.8	97.1	95.8	98.3
95 Hong Kong	99.5	116.6	119.6	106.5	104.5	95.3	114.7	117.3	105.7	103.8
96 Singapore	105.3	105.5	109.7	107.7	119.7	99.8	101.2	106.3	106.1	117.2

97 Islamic Republic of Iran	63.4	72.5	93.2	87.2	107.0	40.5	51.9	71.4	71.6	96.7
98 Iraq	73.8	68.8	93.6	114.6	99.9	44.6	40.8	63.7	108.7	91.7
99 Romania	101.6	111.9	107.2	101.6	98.0	100.7	112.9	106.8	101.3	97.7
High-income oil exporters	**33.7** *m*	**88.1** *m*	**92.6** *m*	**88.9** *m*	**95.0** *m*	**15.3** *m*	**72.1** *m*	**85.8** *m*	**88.3** *m*	**94.1** *m*
	50.3 *u*	**76.0** *u*	**83.9** *u*	**84.5** *u*	**86.8** *u*	**39.7** *u*	**58.8** *u*	**75.0** *u*	**79.1** *u*	**83.4** *u*
	37.6 *w*	**50.1** *w*	**62.3** *w*	**67.5** *w*	**73.3** *w*	**19.8** *w*	**33.7** *w*	**48.5** *w*	**55.9** *w*	**66.2** *w*
100 Saudi Arabia	33.7	45.3	57.5	62.6	69.4	15.3	28.8	42.6	49.5	61.3
101 Kuwait	117.2	88.1	92.6	102.1	96.0	103.8	75.6	85.8	99.6	94.1
102 United Arab Emirates	—	94.6	101.5	88.9	95.0	—	72.1	96.6	88.3	94.9
103 Libya	—	—	—	—	—	—	—	—	—	—
High-income industrial market economies	**103.3** *m*	**102.5** *m*	**102.0** *m*	**100.1** *m*	**100.6** *m*	**102.8** *m*	**102.5** *m*	**101.5** *m*	**100.5** *m*	**100.8** *m*
	104.5 *u*	**101.7** *u*	**102.9** *u*	**101.8** *u*	**101.9** *u*	**103.8** *u*	**101.8** *u*	**103.4** *u*	**102.2** *u*	**101.7** *u*
	104.0 *w*	**103.5** *w*	**101.2** *w*	**101.1** *w*	**101.2** *w*	**102.6** *w*	**103.3** *w*	**101.3** *w*	**100.9** *w*	**101.2** *w*
104 New Zealand	105.8	110.0	106.5	111.0	106.9	104.1	108.9	105.9	110.6	106.3
105 Italy	111.8	108.6	105.3	99.9	96.2	110.4	107.8	104.8	99.7	96.1
106 United Kingdom	92.3	103.5	105.1	102.9	104.2	92.2	103.7	105.3	103.2	104.5
107 Belgium	107.1	103.4	104.5	103.9	97.6	106.3	103.8	110.1	112.7	97.8
108 Austria	105.7	103.9	101.6	98.6	99.0	105.2	103.4	101.4	98.1	98.3
109 Netherlands	104.0	101.6	99.1	100.2	114.0	103.8	102.0	99.6	101.2	115.0
110 France	134.1	116.9	109.4	111.1	108.6	133.4	116.4	113.8	110.0	107.5
111 Australia	116.2	114.6	108.9	111.1	106.0	115.7	114.6	108.5	110.5	105.5
112 Federal Republic of Germany	101.9	100.1	99.6	99.1	97.6	101.8	100.6	99.9	99.0	97.5
113 Finland	91.7	81.9	102.3	96.2	102.8	88.6	79.3	101.1	95.8	102.5
114 Denmark	100.3	95.9	104.2	95.5	98.3	100.5	96.8	103.6	95.4	98.2
115 Japan	99.0	99.5	99.5	101.1	101.7	98.7	99.5	99.6	101.2	101.8
116 Sweden	96.7	94.1	101.1	96.5	97.9	97.8	95.0	101.6	96.6	98.5

— Not available.

Note: *m*, median; *u*, unweighted mean; *w*, weighted mean; and *w'*, weighted mean excluding China and India.

Sources: United Nations 1988; Unesco data.

(Table continues on the following page.)

Table A-4 (continued)

Economy	Total (percent)					Female (percent)				
	1965	1970	1975	1980	1985	1965	1970	1975	1980	1985
117 Canada	105.4	101.1	98.8	103.0	105.5	104.6	100.6	98.8	103.1	104.5
118 Norway	97.3	89.1	100.8	99.5	93.9	97.6	93.4	101.0	99.5	93.8
119 United States	102.6	103.3	99.2	99.1	99.5	100.3	102.9	98.6	98.7	99.7
120 Switzerland	—	—	—	—	—	—	—	—	—	—
Other economies										
121 Angola	41.4	75.3	121.5	158.2	78.4	26.9	53.2	91.6	147.1	71.0
122 Albania	94.1	105.6	115.0	113.2	103.1	88.9	102.3	111.1	110.6	101.9
123 Bulgaria	103.3	100.9	98.8	97.8	102.3	102.6	100.3	98.8	97.7	101.9
124 Czechoslovakia	98.7	96.9	96.4	91.6	98.9	97.2	96.8	96.9	92.3	99.4
125 Cuba	127.9	120.9	124.4	108.4	104.4	126.7	121.0	122.2	105.3	100.9
126 German Democratic Republic	67.3	107.4	105.3	107.6	101.7	66.5	105.6	103.7	106.1	100.7
127 Mongolia	97.8	112.9	107.9	106.4	103.3	97.5	112.9	104.9	104.9	103.9
128 Democratic People's Republic of Korea	—	—	—	—	—	—	—	—	—	—
129 Union of Soviet Socialist Republics	67.4	104.1	97.1	104.0	105.3	67.4	104.1	99.7	107.1	107.9

— Not available.

Note: m, median; u, unweighted mean; w, weighted mean; and w', weighted mean excluding China and India.

Sources: United Nations 1988; Unesco data.

Table A-5. Distribution of Total Primary School Enrollment by Grade, Latest Available Year, 1981–86

Economy	Number of grades in primary system	Year	Distribution (percent) Grade 1	Grade 2	Grade 3	Grade 4	Grade 5	Grade 6	Grade 7	Grade 8	Grade 9
Low-income economies											
1 Ethiopia	6	1986	36.5	17.3	13.8	11.7	10.4	10.3	n.a.	n.a.	n.a.
2 Bhutan	5	1986	41.6	22.1	15.3	12.1	8.9	n.a.	n.a.	n.a.	n.a.
3 Burkina Faso	6	1986	22.6	20.6	19.1	13.3	11.3	13.1	n.a.	n.a.	n.a.
4 Nepal	5	1984	41.7	19.2	15.3	13.3	10.4	n.a.	n.a.	n.a.	n.a.
5 Bangladesh	5	1986	44.9	20.1	15.0	12.0	8.1	n.a.	n.a.	n.a.	n.a.
6 Malawi	8	1986	26.0	18.6	14.2	10.1	7.6	7.7	6.1	9.5	n.a.
7 Zaire	6	1983	25.7	16.5	16.9	15.1	13.7	12.1	n.a.	n.a.	n.a.
8 Mali	6	1986	25.4	20.0	19.0	14.4	11.5	9.8	n.a.	n.a.	n.a.
9 Myanmar	—	—	—	—	—	—	—	—	—	—	—
10 Mozambique	4	1986	37.9	26.5	20.2	15.4	n.a.	n.a.	n.a.	n.a.	n.a.
11 Madagascar	5	1983	33.3	21.1	18.5	14.6	12.5	n.a.	n.a.	n.a.	n.a.
12 Uganda	7	1986	23.8	18.5	16.7	13.6	11.1	9.0	7.3	n.a.	n.a.
13 Burundi	6	1986	25.4	19.1	16.5	14.7	14.1	10.1	n.a.	n.a.	n.a.
14 Tanzania	7	1986	16.6	15.7	14.9	14.5	13.0	12.9	12.4	n.a.	n.a.
15 Togo	6	1986	29.3	19.4	17.4	12.6	11.2	10.1	n.a.	n.a.	n.a.
16 Niger	6	1986	20.5	18.7	16.5	12.9	13.1	18.4	n.a.	n.a.	n.a.
17 Benin	6	1985	24.9	18.9	17.3	13.5	12.5	12.9	n.a.	n.a.	n.a.
18 Somalia	8	1985	17.0	13.6	13.3	12.0	11.8	10.6	10.0	11.7	n.a.
19 Central African Republic	6	1986	23.0	18.9	20.0	14.6	11.6	11.9	n.a.	n.a.	n.a.
20 India	5	1983	30.9	21.1	18.6	15.8	13.5	n.a.	n.a.	n.a.	n.a.
21 Rwanda	8	1986	24.8	17.5	13.9	12.0	10.0	8.4	7.1	6.2	n.a.
22 China	5	1986	22.6	20.0	19.8	19.5	18.2	n.a.	n.a.	n.a.	n.a.

— Not available.
n.a. Not applicable.
Source: Unesco data.

(Table continues on the following page.)

Table A-5 (continued)

Economy	Number of grades in primary system	Year	Disbribution (percent) Grade 1	Grade 2	Grade 3	Grade 4	Grade 5	Grade 6	Grade 7	Grade 8	Grade 9
23 Kenya	8	1985	18.0	14.9	13.7	13.2	11.9	10.8	9.7	7.7	n.a.
24 Zambia	7	1986	15.5	16.5	15.2	14.4	12.7	12.4	13.3	n.a.	n.a.
25 Sierra Leone	—	—	—	—	—	—	—	—	—	—	—
26 Sudan	6	1985	21.5	19.1	17.3	15.5	13.8	12.7	n.a.	n.a.	n.a.
27 Haiti	6	1985	38.3	18.6	15.6	12.0	9.1	6.4	n.a.	n.a.	n.a.
28 Pakistan	5	1986	36.3	18.1	17.8	15.2	12.6	n.a.	n.a.	n.a.	n.a.
29 Lesotho	7	1985	26.3	18.9	15.8	12.7	10.4	8.3	7.5	n.a.	n.a.
30 Ghana	6	1986	21.7	18.1	16.7	15.4	14.3	13.7	n.a.	n.a.	n.a.
31 Sri Lanka	6	1986	18.5	18.0	17.6	16.5	15.3	14.0	n.a.	n.a.	n.a.
32 Mauritania	6	1986	18.6	19.8	16.1	15.2	14.0	16.3	n.a.	n.a.	n.a.
33 Senegal	6	1985	17.6	18.5	18.0	15.9	14.1	16.0	n.a.	n.a.	n.a.
34 Afghanistan	8	1986	21.0	16.7	14.1	12.2	10.8	9.2	8.6	7.5	n.a.
35 Chad	6	1986	40.9	20.8	16.0	9.8	6.5	5.9	n.a.	n.a.	n.a.
36 Guinea	6	1986	20.8	20.4	24.6	12.2	10.4	11.6	n.a.	n.a.	n.a.
37 Democratic Kampuchea	—	—	—	—	—	—	—	—	—	—	—
38 Lao People's Democratic Republic	5	1985	45.1	21.8	14.5	10.4	8.1	n.a.	n.a.	n.a.	n.a.
39 Viet Nam	5	1985	28.7	21.9	18.9	16.6	14.0	n.a.	n.a.	n.a.	n.a.
Lower-middle-income economies											
40 Liberia	6	1984	25.9	20.0	17.7	14.6	11.9	9.9	n.a.	n.a.	n.a.
41 People's Democratic Republic of Yemen	8	1983	18.8	16.6	14.2	12.3	11.6	10.1	8.8	7.6	n.a.
42 Indonesia	6	1986	19.6	17.9	17.7	16.5	15.0	13.3	n.a.	n.a.	n.a.
43 Arab Republic of Yemen	6	1986	24.8	20.8	18.3	14.1	11.1	10.9	n.a.	n.a.	n.a.
44 Philippines	6	1986	21.1	18.8	16.9	15.7	14.6	12.9	n.a.	n.a.	n.a.

45	Morocco	5	1986	22.4	21.8	17.6	18.3	19.9	n.a.	n.a.	n.a.	n.a.
46	Bolivia	8	1986	23.0	17.6	15.0	12.2	10.3	8.5	7.4	6.0	n.a.
47	Zimbabwe	7	1986	15.7	13.8	13.9	14.1	14.7	15.1	12.8	n.a.	n.a.
48	Nigeria	6	1983	20.7	17.7	17.1	16.7	14.9	12.9	n.a.	n.a.	n.a.
49	Dominican Republic	6	1986	31.6	17.3	15.4	13.8	12.1	9.9	n.a.	n.a.	n.a.
50	Papua New Guinea	6	1982	21.9	19.9	16.5	16.3	12.7	12.6	n.a.	n.a.	n.a.
51	Côte d'Ivoire	6	1984	19.3	16.4	15.7	14.3	13.8	20.5	n.a.	n.a.	n.a.
52	Honduras	6	1986	32.0	20.4	16.5	12.7	10.2	8.2	n.a.	n.a.	n.a.
53	Egypt	6	1986	19.1	18.7	16.6	16.8	14.4	14.4	n.a.	n.a.	n.a.
54	Nicaragua	6	1986	35.7	17.3	15.8	13.2	10.9	7.1	n.a.	n.a.	n.a.
55	Thailand	6	1986	18.8	17.0	16.6	16.2	15.8	15.6	n.a.	n.a.	n.a.
56	El Salvador	9	1984	26.0	17.2	13.5	10.8	8.2	7.0	6.7	5.6	5.0
57	Botswana	7	1986	16.5	15.3	14.4	14.2	12.4	12.6	14.5	n.a.	n.a.
59	Cameroon	6	1986	25.3	17.2	18.3	14.0	13.0	12.3	n.a.	n.a.	n.a.
60	Guatemala	6	1986	31.3	20.5	17.1	12.9	9.9	8.4	n.a.	n.a.	n.a.
61	Congo	6	1986	22.7	15.5	20.3	16.6	13.9	11.0	n.a.	n.a.	n.a.
62	Paraguay	6	1986	24.5	20.6	17.8	14.9	12.3	9.9	n.a.	n.a.	n.a.
63	Peru	6	1985	25.7	18.5	16.0	14.6	13.5	11.8	n.a.	n.a.	n.a.
64	Turkey	5	1986	22.1	19.8	19.7	20.9	17.6	n.a.	n.a.	n.a.	n.a.
65	Tunisia	6	1986	17.4	17.9	17.4	16.4	15.6	15.4	n.a.	n.a.	n.a.
66	Ecuador	6	1984	23.5	18.3	16.6	15.4	13.8	12.4	n.a.	n.a.	n.a.
67	Mauritius	6	1986	16.9	17.0	10.1	15.6	14.9	25.6	n.a.	n.a.	n.a.
68	Colombia	5	1986	29.1	21.6	19.0	16.2	14.0	n.a.	n.a.	n.a.	n.a.
69	Chile	8	1986	13.8	13.4	11.4	11.8	13.3	13.2	12.2	10.9	n.a.
70	Costa Rica	6	1986	22.9	19.1	17.0	15.5	13.5	12.1	n.a.	n.a.	n.a.
71	Jordan	6	1985	19.1	17.0	16.0	16.5	16.1	15.2	n.a.	n.a.	n.a.
72	Syrian Arab Republic	6	1986	21.4	18.3	17.0	16.0	14.1	13.2	n.a.	n.a.	n.a.
73	Lebanon	—	—	—	—	—	—	—	—	—	—	—

— Not available.
n.a. Not applicable.
Source: Unesco data.

(*Table continues on the following page.*)

Table A-5 (continued)

Economy	Number of grades in primary system	Year	Disbribution (percent) Grade 1	Grade 2	Grade 3	Grade 4	Grade 5	Grade 6	Grade 7	Grade 8	Grade 9
Upper-middle-income economies											
74 Brazil	8	1985	27.3	18.1	13.6	11.0	11.1	7.9	6.2	4.8	n.a.
75 Malaysia	6	1985	17.4	17.3	17.2	16.5	16.1	15.4	n.a.	n.a.	n.a.
76 South Africa	—	—	—	—	—	—	—	—	—	—	—
77 Mexico	6	1986	21.4	17.8	16.8	15.9	14.8	13.3	n.a.	n.a.	n.a.
78 Uruguay	6	1986	18.9	17.5	17.0	16.6	15.8	14.1	n.a.	n.a.	n.a.
79 Hungary	8	1986	11.9	12.0	12.5	13.2	13.8	14.1	12.1	10.4	n.a.
80 Poland	8	1986	13.7	13.1	13.0	13.3	12.8	12.2	11.3	10.5	n.a.
81 Portugal	6	1985	34.6	0.0	33.9	0.0	16.5	15.0	n.a.	n.a.	n.a.
82 Yugoslavia	4	1985	25.3	24.8	24.9	25.0	n.a.	n.a.	n.a.	n.a.	n.a.
83 Panama	6	1986	21.1	18.4	16.9	15.6	14.6	13.4	n.a.	n.a.	n.a.
84 Argentina	7	1981	19.6	16.5	15.3	14.2	12.9	11.5	10.0	n.a.	n.a.
85 Republic of Korea	6	1986	17.7	15.8	16.1	15.5	16.7	18.1	n.a.	n.a.	n.a.
86 Algeria	6	1986	19.6	18.2	16.9	16.3	14.7	14.3	n.a.	n.a.	n.a.
87 Venezuela	6	1986	22.4	18.5	17.3	15.9	14.0	11.9	n.a.	n.a.	n.a.
88 Gabon	6	1986	31.6	18.2	16.6	11.8	10.4	11.4	n.a.	n.a.	n.a.
89 Greece	6	1985	16.7	16.4	16.3	16.6	16.5	17.5	n.a.	n.a.	n.a.
90 Spain	5	1986	18.3	19.7	19.8	20.1	22.1	n.a.	n.a.	n.a.	n.a.
91 Oman	6	1986	21.3	20.2	18.5	17.1	13.0	10.0	n.a.	n.a.	n.a.
92 Ireland	6	1985	17.3	17.0	16.3	16.2	16.2	16.9	n.a.	n.a.	n.a.
93 Trinidad and Tobago	7	1985	14.6	14.7	14.5	13.4	13.1	11.9	17.8	n.a.	n.a.
94 Israel	8	1986	12.7	12.3	12.7	12.7	13.4	12.8	11.9	11.5	n.a.
95 Hong Kong	6	1984	16.1	16.4	16.6	17.0	17.4	16.5	n.a.	n.a.	n.a.
96 Singapore	6	1984	14.7	16.5	15.3	16.4	18.1	19.0	n.a.	n.a.	n.a.
97 Islamic Republic of Iran	5	1986	25.9	21.4	18.2	17.8	16.7	n.a.	n.a.	n.a.	n.a.

98 Iraq	6	1986	19.7	18.6	17.1	15.7	16.1	12.8	n.a.	n.a.	n.a.
99 Romania	—	—	—	—	—	—	—	—	—	—	—
High-income oil exporters											
100 Saudi Arabia	6	1986	22.5	18.9	17.4	15.9	14.0	11.4	n.a.	n.a.	n.a.
101 Kuwait	4	1986	27.1	24.9	24.7	23.3	n.a.	n.a.	n.a.	n.a.	n.a.
102 United Arab Emirates	6	1986	22.0	19.6	17.4	15.6	13.6	11.8	n.a.	n.a.	n.a.
103 Libya	6	1982	18.4	17.0	16.3	17.9	16.4	14.0	n.a.	n.a.	n.a.
High-income industrial market economies											
104 New Zealand	6	1986	16.3	19.1	15.6	15.9	16.6	16.5	n.a.	n.a.	n.a.
105 Italy	5	1984	17.4	19.1	20.1	21.2	22.2	n.a.	n.a.	n.a.	n.a.
106 United Kingdom	—	—	—	—	—	—	—	—	—	—	—
107 Belgium	6	1986	18.2	17.0	16.5	16.4	16.3	15.6	n.a.	n.a.	n.a.
108 Austria	4	1986	26.0	24.8	24.5	24.7	n.a.	n.a.	n.a.	n.a.	n.a.
109 Netherlands	6	1984	16.9	16.0	15.9	16.3	16.8	18.1	n.a.	n.a.	n.a.
110 France	5	1986	21.9	19.9	19.3	19.5	19.4	n.a.	n.a.	n.a.	n.a.
111 Australia	6	1986	16.6	16.4	16.3	16.4	16.8	17.4	n.a.	n.a.	n.a.
112 Federal Republic of Germany	4	1986	26.2	24.6	24.5	24.7	n.a.	n.a.	n.a.	n.a.	n.a.
113 Finland	6	1986	17.0	16.5	16.9	17.0	16.7	15.9	n.a.	n.a.	n.a.
114 Denmark	6	1986	15.3	15.8	15.9	16.8	18.2	17.9	n.a.	n.a.	n.a.
115 Japan	6	1986	15.2	15.8	16.3	16.7	17.5	18.6	n.a.	n.a.	n.a.
116 Sweden	6	1986	16.1	15.6	16.1	16.4	17.4	18.4	n.a.	n.a.	n.a.
117 Canada	6	1986	17.8	16.8	16.4	16.4	16.2	16.4	n.a.	n.a.	n.a.
118 Norway	6	1986	16.0	16.0	15.8	16.5	17.4	18.5	n.a.	n.a.	n.a.
119 United States	8	1986	13.9	12.9	12.8	12.2	12.4	11.8	12.1	11.9	n.a.
120 Switzerland	6	1986	18.7	18.8	18.7	18.8	14.0	10.9	n.a.	n.a.	n.a.

— Not available.
n.a. Not applicable.
Source: Unesco data.

(Table continues on the following page.)

Table A-5 (continued)

Economy	Number of grades in primary system	Year	Distribution (percent) Grade 1	Grade 2	Grade 3	Grade 4	Grade 5	Grade 6	Grade 7	Grade 8	Grade 9
Other economies											
121 Angola	4	1984	37.7	30.9	19.5	11.9	n.a.	n.a.	n.a.	n.a.	n.a.
122 Albania	8	1986	12.2	12.8	12.6	12.7	13.3	12.5	12.2	11.7	n.a.
123 Bulgaria	8	1986	13.4	12.8	12.8	12.9	12.6	13.4	11.9	10.2	n.a.
124 Czechoslovakia	8	1986	12.1	12.6	12.7	12.8	12.9	12.9	12.7	11.4	n.a.
125 Cuba	6	1986	13.5	16.0	14.7	17.4	18.9	19.4	n.a.	n.a.	n.a.
126 German Democratic Republic	4	1986	26.6	25.8	25.0	22.5	n.a.	n.a.	n.a.	n.a.	n.a.
127 Mongolia	3	1986	35.4	32.4	32.2	n.a.	n.a.	n.a.	n.a.	n.a.	n.a.
128 Democratic People's Republic of Korea	—	—	—	—	—	—	—	—	—	—	—
129 Union of Soviet Socialist Republics	5	1985	24.6	19.4	18.9	18.6	18.5	n.a.	n.a.	n.a.	n.a.

— Not available.
n.a. Not applicable.
Source: Unesco data.

Table A-6. Distribution of Enrollment in Grades 1–5, Latest Available Year, 1982–86

Economy	Year	Distribution (percent)				
		Grade 1	Grade 2	Grade 3	Grade 4	Grade 5
Low-income economies		**28.4 *m***	**21.7 *m***	**19.2 *m***	**15.8 *m***	**13.3 *m***
		30.3 *u*	**21.4 *u***	**19.1 *u***	**15.7 *u***	**13.6 *u***
		31.7 *w*	**20.9 *w***	**18.4 *w***	**15.7 *w***	**13.3 *w***
Excluding China and India		**32.8 *w′***	**20.5 *w′***	**18.1 *w′***	**15.5 *w′***	**13.1 *w′***
1 Ethiopia	1986	40.7	19.3	15.4	13.0	11.6
2 Bhutan	1986	41.6	22.1	15.3	12.1	8.9
3 Burkina Faso	1986	26.0	23.7	22.0	15.3	13.0
4 Nepal	1984	41.7	19.2	15.3	13.3	10.4
5 Bangladesh	1986	44.9	20.1	15.0	12.0	8.1
6 Malawi	1986	33.9	24.3	18.6	13.2	10.0
7 Zaire	1983	29.3	18.8	19.2	17.2	15.5
8 Mali	1986	28.1	22.2	21.0	15.9	12.7
9 Myanmar	—	—	—	—	—	—
10 Mozambique	—	—	—	—	—	—
11 Madagascar	1983	33.3	21.1	18.5	14.6	12.5
12 Uganda	1986	28.4	22.1	20.0	16.3	13.3
13 Burundi	1986	28.3	21.3	18.4	16.3	15.7
14 Tanzania	1986	22.2	21.0	20.0	19.4	17.5
15 Togo	1986	32.6	21.6	19.4	14.0	12.5
16 Niger	1986	25.1	22.9	20.2	15.8	16.0
17 Benin	1985	28.6	21.7	19.9	15.5	14.3
18 Somalia	1985	25.2	20.1	19.6	17.7	17.4
19 Central African Republic	1986	26.1	21.4	22.7	16.6	13.2
20 India	1983	30.9	21.1	18.6	15.8	13.5
21 Rwanda	1986	31.6	22.4	17.8	15.4	12.8

— Not available.

Note: *m*, median; *u*, unweighted mean; *w*, weighted mean; and *w′*, weighted mean excluding China and India.

Source: Unesco data.

(*Table continues on the following page.*)

Table A-6 (continued)

Economy	Year	Distribution (percent)				
		Grade 1	Grade 2	Grade 3	Grade 4	Grade 5
22 China	1986	22.6	20.0	19.8	19.5	18.2
23 Kenya	1985	25.1	20.8	19.1	18.4	16.6
24 Zambia	1986	20.9	22.2	20.5	19.3	17.1
25 Sierra Leone	—	—	—	—	—	—
26 Sudan	1985	24.6	21.9	19.9	17.8	15.8
27 Haiti	1985	40.9	19.9	16.7	12.8	9.7
28 Pakistan	1986	36.3	18.1	17.8	15.2	12.6
29 Lesotho	1985	31.3	22.4	18.8	15.1	12.4
30 Ghana	1986	25.1	21.0	19.4	17.9	16.6
31 Sri Lanka	1986	21.5	20.9	20.5	19.2	17.8
32 Mauritania	1986	22.3	23.6	19.3	18.1	16.7
33 Senegal	1985	20.9	22.0	21.4	18.9	16.7
34 Afghanistan	1986	28.1	22.3	18.8	16.3	14.4
35 Chad	1986	43.5	22.1	17.0	10.5	6.9
36 Guinea	1986	23.5	23.1	27.9	13.7	11.8
37 Democratic Kampuchea	—	—	—	—	—	—
38 Lao People's Democratic Republic	1985	45.1	21.8	14.5	10.4	8.1
39 Viet Nam	1985	28.7	21.9	18.9	16.6	14.0
Lower-middle-income economies		**25.1** *m*	**21.1** *m*	**19.4** *m*	**18.1** *m*	**15.8** *m*
		26.3 *u*	**21.1** *u*	**19.1** *u*	**17.6** *u*	**15.9** *u*
		24.2 *w*	**20.8** *w*	**19.6** *w*	**18.5** *w*	**16.8** *w*
40 Liberia	1984	28.8	22.2	19.6	16.2	13.2
41 People's Democratic Republic of Yemen	1983	25.6	22.5	19.4	16.7	15.8
42 Indonesia	1986	22.6	20.7	20.4	19.0	17.3
43 Arab Republic of Yemen	1986	27.9	23.3	20.6	15.9	12.4
44 Philippines	1986	24.2	21.6	19.4	18.1	16.8

45	Morocco	1986	22.4	21.8	17.6	18.3	19.9
46	Bolivia	1986	29.4	22.6	19.2	15.6	13.2
47	Zimbabwe	1986	21.7	19.2	19.2	19.5	20.4
48	Nigeria	1983	23.8	20.3	19.7	19.2	17.1
49	Dominican Republic	1986	35.1	19.2	17.1	15.3	13.4
50	Papua New Guinea	1982	25.1	22.8	18.9	18.6	14.5
51	Côte d'Ivoire	1984	24.2	20.6	19.7	18.0	17.4
52	Honduras	1986	34.9	22.2	17.9	13.8	11.1
53	Egypt	1986	22.3	21.9	19.4	19.6	16.8
54	Nicaragua	1986	38.4	18.6	17.0	14.2	11.7
55	Thailand	1986	22.3	20.2	19.6	19.2	18.8
56	El Salvador	1984	34.3	22.7	17.8	14.3	10.8
57	Botswana	1986	22.7	21.0	19.8	19.4	17.0
58	Jamaica	1986	20.5	20.1	20.3	20.3	18.8
59	Cameroon	1986	28.8	19.6	20.8	15.9	14.8
60	Guatemala	1986	34.1	22.4	18.6	14.0	10.8
61	Congo	1986	25.5	17.4	22.8	18.6	15.6
62	Paraguay	1986	27.1	22.8	19.8	16.6	13.7
63	Peru	1985	29.1	21.0	18.1	16.5	15.3
64	Turkey	1986	22.1	19.8	19.7	20.9	17.6
65	Tunisia	1986	20.6	21.2	20.5	19.4	18.4
66	Ecuador	1984	26.9	20.9	19.0	17.6	15.7
67	Mauritius	1986	22.7	22.8	13.5	21.0	20.0
68	Colombia	1986	29.1	21.6	19.0	16.2	14.0
69	Chile	1986	21.7	21.1	17.9	18.5	20.8
70	Costa Rica	1986	26.0	21.7	19.3	17.6	15.4
71	Jordan	1985	22.5	20.1	18.9	19.5	19.0
72	Syrian Arab Republic	1986	24.6	21.1	19.6	18.4	16.2
73	Lebanon	—	—	—	—	—	—

— Not available.
Note: m, median; u, unweighted mean; w, weighted mean; and w', weighted mean excluding China and India.
Source: Unesco data.

(Table continues on the following page.)

Table A-6 (*continued*)

Economy		Distribution (percent)				
		Grade 1	*Grade 2*	*Grade 3*	*Grade 4*	*Grade 5*
Upper-middle-income economies		**22.1 *m***	**20.5 *m***	**19.7 *m***	**19.0 *m***	**18.7 *m***
		23.7 *u*	**19.7 *u***	**20.4 *u***	**17.9 *u***	**18.3 *u***
		26.5 *w*	**20.8 *w***	**18.9 *w***	**16.9 *w***	**16.8 *w***
74 Brazil	1985	33.6	22.3	16.8	13.6	13.7
75 Malaysia	1985	20.6	20.5	20.3	19.5	19.1
76 South Africa	—	—	—	—	—	—
77 Mexico	1986	24.6	20.5	19.4	18.4	17.1
78 Uruguay	1986	22.1	20.3	19.8	19.3	18.5
79 Hungary	1986	18.7	18.9	19.7	20.9	21.8
80 Poland	1986	20.7	19.9	19.8	20.1	19.5
81 Portugal	1985	40.7	0.0	39.9	0.0	19.4
82 Yugoslavia	—	—	—	—	—	—
83 Panama	1986	24.4	21.3	19.6	18.0	16.9
84 Argentina	1981	25.0	21.0	19.5	18.1	16.4
85 Republic of Korea	1986	21.7	19.3	19.6	18.9	20.5
86 Algeria	1986	22.9	21.3	19.7	19.0	17.1
87 Venezuela	1986	25.4	21.0	19.7	18.1	15.9
88 Gabon	1986	35.7	20.5	18.8	13.3	11.8
89 Greece	1985	20.2	19.8	19.8	20.1	20.0
90 Spain	1986	18.3	19.7	19.8	20.1	22.1
91 Oman	1986	23.6	22.4	20.6	19.0	14.5
92 Ireland	1985	20.9	20.5	19.7	19.5	19.5
93 Trinidad and Tobago	1985	20.8	20.9	20.6	19.0	18.7
94 Israel	1986	19.9	19.2	19.8	20.0	21.0
95 Hong Kong	1984	19.2	19.6	19.9	20.4	20.8
96 Singapore	1984	18.2	20.4	18.9	20.3	22.3
97 Islamic Republic of Iran	1986	25.9	21.4	18.2	17.8	16.7
98 Iraq	1986	22.6	21.3	19.6	18.0	18.5

99	Romania	—	—	—	—	—	—
High-income oil exporters			**25.0 *m* 23.9 *u* 24.2 *w***	**21.3 *m* 21.1 *u* 20.9 *w***	**19.6 *m* 19.4 *u* 19.4 *w***	**18.0 *m* 18.8 *u* 18.8 *w***	**15.8 *m* 16.8 *u* 16.7 *w***
100	Saudi Arabia	1986	25.4	21.3	19.6	18.0	15.8
101	Kuwait	—	—	—	—	—	—
102	United Arab Emirates	1986	25.0	22.2	19.7	17.7	15.4
103	Libya	1982	21.4	19.7	18.9	20.8	19.1
High-income industrial market economies			**20.2 *m* 20.1 *u* 20.5 *w***	**19.8 *m* 20.0 *u* 19.8 *w***	**19.7 *m* 19.7 *u* 19.8 *w***	**20.0 *m* 20.0 *u* 19.7 *w***	**20.1 *m* 20.2 *u* 20.1 *w***
104	New Zealand	1986	19.5	22.9	18.7	19.0	19.9
105	Italy	1984	17.4	19.1	20.1	21.2	22.2
106	United Kingdom	—	—	—	—	—	—
107	Belgium	1986	21.6	20.1	19.5	19.5	19.3
108	Austria	—	—	—	—	—	—
109	Netherlands	1984	20.7	19.5	19.4	19.9	20.6
110	France	1986	21.9	19.9	19.3	19.5	19.4
111	Australia	1986	20.1	19.8	19.8	19.9	20.3
112	Federal Republic of Germany	—	—	—	—	—	—
113	Finland	1986	20.2	19.7	20.1	20.3	19.8
114	Denmark	1986	18.6	19.3	19.4	20.5	22.2
115	Japan	1986	18.7	19.4	20.0	20.5	21.5
116	Sweden	1986	19.7	19.1	19.7	20.1	21.3
117	Canada	1986	21.3	20.1	19.6	19.6	19.4
118	Norway	1986	19.6	19.6	19.3	20.2	21.3
119	United States	1986	21.7	20.1	19.9	19.0	19.3
120	Switzerland	1986	21.0	21.1	21.0	21.1	15.7

— Not available.
Note: *m*, median; *u*, unweighted mean; *w*, weighted mean; and *w′*, weighted mean excluding China and India.
Source: Unesco data.

(Table continues on the following page.)

Table A-6 (*continued*)

Economy	Year	Distribution (percent)				
		Grade 1	*Grade 2*	*Grade 3*	*Grade 4*	*Grade 5*
Other economies						
121 Angola	—	—	—	—	—	—
122 Albania	1986	19.2	20.1	19.8	20.0	20.9
123 Bulgaria	1986	20.7	19.8	19.9	20.0	19.5
124 Czechoslovakia	1986	19.2	20.0	20.1	20.3	20.4
125 Cuba	1986	16.8	19.9	18.3	21.6	23.4
126 German Democratic Republic	—	—	—	—	—	—
127 Mongolia	—	—	—	—	—	—
128 Democratic People's Republic of Korea	—	—	—	—	—	—
129 Union of Soviet Socialist Republics	1985	24.6	19.4	18.9	18.6	18.5

— Not available.

Note: m, median; u, unweighted mean; w, weighted mean; and w', weighted mean excluding China and India.

Source: Unesco data.

Table A-7. Ratio of Female to Male Students in Primary School by Grade, Latest Available Year, 1980–86

Economy	Number of grades in primary system	Year	Ratio								
			Grade 1	Grade 2	Grade 3	Grade 4	Grade 5	Grade 6	Grade 7	Grade 8	Grade 9
Low-income economies			**0.75 *m***	**0.71 *m***	**0.68 *m***	**0.66 *m***	**0.65 *m***	**0.64 *m***	**0.73 *m***	**0.60 *m***	**n.a.**
			0.75 *u*	**0.73 *u***	**0.71 *u***	**0.70 *u***	**0.69 *u***	**0.69 *u***	**0.85 *u***	**0.62 *u***	**n.a.**
			0.70 *w*	**0.70 *w***	**0.66 *w***	**0.65 *w***	**0.63 *w***	**0.76 *w***	**0.83 *w***	**0.69 *w***	**n.a.**
Excluding China and India			**0.75 *w′***	**0.76 *w′***	**0.71 *w′***	**0.70 *w′***	**0.71 *w′***	**0.76 *w′***	**0.83 *w′***	**0.69 *w′***	**n.a.**
1 Ethiopia	6	1986	0.65	0.60	0.59	0.61	0.61	0.67	n.a.	n.a.	n.a.
2 Bhutan	5	1986	0.58	0.53	0.55	0.48	0.47	n.a.	n.a.	n.a.	n.a.
3 Burkina Faso	6	1986	0.59	0.59	0.59	0.59	0.59	0.56	n.a.	n.a.	n.a.
4 Nepal	5	1984	0.45	0.40	0.40	0.37	0.37	n.a.	n.a.	n.a.	n.a.
5 Bangladesh	5	1986	0.76	0.64	0.60	0.57	0.52	n.a.	n.a.	n.a.	n.a.
6 Malawi	8	1986	0.91	0.85	0.83	0.81	0.75	0.75	0.68	0.45	n.a.
7 Zaire	6	1983	0.77	0.94	0.69	0.70	0.70	0.66	n.a.	n.a.	n.a.
8 Mali	6	1986	0.60	0.59	0.60	0.61	0.58	0.55	n.a.	n.a.	n.a.
9 Myanmar	—	—	—	—	—	—	—	—	—	—	—
10 Mozambique	4	1986	0.87	0.78	0.73	0.66	n.a.	n.a.	n.a.	n.a.	n.a.
11 Madagascar	5	1982	0.91	0.88	0.82	0.87	0.82	n.a.	n.a.	n.a.	n.a.
12 Uganda	7	1986	0.90	0.88	0.85	0.83	0.79	0.73	0.60	n.a.	n.a.
13 Burundi	6	1986	0.83	0.78	0.76	0.73	0.66	0.63	n.a.	n.a.	n.a.
14 Tanzania	7	1986	0.97	0.98	1.00	1.05	1.00	1.02	0.99	n.a.	n.a.
15 Togo	6	1986	0.68	0.67	0.65	0.61	0.55	0.43	n.a.	n.a.	n.a.
16 Niger	6	1986	0.55	0.57	0.55	0.56	0.58	0.55	n.a.	n.a.	n.a.
17 Benin	6	1985	0.52	0.52	0.51	0.52	0.49	0.45	n.a.	n.a.	n.a.
18 Somalia	8	1985	0.54	0.53	0.51	0.46	0.47	0.48	0.53	0.60	n.a.
19 Central African Republic	6	1986	0.70	0.65	0.64	0.61	0.56	0.50	n.a.	n.a.	n.a.
20 India	5	1983	0.67	0.67	0.63	0.61	0.58	n.a.	n.a.	n.a.	n.a.

— Not available.
n.a. Not applicable.
Note: *m*, median; *u*, unweighted mean; *w*, weighted mean; and *w′*, weighted mean excluding China and India.
Source: Unesco data.

(Table continues on the following page.)

Table A-7 (continued)

Economy	Number of grades in primary system	Year	Ratio								
			Grade 1	Grade 2	Grade 3	Grade 4	Grade 5	Grade 6	Grade 7	Grade 8	Grade 9
21 Rwanda	8	1986	0.99	0.98	0.98	1.01	1.00	0.97	0.88	0.83	n.a.
22 China	5	1986	0.86	0.85	0.83	0.80	0.76	n.a.	n.a.	n.a.	n.a.
23 Kenya	8	1985	0.94	0.93	0.94	0.97	0.98	0.95	0.89	0.79	n.a.
24 Zambia	7	1986	0.98	0.96	0.94	0.92	0.88	0.84	0.73	n.a.	n.a.
25 Sierra Leone	—	—	—	—	—	—	—	—	—	—	—
26 Sudan	6	1985	0.65	0.66	0.67	0.68	0.70	0.74	n.a.	n.a.	n.a.
27 Haiti	6	1985	0.86	0.90	0.86	0.89	0.90	0.85	n.a.	n.a.	n.a.
28 Pakistan	5	1986	0.55	0.61	0.46	0.39	0.42	n.a.	n.a.	n.a.	n.a.
29 Lesotho	7	1985	1.03	1.08	1.22	1.34	1.54	1.74	1.88	n.a.	n.a.
30 Ghana	6	1984	0.85	0.80	0.77	0.77	0.74	0.71	n.a.	n.a.	n.a.
31 Sri Lanka	6	1986	0.94	0.93	0.92	0.93	0.93	0.95	n.a.	n.a.	n.a.
32 Mauritania	6	1986	0.74	0.71	0.68	0.64	0.62	0.56	n.a.	n.a.	n.a.
33 Senegal	6	1985	0.74	0.72	0.68	0.67	0.65	0.59	n.a.	n.a.	n.a.
34 Afghanistan	8	1986	0.58	0.52	0.50	0.50	0.47	0.45	0.45	0.44	n.a.
35 Chad	6	1986	0.47	0.40	0.34	0.30	0.31	0.25	n.a.	n.a.	n.a.
36 Guinea	6	1986	0.50	0.50	0.38	0.46	0.44	0.39	n.a.	n.a.	n.a.
37 Democratic Kampuchea	—	—	—	—	—	—	—	—	—	—	—
38 Lao People's Democratic Republic	5	1985	0.84	0.77	0.82	0.84	0.74	n.a.	n.a.	n.a.	n.a.
39 Viet Nam	5	1985	0.88	0.91	0.93	0.90	0.95	n.a.	n.a.	n.a.	n.a.
Lower-middle-income economies			**0.91 *m***	**0.93 *m***	**0.92 *m***	**0.92 *m***	**0.92 *m***	**0.92 *m***	**0.91 *m***	**0.88 *m***	**1.02 *m***
			0.86 *u*	**0.87 *u***	**0.86 *u***	**0.86 *u***	**0.86 *u***	**0.85 *u***	**0.87 *u***	**0.76 *u***	**1.02 *u***
			0.88 *w*	**0.88 *w***	**0.88 *w***	**0.88 *w***	**0.87 *w***	**0.85 *w***	**0.88 *w***	**0.89 *w***	**1.02 *w***
40 Liberia	6	1980	0.63	0.57	0.54	0.50	0.50	0.46	n.a.	n.a.	n.a.
41 People's Democratic Republic of Yemen	8	1983	0.43	0.37	0.36	0.33	0.33	0.34	0.34	0.29	n.a.

42	Indonesia	6	1986	0.93	0.93	0.93	0.93	0.93	0.91	n.a.	n.a.	n.a.
43	Arab Republic of Yemen	6	1986	0.36	0.32	0.28	0.23	0.20	0.15	n.a.	n.a.	n.a.
44	Philippines	6	1986	0.92	0.92	0.93	0.94	0.98	0.99	n.a.	n.a.	n.a.
45	Morocco	5	1986	0.66	0.64	0.59	0.59	0.61	n.a.	n.a.	n.a.	n.a.
46	Bolivia	8	1986	0.97	1.02	0.84	0.82	0.92	0.73	0.79	0.76	n.a.
47	Zimbabwe	7	1986	0.98	0.97	0.98	0.98	0.97	0.94	0.84	n.a.	n.a.
48	Nigeria	6	1983	0.81	0.80	0.78	0.79	0.77	0.75	n.a.	n.a.	n.a.
49	Dominican Republic	6	1986	0.88	0.94	0.99	1.00	1.04	1.08	n.a.	n.a.	n.a.
50	Papua New Guinea	6	1982	0.81	0.75	0.75	0.73	0.71	0.69	n.a.	n.a.	n.a.
51	Côte d'Ivoire	6	1984	0.78	0.77	0.76	0.72	0.70	0.53	n.a.	n.a.	n.a.
52	Honduras	6	1986	0.92	1.05	1.00	1.04	1.06	1.08	n.a.	n.a.	n.a.
53	Egypt	6	1986	0.81	0.80	0.79	0.76	0.75	0.72	n.a.	n.a.	n.a.
54	Nicaragua	6	1986	0.96	1.04	1.10	1.18	1.22	1.31	n.a.	n.a.	n.a.
55	Thailand	6	1980	0.92	0.92	0.93	0.94	0.95	0.93	n.a.	n.a.	n.a.
56	El Salvador	9	1984	0.96	0.97	0.99	0.99	1.01	1.04	1.01	1.01	1.02
57	Botswana	7	1986	0.99	1.02	1.03	1.02	1.10	1.17	1.26	n.a.	n.a.
58	Jamaica	6	1986	0.94	0.93	0.96	0.99	0.91	1.10	n.a.	n.a.	n.a.
59	Cameroon	6	1986	0.84	0.86	0.84	0.86	0.85	0.82	n.a.	n.a.	n.a.
60	Guatemala	6	1986	0.87	0.84	0.80	0.79	0.78	0.80	n.a.	n.a.	n.a.
61	Congo	6	1986	0.74	0.96	0.92	0.95	0.97	0.98	n.a.	n.a.	n.a.
62	Paraguay	6	1986	0.93	0.90	0.92	0.92	0.92	0.94	n.a.	n.a.	n.a.
63	Peru	6	1985	0.96	0.96	0.94	0.92	0.88	0.87	n.a.	n.a.	n.a.
64	Turkey	5	1986	0.89	0.90	0.89	0.90	0.89	n.a.	n.a.	n.a.	n.a.
65	Tunisia	6	1986	0.87	0.84	0.81	0.80	0.76	0.74	n.a.	n.a.	n.a.
66	Ecuador	6	1983	0.96	0.95	0.97	0.98	0.97	0.96	n.a.	n.a.	n.a.
67	Mauritius	6	1986	0.98	0.97	1.01	0.98	0.96	0.97	n.a.	n.a.	n.a.
68	Colombia	5	1986	0.95	0.94	1.07	1.15	0.94	n.a.	n.a.	n.a.	n.a.
69	Chile	8	1986	0.93	0.95	0.92	0.92	0.92	0.96	0.97	1.00	n.a.

— Not available.
n.a. Not applicable.
Note: m, median; u, unweighted mean; w, weighted mean; and w', weighted mean excluding China and India.
Source: Unesco data.

(*Table continues on the following page.*)

Table A-7 (continued)

Economy	Number of grades in primary system	Year	Ratio								
			Grade 1	Grade 2	Grade 3	Grade 4	Grade 5	Grade 6	Grade 7	Grade 8	Grade 9
70 Costa Rica	6	1986	0.91	0.93	0.94	0.95	0.95	0.97	n.a.	n.a.	n.a.
71 Jordan	6	1985	0.86	0.95	0.92	0.93	0.92	0.91	n.a.	n.a.	n.a.
72 Syrian Arab Republic	6	1986	0.91	0.89	0.87	0.86	0.83	0.78	n.a.	n.a.	n.a.
73 Lebanon	—	—	—	—	—	—	—	—	—	—	—
Upper-middle-income economies			**0.93 *m***	**0.94 *m***	**0.94 *m***	**0.95 *m***	**0.94 *m***	**0.95 *m***	**0.99 *m***	**0.98 *m***	**n.a.**
			0.92 *u*	**0.93 *u***	**0.93 *u***	**0.92 *u***	**0.91 *u***	**0.92 *u***	**1.00 *u***	**0.98 *u***	**n.a.**
			0.90 *w*	**0.90 *w***	**0.91 *w***	**0.90 *w***	**0.89 *w***	**0.93 *w***	**1.01 *w***	**0.98 *w***	**n.a.**
74 Brazil	—	—	—	—	—	—	—	—	—	—	—
75 Malaysia	6	1985	0.95	0.93	0.94	0.95	0.94	0.95	n.a.	n.a.	n.a.
76 South Africa	—	—	—	—	—	—	—	—	—	—	—
77 Mexico	6	1986	0.91	0.94	0.95	0.96	0.96	0.98	n.a.	n.a.	n.a.
78 Uruguay	6	1986	0.93	0.94	0.93	0.95	0.97	1.01	n.a.	n.a.	n.a.
79 Hungary	8	1986	0.95	0.95	0.95	0.95	0.94	0.95	0.97	0.98	n.a.
80 Poland	—	—	—	—	—	—	—	—	—	—	—
81 Portugal	6	1985	0.88	—	0.93	—	0.90	0.93	n.a.	n.a.	n.a.
82 Yugoslavia	4	1985	0.93	0.94	0.94	0.95	n.a.	n.a.	n.a.	n.a.	n.a.
83 Panama	6	1986	0.89	0.89	0.91	0.92	0.94	0.97	n.a.	n.a.	n.a.
84 Argentina	7	1981	0.93	0.94	0.96	0.97	0.98	1.01	1.04	n.a.	n.a.
85 Republic of Korea	6	1986	0.94	0.94	0.95	0.93	0.94	0.94	n.a.	n.a.	n.a.
86 Algeria	6	1986	0.83	0.82	0.80	0.78	0.76	0.71	n.a.	n.a.	n.a.
87 Venezuela	6	1986	0.89	0.93	0.95	0.98	1.00	1.06	n.a.	n.a.	n.a.
88 Gabon	6	1986	0.98	1.01	1.02	1.01	0.99	0.93	n.a.	n.a.	n.a.
89 Greece	6	1985	0.94	0.93	0.95	0.94	0.94	0.93	n.a.	n.a.	n.a.
90 Spain	5	1986	0.95	0.94	0.95	0.95	0.94	n.a.	n.a.	n.a.	n.a.
91 Oman	6	1986	0.93	0.95	0.90	0.74	0.68	0.62	n.a.	n.a.	n.a.
92 Ireland	6	1985	0.96	0.96	0.95	0.94	0.95	0.96	n.a.	n.a.	n.a.
93 Trinidad and Tobago	7	1985	0.99	0.99	0.96	0.97	0.98	1.03	1.02	n.a.	n.a.

94 Israel	8	1986	0.97	0.97	0.98	0.97	0.97	0.97	0.96	0.98	n.a.
95 Hong Kong	6	1984	0.93	0.90	0.90	0.88	0.93	0.93	n.a.	n.a.	n.a.
96 Singapore	6	1984	0.92	0.92	0.93	0.91	0.92	0.92	n.a.	n.a.	n.a.
97 Islamic Republic of Iran	5	1986	0.82	0.81	0.80	0.76	0.70	n.a.	n.a.	n.a.	n.a.
98 Iraq	6	1986	0.87	0.85	0.83	0.81	0.76	0.70	n.a.	n.a.	n.a.
99 Romania	—	—	—	—	—	—	—	—	—	—	—
High-income oil exporters			**0.94 *m***	**0.94 *m***	**0.94 *m***	**0.91 *m***	**0.86 *m***	**0.82 *m***	**n.a.**	**n.a.**	**n.a.**
			0.92 *u*	**0.91 *u***	**0.91 *u***	**0.89 *u***	**0.84 *u***	**0.85 *u***	**n.a.**	**n.a.**	**n.a.**
			0.88 *w*	**0.86 *w***	**0.87 *w***	**0.85 *w***	**0.80 *w***	**0.78 *w***	**n.a.**	**n.a.**	**n.a.**
100 Saudi Arabia	6	1986	0.84	0.81	0.83	0.80	0.75	0.74	n.a.	n.a.	n.a.
101 Kuwait	4	1986	0.95	0.96	0.94	0.93	n.a.	n.a.	n.a.	n.a.	n.a.
102 United Arab Emirates	6	1986	0.94	0.94	0.94	0.92	0.93	0.98	n.a.	n.a.	n.a.
103 Libya	6	1982	0.94	0.93	0.94	0.90	0.86	0.82	n.a.	n.a.	n.a.
High-income industrial market economies			**0.95 *m***	**0.95 *m***	**0.95 *m***	**0.96 *m***	**0.95 *m***	**0.96 *m***	**0.96 *m***	**0.93 *m***	**n.a.**
			0.94 *u*	**0.95 *u***	**0.95 *u***	**0.96 *u***	**0.96 *u***	**0.96 *u***	**0.96 *u***	**0.93 *u***	**n.a.**
			0.92 *w*	**0.96 *w***	**0.94 *w***	**0.96 *w***	**0.95 *w***	**0.93 *w***	**0.96 *w***	**0.93 *w***	**n.a.**
104 New Zealand	6	1986	0.94	0.93	0.98	0.96	0.95	0.95	n.a.	n.a.	n.a.
105 Italy	5	1984	0.94	0.95	0.95	0.95	0.95	n.a.	n.a.	n.a.	n.a.
106 United Kingdom	—	—	—	—	—	—	—	—	—	—	—
107 Belgium	6	1986	0.95	0.95	0.97	0.97	0.96	0.97	n.a.	n.a.	n.a.
108 Austria	4	1986	0.93	0.94	0.94	0.95	n.a.	n.a.	n.a.	n.a.	n.a.
109 Netherlands	6	1984	0.95	0.96	0.98	0.99	0.99	1.00	n.a.	n.a.	n.a.
110 France	5	1986	0.92	0.94	0.94	0.95	0.94	n.a.	n.a.	n.a.	n.a.
111 Australia	6	1986	0.93	0.94	0.95	0.95	0.95	0.95	n.a.	n.a.	n.a.
112 Federal Republic of Germany	4	1986	0.96	0.96	0.96	0.96	n.a.	n.a.	n.a.	n.a.	n.a.
113 Finland	6	1986	0.96	0.94	0.96	0.96	0.95	0.96	n.a.	n.a.	n.a.

— Not available.

n.a. Not applicable.

Note: *m*, median; *u*, unweighted mean; *w*, weighted mean; and *w′*, weighted mean excluding China and India.

Source: Unesco data.

(*Table continues on the following page.*)

Table A-7 (continued)

Economy	Number of grades in primary system	Year	Ratio								
			Grade 1	Grade 2	Grade 3	Grade 4	Grade 5	Grade 6	Grade 7	Grade 8	Grade 9
114 Denmark	6	1986	0.96	0.95	0.96	0.95	0.97	0.96	n.a.	n.a.	n.a.
115 Japan	6	1986	0.95	0.95	0.95	0.95	0.95	0.95	n.a.	n.a.	n.a.
116 Sweden	6	1983	0.95	0.96	0.95	0.96	0.95	0.95	n.a.	n.a.	n.a.
117 Canada	6	1986	0.91	0.93	0.94	0.94	0.94	0.94	n.a.	n.a.	n.a.
118 Norway	6	1986	0.95	0.96	0.95	0.96	0.97	0.97	n.a.	n.a.	n.a.
119 United States	8	1986	0.90	0.98	0.92	0.98	0.95	0.91	0.96	0.93	n.a.
120 Switzerland	6	1986	0.98	0.96	0.96	0.98	0.96	0.97	n.a.	n.a.	n.a.
Other economies											
121 Angola	4	1981	0.94	0.94	0.80	0.66	n.a.	n.a.	n.a.	n.a.	n.a.
122 Albania	8	1986	0.93	0.86	0.93	0.91	0.89	0.89	0.90	0.92	n.a.
123 Bulgaria	8	1986	0.94	0.95	0.96	0.93	0.95	0.93	0.95	0.91	n.a.
124 Czechoslovakia	8	1986	0.95	0.96	0.96	0.96	0.96	0.97	0.97	0.99	n.a.
125 Cuba	6	1986	0.93	0.84	0.90	0.88	0.91	0.91	n.a.	n.a.	n.a.
126 German Democratic Republic	4	1986	0.93	0.93	0.94	0.93	n.a.	n.a.	n.a.	n.a.	n.a.
127 Mongolia	3	1982	0.82	1.00	1.21	n.a.	n.a.	n.a.	n.a.	n.a.	n.a.
128 Democratic People's Republic of Korea	—	—	—	—	—	—	—	—	—	—	—
129 Union of Soviet Socialist Republics	—	—	—	—	—	—	—	—	—	—	—

— Not available.
n.a. Not applicable.
Note: m, median; u, unweighted mean; w, weighted mean; and w', weighted mean excluding China and India.
Source: Unesco data.

Table A-8. Number of Primary Schools and Student-School Ratio, Selected Years, 1965–85

Economy	Total primary schools (thousands)					Student-school ratio				
	1965	1970	1975	1980	1985	1965	1970	1975	1980	1985
Low-income economies	**2,180.1**	**1,530.8**	**1,732.1**	**1,648.6**	**1,610.9**	**169.0 *m***	**181.0 *m***	**209.0 *m***	**230.0 *m***	**236.0 *m***
						160.6 *u*	**209.3 *u***	**209.3 *u***	**257.9 *u***	**268.9 *u***
						84.8 *w*	**124.0 *w***	**144.4 *w***	**165.7 *w***	**168.3 *w***
Excluding China and India	**107.1**	**161.3**	**185.3**	**245.8**	**250.5**	**147.8 *w′***	**170.4 *w′***	**177.5 *w′***	**214.3 *w′***	**201.6 *w′***
1 Ethiopia	1.6	2.3	3.7	5.8	7.9	237.0	285.0	293.0	367.0	310.0
2 Bhutan	—	—	—	—	0.1	—	—	—	—	454.0
3 Burkina Faso	0.6	0.6	0.7	0.9	1.8	150.0	176.0	202.0	224.0	195.0
4 Nepal	5.7	—	8.3	10.1	11.9	68.0	—	65.0	106.0	153.0
5 Bangladesh	27.6	29.1	39.9	43.9	44.2	155.0	182.0	209.0	188.0	202.0
6 Malawi	2.5	2.0	2.1	2.3	2.5	135.0	181.0	306.0	352.0	377.0
7 Zaire	9.9	4.8	—	—	—	209.0	643.0	—	—	—
8 Mali	0.7	0.8[a]	1.1	1.2	1.3	200.0	271.0	229.0	243.0	225.0
9 Myanmar	13.5	17.4	18.7	22.0	—	167.0	183.0	186.0	189.0	—
10 Mozambique	2.9	4.1	5.9[b]	5.7	—	122.0	121.0	216.0	243.0	—
11 Madagascar	3.8	5.8	8.0	13.6	14.0[c]	178.0	162.0	151.0	127.0	116.0
12 Uganda	2.6	2.8	3.5	4.3	6.7[d]	219.0	257.0	278.0	301.0	—
13 Burundi	—	1.0	0.6	0.7	1.0	—	182.0	216.0	251.0	386.0
14 Tanzania	3.7	4.2	5.8	9.8	10.2	208.0	204.0	275.0	344.0	311.0
15 Togo	0.8	0.9	1.4	2.2	2.3	195.0	254.0	259.0	230.0	201.0
16 Niger	0.5	0.7	1.2	1.7	1.9	124.0	127.0	119.0	135.0	145.0
17 Benin	0.7	0.9	1.3	2.3	2.7	179.0	173.0	200.0	165.0	165.0
18 Somalia	0.3	0.2	0.7	1.4	1.2	96.0	163.0	282.0	194.0	162.0
19 Central African Republic	0.7	0.8	0.7	0.8	1.0	183.0	220.0	316.0	308.0	310.0
20 India	391.1	408.4	453.5	485.5	528.1	129.0	140.0	145.0	152.0	164.0

— Not available.

Note: *m*, median; *u*, unweighted mean; *w*, weighted mean; and *w′*, weighted mean excluding China and India.

a. 1969 data. b. 1976 data. c. 1984 data. d. 1986 data. e. 1979 data. f. 1974 data.

Source: Unesco data.

(*Table continues on the following page.*)

Table A-8 (*continued*)

Economy	Total primary schools (thousands) 1965	1970	1975	1980	1985	Student-school ratio 1965	1970	1975	1980	1985
21 Rwanda	—	2.0	1.7	1.6	1.6	—	210.0	236.0	441.0	523.0
22 China	1,681.9	961.1	1,093.3	917.3	832.3	69.0	110.0	138.0	159.0	161.0
23 Kenya	5.1	6.1	8.2	10.3	12.9	204.0	234.0	351.0	381.0	365.0
24 Zambia	2.0	2.6	2.7	2.8	3.1	205.0	267.0	323.0	372.0	435.0
25 Sierra Leone	0.9	1.0	1.1	1.2	—	140.0	166.0	187.0	263.0	—
26 Sudan	2.5	4.1	4.7	6.0	6.8	171.0	201.0	249.0	244.0	256.0
27 Haiti	1.9	2.2	2.8	3.3	3.7	149.0	167.0	174.0	195.0	236.0
28 Pakistan	32.9	43.7	52.8	59.2	86.1	96.0	91.0	99.0	92.0	90.0
29 Lesotho	—	1.4	1.1	1.1	1.1[c]	—	131.0	202.0	223.0	270.0
30 Ghana	10.4	10.8	7.0	7.8	9.0	136.0	131.0	165.0	177.0	167.0
31 Sri Lanka	—	—	8.6[b]	8.8	9.3	—	—	161.0	237.0	241.0
32 Mauritania	—	—	—	0.6	0.9	—	—	—	151.0	157.0
33 Senegal	1.2	—	—	1.7	2.3	182.0	—	—	247.0	254.0
34 Afghanistan	1.9	3.0	3.4	3.8	0.8	188.0	180.0	231.0	294.0	726.0
35 Chad	0.6	0.8[a]	0.8[b]	—	1.2	273.0	230.0	264.0	—	281.0
36 Guinea	1.6	2.0	2.1	2.6	2.3	107.0	96.0	95.0	99.0	120.0
37 Democratic Kampuchea	3.9	1.5	—	—	—	205.0	225.0	—	—	—
38 Lao People's Democratic Republic	2.7	3.3	4.9[c]	6.3	8.0	60.0	74.0	85.0	76.0	65.0
39 Viet Nam	—	—	—	6.3[e]	12.5	—	—	—	1,258.0	650.0
Lower-middle-income economies	**189.9**	**243.9**	**354.6**	**384.6**	**458.9**	**166.0 *m***	**175.0 *m***	**229.0 *m***	**240.0 *m***	**221.0 *m***
						205.0 *u*	**235.5 *u***	**235.5 *u***	**265.1 *u***	**265.1 *u***
						172.8 *w*	**189.8 *w***	**199.4 *w***	**213.7 *w***	**209.7 *w***
40 Liberia	0.8	0.9	1.1[f]	1.6	—	105.0	134.0	136.0	92.0	—
41 People's Democratic Republic of Yemen	0.3	0.9	1.0	0.9	—	168.0	149.0	229.0	297.0	—
42 Indonesia	51.4	64.0	72.8	128.9	168.6	227.0	232.0	244.0	198.0	177.0

43 Arab Republic of Yemen	1.0	0.8	2.1	3.1	5.8	69.0	110.0	121.0	141.0	169.0
44 Philippines	35.1	—	30.8	30.6	33.1	166.0	—	247.0	263.0	270.0
45 Morocco	1.4	1.5	1.9	2.3	3.6	797.0	784.0	815.0	944.0	633.0
46 Bolivia	6.9	8.2	9.5	—	8.0[c]	72.0	83.0	90.0	—	148.0
47 Zimbabwe	3.1	3.8	3.6	3.2	4.2	218.0	194.0	240.0	386.0	527.0
48 Nigeria	15.0	14.9	21.2	36.5	35.4	194.0	236.0	291.0	377.0	365.0
49 Dominican Republic	—	5.2	5.5	4.6	6.3	—	147.0	166.0	240.0	194.0
50 Papua New Guinea	3.3	1.6	1.8	2.1	—	42.0	119.0	132.0	143.0	—
51 Côte d'Ivoire	1.8	2.3	2.9	4.8	6.0[c]	197.0	219.0.	232.0	213.0	197.0
52 Honduras	3.8	4.1	4.6	5.5	6.3[c]	75.0	93.0	100.0	109.0	117.0
53 Egypt	8.0	8.6	10.3	12.1	14.1[d]	437.0	441.0	406.0	385.0	441.0
54 Nicaragua	2.1	2.1	2.3	4.4	4.0	98.0	136.0	148.0	107.0	140.0
55 Thailand	26.0	—	42.2	—	32.4	178.0	—	158.0	—	221.0
56 El Salvador	2.7	2.8	3.1	3.2	2.6[c]	141.0	182.0	245.0	261.0	340.0
57 Botswana	0.2	0.3	0.3	0.4	0.5	331.0	277.0	388.0	430.0	447.0
58 Jamaica	0.7	0.7	0.9	0.9	0.8[d]	463.0	537.0	413.0	399.0	—
59 Cameroon	4.1	4.1[a]	4.5	5.0	5.9	181.0	227.0	250.0	276.0	289.0
60 Guatemala	4.2	5.3	6.1	7.0	8.0	91.0	95.0	103.0	115.0	127.0
61 Congo	0.8	0.9	1.0	1.3	1.6	233.0	268.0	319.0	301.0	297.0
62 Paraguay	2.7	3.0	—	—	3.9	132.0	141.0	—	—	146.0
63 Peru	18.5	18.4	19.7	20.8	24.3	103.0	127.0	144.0	152.0	153.0
64 Turkey	30.4	38.2	42.0	45.5	47.6	131.0	131.0	130.0	124.0	139.0
65 Tunisia	2.0	2.2	2.3	2.7	3.4	367.0	425.0	406.0	390.0	380.0
66 Ecuador	6.7	7.7	9.5	11.5	—	119.0	132.0	128.0	133.0	—
67 Mauritius	0.3	0.3	—	0.3	0.3	448.0	501.0	—	429.0	469.0
68 Colombia	23.6	27.1	33.2	33.6	34.0	96.0	121.0	118.0	124.0	119.0
69 Chile	7.2	7.4	8.5	8.2[e]	8.6	212.0	276.0	270.0	273.0	240.0
70 Costa Rica	2.1	2.5	2.8	2.9	3.1	132.0	140.0	129.0	120.0	117.0

— Not available.

Note: *m*, median; *u*, unweighted mean; *w*, weighted mean; and *w'*, weighted mean excluding China and India.

a. 1969 data. b. 1976 data. c. 1984 data. d. 1986 data. e. 1979 data. f. 1974 data.

Source: Unesco data.

(Table continues on the following page.)

Table A-8 (continued)

Economy	Total primary schools (thousands)					Student-school ratio				
	1965	1970	1975	1980	1985	1965	1970	1975	1980	1985
71 Jordan	1.2	0.9	1.2	1.1	1.2	246.0	308.0	322.0	413.0	442.0
72 Syrian Arab Republic	4.6	5.5	7.0	7.8	9.0	153.0	168.0	182.0	199.0	226.0
73 Lebanon	2.5	1.8	—	2.1[e]	2.1[c]	142.0	242.0	—	185.0	157.0
Upper-middle-income economies	**333.6**	**372.2**	**423.8**	**428.4**	**478.8**	**140.9 *m* 243.5 *u* 170.5 *w***	**152.4 *m* 267.8 *u* 193.0 *w***	**157.6 *m* 267.8 *u* 197.0 *w***	**166.5 *m* 298.5 *u* 212.5 *w***	**180.8 *m* 304.8 *u* 215.0 *w***
74 Brazil	124.5	146.1	188.3	201.9	187.3	80.0	88.0	104.0	112.0	132.0
75 Malaysia	—	6.4	6.4	6.4	6.7	—	263.0	296.0	314.0	328.0
76 South Africa	—	—	—	—	—	—	—	—	—	—
77 Mexico	39.1	46.0	55.6	76.2	76.7	177.0	201.0	206.0	192.0	197.0
78 Uruguay	2.3	2.3	2.3	2.3	2.4	146.0	154.0	140.0	144.0	148.0
79 Hungary	6.0	5.5	4.5	3.6	3.5	236.0	203.0	234.0	323.0	371.0
80 Poland	26.6	26.1	14.7	12.6	16.3	195.0	201.0	293.0	331.0	295.0
81 Portugal	17.5	17.0	13.1	12.5	12.7	51.0	58.0	92.0	99.0	97.0
82 Yugoslavia	14.1	14.0	13.4	12.7	12.1	117.0	113.0	112.0	113.0	120.0
83 Panama	1.6	1.8	2.2	2.3	2.5	127.0	142.0	152.0	147.0	136.0
84 Argentina	19.0	19.8	20.6	20.8[e]	20.7	164.0	171.0	173.0	184.0	222.0
85 Republic of Korea	5.1	6.0	6.4	6.5	6.5	969.0	958.0	875.0	870.0	747.0
86 Algeria	4.4	6.1	7.8	9.3	11.4	309.0	309.0	341.0	335.0	305.0
87 Venezuela	10.9	10.5	11.9	12.3[e]	13.2	133.0	169.0	177.0	200.0	210.0
88 Gabon	0.6	0.7	0.7	0.9	0.9	132.0	144.0	184.0	172.0	204.0
89 Greece	10.3	9.5	9.6	9.5	8.7	94.0	96.0	97.0	95.0	100.2
90 Spain	—	—	30.7[f]	—	18.9	—	—	120.0	—	184.0
91 Oman	0.0	0.0	0.2	0.2	0.3	—	—	273.0	460.0	592.0
92 Ireland	5.0	4.1	3.6	3.4	3.3	100.0	97.0	112.0.	124.0	127.0
93 Trinidad and Tobago	0.5	0.5	0.5	0.5	0.5	424.0	451.0	398.0	334.0	337.0

94 Israel	1.4	1.6	1.5	1.6	1.6	318.0	299.0	357.0	389.0	437.0
95 Hong Kong	1.7	1.5	1.1	0.8	0.8[c]	351.0	493.0	584.0	675.0	670.0
96 Singapore	0.5	0.4	0.4	0.3	0.3[c]	714.0	909.0	821.0	972.0	962.0
97 Islamic Republic of Iran	23.0	25.8	36.7	39.2	50.4	112.0	132.0	122.0	122.0	135.0
98 Iraq	4.5	5.6	7.6	11.3	8.1	214.0	196.0	234.0	231.0	348.0
99 Romania	15.0	14.9	14.7	14.4	14.1	195.0	193.0	197.0	225.0	215.0
High-income oil exporters	**2.0**	**3.3**	**5.8**	**8.7**	**8.1**	**329.0 *m***	**269.0 *m***	**418.5 *m***	**349.0 *m***	**374.5 *m***
						340.7 *u*	**409.5 *u***	**409.5 *u***	**401.5 *u***	**374.5 *u***
						251.5 *w*	**257.1 *w***	**242.5 *w***	**210.0 *w***	**187.3 *w***
100 Saudi Arabia	1.3	1.9	3.5	5.7	7.8	197.0	222.0	194.0	163.0	172.0
101 Kuwait	0.1	0.1	0.2	0.2	0.3	496.0	755.0	559.0	745.0	577.0
102 United Arab Emirates	—	—	0.1[c]	0.2	—	—	—	607.0	443.0	—
103 Libya	0.6	1.3	2.0	2.6	—	329.0	269.0	278.0	255.0	—
High-income industrial market economies	**220.5**	**291.6**	**270.4**	**185.9**	**183.3**	**157.0 *m***	**178.0 *m***	**157.0 *m***	**152.0 *m***	**139.5 *m***
						178.4 *u*	**188.7 *u***	**188.7 *u***	**179.9 *u***	**159.3 *u***
						213.2 *w*	**235.2 *w***	**236.9 *w***	**229.9 *w***	**168.3 *w***
104 New Zealand	2.6	2.6	2.5	2.3	—	141.0	154.0	157.0	166.0	—
105 Italy	44.7	37.1	33.2	30.3	27.7	101.0	131.0	146.0	146.0	134.0
106 United Kingdom	29.9	29.5	27.0	26.5	24.8	173.0	197.0	212.0	185.0	173.0
107 Belgium	9.0	—	7.8	5.0	4.4	109.0	—	121.0	168.0	166.0
108 Austria	4.3	4.0	3.6	3.5	3.8	110.0	133.0	139.0	114.0	90.0
109 Netherlands	7.9	8.2	8.6	8.7	8.4	178.0	178.0	169.0	153.0	175.0
110 France	72.1	63.5	55.9	51.4	47.9	77.0	78.0	82.0	90.0	86.0
111 Australia	9.0	8.4	8.0	—	8.0[d]	173.0	216.0	204.0	—	—
112 Federal Republic of Germany	30.2	21.5	18.1	18.4	19.6	184.0	185.0	216.0	151.0	116.0

— Not available.

Note: *m*, median; *u*, unweighted mean; *w*, weighted mean; and *w′*, weighted mean excluding China and India.

a. 1969 data. b. 1976 data. c. 1984 data. d. 1986 data. e. 1979 data. f. 1974 data.

Source: Unesco data.

(*Table continues on the following page.*)

Table A-8 (continued)

Economy	Total primary schools (thousands) 1965	1970	1975	1980	1985	Student-school ratio 1965	1970	1975	1980	1985
113 Finland	5.8	4.5	4.4[f]	4.2	—	79.0	86.0	92.0	89.0	—
114 Denmark	1.7	2.4	2.2[f]	3.3	2.6	259.0	185.0	222.0	189.0	155.0
115 Japan	26.2	25.0	24.7	24.9	25.0	379.0	385.0	420.0	475.0	444.0
116 Sweden	—	—	4.8[f]	4.9	4.8[c]	—	—	144.0	136.0	131.0
117 Canada	—	—	—	—	15.6	—	—	—	—	145.0
118 Norway	3.5	3.1	3.4	3.5	3.5	118.0	124.0	115.0	111.0	96.0
119 United States	78.0	81.8	77.6	80.5[e]	—	416.0	415.0	392.0	346.0	—
120 Switzerland	—	—	—	—	—	—	—	—	—	—
Other economies										
121 Angola	2.6	4.4	—	6.1	—	84.0	99.0	—	213.0	—
122 Albania	3.3	1.4	—	1.6	1.6	106.0	355.0	—	345.0	340.0
123 Bulgaria	4.8	3.9	3.4	3.2	3.0	233.0	269.0	288.0	311.0	360.0
124 Czechoslovakia	11.3	10.8	9.3	6.8	6.3	197.0	182.0	202.0	280.0	329.0
125 Cuba	14.1	15.2	14.9	12.2	10.2	87.0	101.0	121.0	120.0	106.0
126 German Democratic Republic	8.1	6.6	5.6[b]	5.6	5.6	281.0	187.0	189.0	152.0	154.0
127 Mongolia	—	—	0.2[f]	0.1[e]	—	—	—	640.0	1,413.0	—
128 Democratic People's Republic of Korea	—	—	4.7[b]	—	—	—	—	545.0	—	—
129 Union of Soviet Socialist Republics	188.7	172.6	147.1	130.0	128.0	133.0	149.0	145.0	167.0	184.0

— Not available.

Note: *m*, median; *u*, unweighted mean; *w*, weighted mean; and *w′*, weighted mean excluding China and India.

a. 1969 data. b. 1976 data. c. 1984 data. d. 1986 data. e. 1979 data. f. 1974 data.

Source: Unesco data.

Table A-9. Number of Primary School Teachers and Student-Teacher Ratio, Selected Years, 1965–85

Economy	Total primary teachers (thousands)					Student-teacher ratio				
	1965	1970	1975	1980	1985	1965	1970	1975	1980	1985
Low-income economies	**5,636.1**	**5,712.2**	**7,901.4**	**8,542.8**	**8,971.4**	**42.5 *m***	**42.0 *m***	**41.0 *m***	**41.5 *m***	**38.0 *m***
						44.4 *u*	**42.1 *u***	**43.9 *u***	**43.8 *u***	**41.5 *u***
						34.6 *w*	**34.8 *w***	**33.4 *w***	**32.6 *w***	**32.0 *w***
Excluding China and India	**571.9**	**724.0**	**1,139.3**	**1,325.4**	**1,727.1**	**49.9 *w′***	**50.0 *w′***	**41.6 *w′***	**44.0 *w′***	**38.6 *w′***
1 Ethiopia	9.1	13.5	24.5	33.3	50.9	41.0	48.0	44.0	64.0	48.0
2 Bhutan	0.4	0.4	0.6[a]	1.0[a]	1.2[a]	22.0	22.0	26.0	31.0	37.0
3 Burkina Faso	1.9	2.4	3.0	3.7	5.7	47.0	44.0	47.0	54.0	62.0
4 Nepal	12.4	18.0	18.9	27.8	51.3	25.0	22.0	29.0	38.0	35.0
5 Bangladesh	95.7	113.7	164.7	153.9	189.9	45.0	46.0	51.0	54.0	47.0
6 Malawi	8.5	8.4	10.6	12.5	15.4	40.0	43.0	61.0	65.0	61.0
7 Zaire	55.2	72.5	84.4[a]	99.9[a]	121.2[a]	37.0	43.0	42.0	42.0	41.0
8 Mali	3.1	5.1	6.2	6.9	8.6	46.0	40.0	41.0	42.0	34.0
9 Myanmar	42.2	68.2	66.3	80.3	111.6[a]	53.0	47.0	52.0	52.0	45.0
10 Mozambique	4.6	7.2	11.5[a]	16.1	19.4	78.0	69.0	85.0	81.0	64.0
11 Madagascar	9.5	14.4	20.1	39.5	42.1[a]	71.0	65.0	60.0	44.0	38.0
12 Uganda	16.3	21.5	28.7	38.4	65.0[a]	35.0	34.0	34.0	34.0	31.0
13 Burundi	3.6	5.0	4.2	5.0	6.9	40.0	37.0	31.0	35.0	56.0
14 Tanzania	14.7	18.3	29.7	81.4	92.6	52.0	47.0	54.0	41.0	34.0
15 Togo	3.1	3.9	6.1	9.2	10.0	50.0	58.0	60.0	55.0	46.0
16 Niger	1.5	2.3	3.6	5.5	7.4	42.0	39.0	39.0	41.0	37.0
17 Benin	3.1	3.8	4.9	8.0	13.5	41.0	41.0	53.0	48.0	33.0
18 Somalia	1.1	1.0	3.5	8.1	9.7	26.0	33.0	57.0	33.0	20.0
19 Central African Republic	2.4	2.8	3.3	4.1	4.7	54.0	64.0	67.0	60.0	66.0
20 India	1,207.2	1,376.2	1,559.1	1,718.0	1,867.5	42.0	41.0	42.0	43.0	46.0

— Not available.

Note: *m*, median; *u*, unweighted mean; *w*, weighted mean; and *w′*, weighted mean excluding China and India.

a. Estimate.

Sources: Unesco data; Unesco 1988.

(*Table continues on the following page.*)

Table A-9 (*continued*)

Economy	*Total primary teachers (thousands)* 1965	1970	1975	1980	1985	*Student-teacher ratio* 1965	1970	1975	1980	1985
21 Rwanda	4.9	7.0	8.0	11.9	14.9	67.0	60.0	50.0	59.0	56.0
22 China	3,857.0	3,612.0	5,203.0	5,499.4	5,376.8	30.0	29.0	29.0	27.0	25.0
23 Kenya	30.6	41.5	86.1	102.5	138.4	34.0	34.0	33.0	38.0	34.0
24 Zambia	8.0	14.9	18.1	21.5	27.3	51.0	47.0	48.0	49.0	49.0
25 Sierra Leone	4.0	5.1	6.4	9.5	11.2[a]	32.0	32.0	32.0	33.0	33.0
26 Sudan	9.0	17.7	31.7	43.5	50.1	48.0	47.0	37.0	34.0	35.0
27 Haiti	6.2	7.8	11.8	14.6	23.2	46.0	47.0	41.0	44.0	38.0
28 Pakistan	67.0	85.7	116.8	131.1	165.3[a]	42.0	41.0	40.0	37.0	40.0
29 Lesotho	2.9	4.0	4.2	5.1	6.0[a]	57.0	46.0	53.0	48.0	52.0
30 Ghana	44.1	48.0	39.2	50.2	69.3	32.0	30.0	30.0	29.0	23.0
31 Sri Lanka	67.2	55.7	45.6	—	70.1	33.0	30.0	30.0	—	32.0
32 Mauritania	1.0	1.3	1.4	2.2	2.8	20.0	24.0	35.0	41.0	51.0
33 Senegal	5.1	5.8	7.6	9.2	12.6	43.0	45.0	41.0	46.0	46.0
34 Afghanistan	6.7	13.1	21.1	35.4	15.6	53.0	41.0	37.0	32.0	37.0
35 Chad	2.0	2.8	2.8	3.9	4.8	83.0	68.0	77.0	64.0	71.0
36 Guinea	4.0	4.4	5.0	7.2	7.6	43.0	44.0	40.0	36.0	36.0
37 Democratic Kampuchea	16.5	20.0	21.9	22.8	24.0	48.0	17.0	22.0	22.0	22.0
38 Lao People's Democratic Republic	4.3	6.8	11.8	16.1	21.0	37.0	36.0	27.0	30.0	25.0
39 Viet Nam	—	—	205.0	204.1	235.8	—	—	36.0	39.0	34.0
Lower-middle-income economies	**1,282.5**	**1,785.5**	**2,250.9**	**2,936.6**	**3,542.5**	**36.0 *m***	**36.5 *m***	**35.0 *m***	**34.0 *m***	**31.5 *m***
						38.1 *u*	**37.7 *u***	**35.7 *u***	**34.5 *u***	**33.4 *u***
						37.0 *w*	**33.0 *w***	**32.0 *w***	**31.9 *w***	**29.0 *w***
40 Liberia	2.6	3.4	2.8	3.5	4.5	32.0	36.0	38.0	41.0	29.0
41 People's Democratic Republic of Yemen	1.6	4.3	6.7	10.1	12.3[a]	31.0	31.0	34.0	27.0	26.0
42 Indonesia	282.0	514.0	603.3	787.4	1,181.8	41.0	29.0	29.0	32.0	25.0

43	Arab Republic of Yemen	1.2	1.7	6.9	11.7	19.4	56.0	51.0	37.0	39.0	50.0
44	Philippines	185.1	243.8	261.8	272.9	289.3	31.0	29.0	29.0	31.0	31.0
45	Morocco	28.9	34.3	38.7	58.7	81.9	39.0	34.0	40.0	37.0	28.0
46	Bolivia	17.8	25.5	38.7	48.9	47.3[a]	28.0	27.0	22.0	20.0	25.0
47	Zimbabwe	16.9	18.4	21.6	28.1	56.1	40.0	40.0	40.0	44.0	40.0
48	Nigeria	87.1	103.2	177.2	369.6	292.8	33.0	34.0	35.0	37.0	44.0
49	Dominican Republic	10.5	13.8	17.9	22.1[a]	28.0	53.0	55.0	51.0	50.0	44.0
50	Papua New Guinea	7.5	6.4	7.5	9.5	11.2[a]	19.0	30.0	32.0	31.0	31.0
51	Côte d'Ivoire	7.5	11.2	15.4	26.5	33.5[a]	47.0	45.0	44.0	39.0	36.0
52	Honduras	9.9	10.8	13.0	16.4	19.9[a]	29.0	35.0	35.0	37.0	38.0
53	Egypt	89.2	99.9	121.5	144.4[a]	207.2	39.0	38.0	34.0	32.0	30.0
54	Nicaragua	5.8	7.4	8.5	13.3	16.9	35.0	39.0	40.0	35.0	33.0
55	Thailand	133.7	162.5	239.1	326.1	369.8	35.0	35.0	28.0	23.0	19.0
56	El Salvador	11.1	13.6	17.3	17.4	21.5[a]	34.0	37.0	44.0	48.0	42.0
57	Botswana	1.7	2.3	3.5	5.3	7.0	40.0	36.0	33.0	32.0	32.0
58	Jamaica	4.6	7.6	9.5	9.0	9.8[a]	57.0	47.0	39.0	40.0	35.0
59	Cameroon	15.3	19.4	22.2	26.8	33.6	48.0	48.0	51.0	52.0	51.0
60	Guatemala	11.8	14.1	18.1	23.8	27.8	33.0	36.0	35.0	34.0	37.0
61	Congo	3.1	3.9	5.4	7.2	7.7	60.0	62.0	59.0	54.0	61.0
62	Paraguay	11.8	13.1	15.4	18.9	22.8	30.0	32.0	29.0	27.0	25.0
63	Peru	53.0	66.0	72.6	84.4	106.6	36.0	35.0	39.0	37.0	35.0
64	Turkey	86.9	132.6	171.0	212.5	212.7	46.0	38.0	32.0	27.0	31.0
65	Tunisia	13.2	19.7	23.3	27.4	40.9	56.0	47.0	40.0	39.0	32.0
66	Ecuador	21.4	26.6	32.3	42.4	51.4[a]	36.0	37.0	38.0	36.0	33.0
67	Mauritius	4.0	4.7	5.8	6.4	6.5	34.0	32.0	26.0	20.0	22.0
68	Colombia	63.3	86.0	122.0	136.4	132.9	36.0	38.0	32.0	31.0	30.0
69	Chile	41.7	52.4	65.8	64.7[a]	62.5[a]	37.0	39.0	35.0	34.0	33.0
70	Costa Rica	10.3	11.7	12.4	12.6	11.5	27.0	30.0	29.0	28.0	31.0
71	Jordan	7.7	7.2	11.1	14.3	17.0	38.0	39.0	35.0	32.0	31.0

— Not available.

Note: m, median; u, unweighted mean; w, weighted mean; and w', weighted mean excluding China and India.

a. Estimate.

Sources: Unesco data; Unesco 1988.

(Table continues on the following page.)

Table A-9 (*continued*)

Economy	Total primary teachers (thousands) 1965	1970	1975	1980	1985	Student-teacher ratio 1965	1970	1975	1980	1985
72 Syrian Arab Republic	19.5	25.1	37.6	55.3	78.4	36.0	37.0	34.0	28.0	26.0
73 Lebanon	14.8	18.9	25.0[a]	22.6	20.0[a]	23.0	23.0	17.0	18.0	18.0
Upper-middle-income economies	**1,727.7**	**2,233.1**	**2,828.3**	**3,111.1**	**3,778.1**	**31.0 *m***	**31.0 *m***	**29.0 *m***	**27.0 *m***	**24.0 *m***
						32.6 *u*	**31.4 *u***	**28.8 *u***	**26.9 *u***	**24.8 *u***
						32.0 *w*	**30.1 *w***	**27.0 *w***	**27.0 *w***	**24.0 *w***
74 Brazil	351.5	619.0	896.7	884.3	1040.6	28.0	25.0	22.0	26.0	24.0
75 Malaysia	50.3	54.4	59.3	73.7	91.4	29.0	31.0	32.0	27.0	24.0
76 South Africa	75.0	80.5	157.7	160.3	185.0[a]	41.0	50.0	29.0	27.0	24.0
77 Mexico	148.3	201.5	255.9	375.2	449.8	47.0	46.0	45.0	39.0	34.0
78 Uruguay	10.8	12.0	13.6	14.8	14.2	31.0	29.0	24.0	22.0	25.0
79 Hungary	62.2	63.1	66.9	75.4	88.1	23.0	18.0	16.0	15.0	15.0
80 Poland	182.8	211.5	191.2	194.0	307.8	28.0	25.0	23.0	21.0	16.0
81 Portugal	28.0	29.6	59.5	68.7	73.3	32.0	34.0	20.0	18.0	17.0
82 Yugoslavia	52.6	58.4	60.9	59.4	61.3	31.0	27.0	25.0	24.0	24.0
83 Panama	6.8	9.4	12.5	12.4	13.4	30.0	27.0	27.0	27.0	25.0
84 Argentina	153.7	175.9	196.0	193.6[a]	229.7	20.0	19.0	18.0	20.0	20.0
85 Republic of Korea	79.2	101.1	108.1	119.1	126.8	62.0	57.0	52.0	48.0	38.0
86 Algeria	31.2	47.2	65.0	88.5	125.0	43.0	40.0	41.0	35.0	28.0
87 Venezuela	42.6	50.8	69.5	92.6	108.1	34.0	35.0	30.0	27.0	26.0
88 Gabon	2.0	2.2	2.7	3.4	4.0	39.0	46.0	48.0	45.0	46.0
89 Greece	27.4	29.3	31.0	37.3	38.0	35.0	31.0	30.0	24.0	23.0
90 Spain	99.4	115.6	126.0	127.7	137.8	34.0	34.0	29.0	28.0	25.0
91 Oman	0.0	0.2	2.1	4.0	6.7	—	18.0	27.0	23.0	27.0
92 Ireland	11.2	11.8	13.1	14.6	15.7	34.0	34.0	31.0	29.0	27.0
93 Trinidad and Tobago	5.6	6.4	6.5	7.0	7.6	38.0	35.0	31.0	24.0	22.0
94 Israel	22.2	27.8	35.7	45.9	48.2	20.0	17.0	15.0	14.0	14.0
95 Hong Kong	20.3	22.4	20.7	17.9	19.7[a]	29.0	33.0	31.0	30.0	27.0

96 Singapore	12.2	12.3	10.8	9.5	10.7[a]	29.0	30.0	30.0	31.0	27.0
97 Islamic Republic of Iran	81.2	105.3	152.1	181.0[a]	309.7	32.0	32.0	29.0	27.0	22.0
98 Iraq	42.9	49.8	69.8	94.0	118.4	22.0	22.0	25.0	28.0	24.0
99 Romania	128.3	135.6	145.0	156.8	147.1	23.0	21.0	20.0	21.0	21.0
High-income oil exporters	**20.4**	**34.3**	**68.7**	**101.4**	**143.0**	**22.0 *m***	**25.5 *m***	**19.0 *m***	**18.0 *m***	**18.0 *m***
						24.7 *u*	**25.0 *u***	**19.0 *u***	**17.3 *u***	**17.8 *u***
						25.0 *w*	**26.0 *w***	**20.0 *w***	**18.0 *w***	**17.0 *w***
100 Saudi Arabia	11.4	17.4	34.5	50.5	83.4	22.0	24.0	20.0	18.0	16.0
101 Kuwait	2.6	3.6	6.4	8.0	9.6	21.0	21.0	18.0	19.0	18.0
102 United Arab Emirates	—	1.0	3.5	6.3	8.5	—	27.0	15.0	14.0	18.0
103 Libya	6.4	12.3	24.3	36.6	41.5	31.0	28.0	23.0	18.0	19.0
High-income industrial market economies	**2,548.7**	**2,827.3**	**3,081.0**	**3,121.5**	**3,045.1**	**25.0 *m***	**23.0 *m***	**20.0 *m***	**18.0 *m***	**17.0 *m***
						24.5 *u*	**23.5 *u***	**20.8 *u***	**18.6 *u***	**17.5 *u***
						28.0 *w*	**26.0 *w***	**23.0 *w***	**21.0 *w***	**20.0 *w***
104 New Zealand	14.6	16.4	17.0	19.3	16.5	25.0	24.0	23.0	20.0	20.0
105 Italy	207.2	224.6	255.3	273.7	258.0[a]	22.0	22.0	19.0	16.0	14.0
106 United Kingdom	198.7	233.6	269.6	246.3	229.5	26.0	25.0	21.0	20.0	19.0
107 Belgium	47.5	51.7	48.6	46.4	44.2	20.0	19.0	19.0	18.0	17.0
108 Austria	21.4	24.0	25.5	27.5	32.8	22.0	22.0	20.0	15.0	10.0
109 Netherlands	45.0	49.2	52.7	57.5	87.8	31.0	30.0	28.0	23.0	17.0
110 France	217.5	215.1	242.2	242.7	219.4	25.0	23.0	19.0	19.0	19.0
111 Australia	55.2	60.9	72.1	85.1	91.5	28.0	28.0	23.0	20.0	17.0
112 Federal Republic of Germany	137.7	155.8[a]	231.5[a]	165.1[a]	133.5	26.0	25.0	17.0	17.0	17.0
113 Finland	19.9	17.4	24.5	25.9	26.0[a]	23.0	22.0	19.0	14.0	15.0
114 Denmark	27.2	27.7	40.9	36.4	34.7	16.0	16.0	12.0	12.0	12.0

— Not available.

Note: *m*, median; *u*, unweighted mean; *w*, weighted mean; and *w*′, weighted mean excluding China and India.

a. Estimate.

Sources: Unesco data; Unesco 1988.

(*Table continues on the following page.*)

Table A-9 (continued)

Economy	Total primary teachers (thousands)					Student-teacher ratio				
	1965	1970	1975	1980	1985	1965	1970	1975	1980	1985
115 Japan	345.9	364.9	402.6	471.0	464.2	29.0	26.0	26.0	25.0	24.0
116 Sweden	30.0	30.8	34.2	40.5[a]	38.3[a]	20.0	20.0	20.0	16.0	16.0
117 Canada	138.7	159.9	119.3	119.6[a]	132.6	26.0	23.0	20.0	18.0	17.0
118 Norway	19.2	19.7	23.4	23.4	20.5	21.0	20.0	17.0	17.0	16.0
119 United States	1,004.0	1,157.1[a]	1,201.0	1,223.1	1,200.5	32.0	29.0	25.0	22.0	22.0
120 Switzerland	19.0	18.5[a]	20.6[a]	18.0[a]	15.1[a]	24.0	25.0	25.0	25.0	25.0
Other economies										
121 Angola	4.8	9.8	20.1[a]	32.7[a]	19.2[a]	45.0	44.0	41.0	40.0	39.0
122 Albania	13.0	18.9	24.1[a]	26.0	27.2	27.0	26.0	24.0	21.0	20.0
123 Bulgaria	49.4	47.8	48.4	51.6	61.2	23.0	22.0	20.0	19.0	18.0
124 Czechoslovakia	96.0	97.7	95.6	90.4	96.4	23.0	20.0	20.0	21.0	22.0
125 Cuba	41.9	56.6	77.5	84.0	77.1	29.0	27.0	23.0	17.0	14.0
126 German Democratic Republic	72.2	63.2	53.7	53.4	58.4	20.0	20.0	20.0	16.0	15.0
127 Mongolia	3.3	4.8	4.2[a]	4.6[a]	4.8[a]	32.0	30.0	31.0	32.0	32.0
128 Democratic People's Republic of Korea	—	—	—	—	—	—	—	—	—	—
129 Union of Soviet Socialist Republics	1,251.0	1,289.9	1,124.5	1,206.3	1,387.4	20.0	20.0	19.0	18.0	17.0

— Not available.

Note: *m*, median; *u*, unweighted mean; *w*, weighted mean; and *w'*, weighted mean excluding China and India.

a. Estimate.

Sources: Unesco data; Unesco 1988.

Table A-10. Students Enrolled in Private Primary Schools, Selected Years, 1965–85

(percentage of total enrollment)

Economy	1965	1970	1975	1980	1985
Low-income economies	**11.5 *m***	**6.1 *m***	**6.2 *m***	**3.0 *m***	**1.2 *m***
	25.0 *u*	**18.0 *u***	**18.6 *u***	**10.7 *u***	**5.5 *u***
	2.7 *w*	**1.5 *w***	**1.2 *w***	**1.5 *w***	**1.4 *w***
Excluding China and India	**26.1 *w′***	**11.5 *w′***	**8.2 *w′***	**8.4 *w′***	**5.1 *w′***
1 Ethiopia	25.2	28.1	28.6	15.6	11.0
2 Bhutan	—	—	—	—	—
3 Burkina Faso	34.6	3.5	7.2	8.4	8.7
4 Nepal	—	—	—	—	1.2[a]
5 Bangladesh	6.1	5.8	4.1	14.6	11.0
6 Malawi	76.9	11.2	10.2	6.8	5.5
7 Zaire	90.7	—	—	—	—
8 Mali	7.4	5.9[b]	6.0	4.2	4.2
9 Myanmar	—	—	—	—	—
10 Mozambique	—	—	—	0.0	0.0
11 Madagascar	27.2	24.4	23.3	12.7	13.2[a]
12 Uganda	—	—	—	—	—
13 Burundi	95.7	93.9	92.4[c]	4.3	1.2
14 Tanzania	7.8	3.3	3.7	0.2	0.3
15 Togo	39.7	33.9	28.6	23.3	23.4
16 Niger	6.0	6.1	4.6	3.0	2.7
17 Benin	42.2	34.2	6.4	3.4	0.0
18 Somalia	19.4	25.1	0.0	0.0	0.0
19 Central African Republic	0.0	0.0	0.0	0.0	0.0

— Not available.

Note: *m*, median; *u*, unweighted mean; *w*, weighted mean; and *w′*, weighted mean excluding China and India.

a. 1984 data. b. 1969 data. c. 1974 data. d. 1979 data. e. 1986 data. f. 1976 data. g. 1981 data.

Source: Unesco data.

(Table continues on the following page.)

Table A-10 (*continued*)

Economy	1965	1970	1975	1980	1985
20 India	—	—	—	—	—
21 Rwanda	—	—	0.0[c]	—	0.6
22 China	0.0	0.0	0.0	0.0	0.0
23 Kenya	4.1	3.4	0.0	—	—
24 Zambia	—	27.4	24.4	0.6	0.3
25 Sierra Leone	—	—	78.2	—	—
26 Sudan	2.4	4.3	2.0	2.5	2.3
27 Haiti	26.3	37.7	42.5	56.8	58.5
28 Pakistan	—	—	0.0	0.0	0.0
29 Lesotho	95.8	100.0	100.0	100.0[d]	—
30 Ghana	—	2.1	—	—	6.3[e]
31 Sri Lanka	—	—	—	1.3	1.4
32 Mauritania	0.0	0.0	—	—	0.1[e]
33 Senegal	13.1	12.4	12.3	11.0	8.8
34 Afghanistan	0.0	0.0	0.0	0.0	0.0
35 Chad	12.4	8.0	9.7	—	5.5
36 Guinea	0.0	0.0	0.0	0.0	0.0
37 Democratic Kampuchea	5.4	3.3	—	—	—
38 Lao People's Democratic Republic	10.6	11.3	—	0.0	0.0
39 Viet Nam	—	—	—	—	0.0
Lower-middle-income economies	**13.1** ***m***	**14.2** ***m***	**10.0** ***m***	**7.8** ***m***	**9.0** ***m***
	20.1 ***u***	**20.1** ***u***	**14.7** ***u***	**14.9** ***u***	**16.2** ***u***
	16.3 ***w***	**19.8** ***w***	**10.9** ***w***	**14.2** ***w***	**14.1** ***w***
40 Liberia	25.1	34.3	34.6[c]	34.6[d]	33.4[a]
41 People's Democratic Republic of Yemen	10.1	3.8	—	—	—
42 Indonesia	12.0	33.2[b]	—	20.7	17.3
43 Arab Republic of Yemen	—	4.7[b]	1.0	4.8	7.5
44 Philippines	4.5	4.9	5.3	5.6[d]	6.0

45	Morocco	6.4	4.5	4.7	3.0	3.4
46	Bolivia	25.9	15.8[b]	8.8	—	7.7[a]
47	Zimbabwe	—	86.6	86.6	83.5	87.6
48	Nigeria	76.2	37.7	—	—	—
49	Dominican Republic	7.3	11.6	12.2	17.8	24.1
50	Papua New Guinea	66.3	63.0	—	2.0	—
51	Côte d'Ivoire	27.8	21.8	18.6	14.0	11.2[a]
52	Honduras	6.6	5.9	5.2	5.3	5.1
53	Egypt	12.9	10.9	4.8	5.0	4.8
54	Nicaragua	15.7	15.1	13.3	11.8	13.3
55	Thailand	13.1	14.2	11.1	8.4	9.0
56	El Salvador	4.1	4.8	6.6	7.2	8.1[a]
57	Botswana	4.4	5.2	5.4	4.7	2.8
58	Jamaica	—	—	5.1	3.7	—
59	Cameroon	61.0	53.6	42.9	36.3	33.8
60	Guatemala	19.1	15.5	13.6	14.2	13.7
61	Congo	0.0	0.0	0.0	0.0	0.0
62	Paraguay	10.4	12.9	—	—	13.7
63	Peru	14.1	14.5	12.9	13.1	14.4
64	Turkey	0.7	0.5	0.4	0.4[d]	0.4
65	Tunisia	2.3	1.3[b]	1.3	0.9	0.4
66	Ecuador	18.4	17.9	16.9	15.9	—
67	Mauritius	33.6	29.3	27.8	25.9	23.2
68	Colombia	14.0	13.4	15.2	14.5	13.5
69	Chile	27.4	22.8	18.3	20.2	31.8
70	Costa Rica	3.8	3.5	3.7	2.6	3.5
71	Jordan	28.4	30.6	29.8	6.3	7.8
72	Syrian Arab Republic	10.2	6.1	5.2	4.8	4.5

— Not available.

Note: m, median; u, unweighted mean; w, weighted mean; and w', weighted mean excluding China and India.

a. 1984 data. b. 1969 data. c. 1974 data. d. 1979 data. e. 1986 data. f. 1976 data. g. 1981 data.

Source: Unesco data.

(Table continues on the following page.)

Table A-10 (*continued*)

Economy	*1965*	*1970*	*1975*	*1980*	*1985*
73 Lebanon	61.3	61.9	—	61.1	68.5[g]
Upper-middle-income economies	**6.4 *m***	**7.2 *m***	**6.8 *m***	**6.4 *m***	**6.3 m**
	20.7 *u*	**19.4 *u***	**20.1 *u***	**19.4 *u***	**19.0 *u***
	9.7 *w*	**9.3 *w***	**10.5 *w***	**9.4 *w***	**9.0 *w***
74 Brazil	11.0	7.5	12.9	12.8	12.1
75 Malaysia	—	—	—	—	—
76 South Africa	—	—	—	—	—
77 Mexico	9.5	7.8	6.0	4.9	5.0
78 Uruguay	18.1	18.5	17.2	16.4	15.4
79 Hungary	0.0	0.0	0.0	0.0	0.0
80 Poland	0.0	0.0	0.0	0.0	0.0
81 Portugal	5.1	5.3	4.8	7.0	6.7[a]
82 Yugoslavia	0.0	0.0	0.0	0.0	0.0
83 Panama	5.1	5.4	5.0	6.3	7.5
84 Argentina	13.8	15.9	17.3	17.8[d]	18.6
85 Republic of Korea	0.5	1.1	1.2	1.3	1.5
86 Algeria	1.9	1.9	0.8	0.0	0.0
87 Venezuela	13.0	11.7	11.1	11.1	11.5
88 Gabon	52.8	49.4	45.1	39.0	33.4
89 Greece	6.4	7.1	7.8	6.4	5.8
90 Spain	24.3	27.9	37.1	35.5	34.2
91 Oman	—	0.0	0.3	0.3	1.2
92 Ireland	131.7	130.1	100.0	100.0	100.0
93 Trinidad and Tobago	—	—	73.7[f]	72.9	72.1[a]
94 Israel	—	—	—	—	—
95 Hong Kong	92.4	92.5	93.4	94.2	93.2[a]
96 Singapore	39.8	35.7	—	—	—
97 Islamic Republic of Iran	6.2	7.2	7.6	0.0	0.0
98 Iraq	2.2	2.2	0.0	0.0	0.0

Economy					
99 Romania	0.0	0.0	0.0	0.0	0.0
High-income oil exporters	**6.5 *m***	**2.1 *m***	**6.8 *m***	**9.4 *m***	**14.4 *m***
	5.8 *u*	**8.9 *u***	**8.2 *u***	**9.0 *u***	**14.3 *u***
	5.2 *w*	**3.4 *w***	**3.7 *w***	**3.5 *w***	**5.3 *w***
100 Saudi Arabia	6.5	2.1	2.6	2.8	3.2
101 Kuwait	8.0	24.0	17.5	16.0	28.3
102 United Arab Emirates	—	—	10.9	17.1	25.6
103 Libya	2.8	0.5	1.7	0.0	0.0
High-income industrial market economies	**6.3 *m***	**7.0 *m***	**6.5 *m***	**7.2 *m***	**4.2 *m***
	14.9 *u*	**15.6 *u***	**13.8 *u***	**13.7 *u***	**13.1 *u***
	12.6 *w*	**10.2 *w***	**11.3 *w***	**9.9 *w***	**10.2 *w***
104 New Zealand	15.8	12.7	9.0	8.5	2.0
105 Italy	8.4	7.0	6.8	7.2	7.7
106 United Kingdom	4.3	—	—	4.0	4.5
107 Belgium	53.7	54.0	51.6	53.0	54.6
108 Austria	2.6	—	3.1	3.4	3.9
109 Netherlands	73.8	72.5	70.2	68.3	68.5
110 France	14.6	13.9	14.1	14.5	15.1
111 Australia	24.2	21.9	—	19.7	23.4
112 Federal Republic of Germany	0.6	0.5[b]	—	—	1.6
113 Finland	—	—	—	—	—
114 Denmark	6.3	5.9[b]	6.5	8.8	9.0
115 Japan	0.5	0.6	0.6	0.5	0.5
116 Sweden	0.0	0.0	0.4	0.6	0.7[e]
117 Canada	2.4	1.8	2.0	2.7[g]	3.2
118 Norway	0.4	0.4	0.4	0.6	0.8
119 United States	16.4	11.5	12.8	11.1	11.4

— Not available.

Note: *m*, median; *u*, unweighted mean; *w*, weighted mean; and *w′*, weighted mean excluding China and India.

a. 1984 data. b. 1969 data. c. 1974 data. d. 1979 data. e. 1986 data. f. 1976 data. g. 1981 data.

Source: Unesco data.

(Table continues on the following page.)

Table A-10 (continued)

Economy	1965	1970	1975	1980	1985
120 Switzerland	—	—	1.8[f]	2.2	2.2
Other economies					
121 Angola	19.3	—	—	0.0	0.0[a]
122 Albania	0.0	0.0	—	0.0	0.0
123 Bulgaria	0.0	0.0	0.0	0.0	0.0
124 Czechoslovakia	0.0	0.0	0.0	0.0	0.0
125 Cuba	0.0	0.0	0.0	0.0	0.0
126 German Democratic Republic	0.0	0.0	0.0	0.0	0.0
127 Mongolia	0.0	0.0	0.0	0.0	0.0[a]
128 Democratic People's Republic of Korea	—	—	—	—	—
129 Union of Soviet Socialist Republics	0.0	0.0	0.0	0.0	0.0

— Not available.
Note: *m*, median; *u*, unweighted mean; *w*, weighted mean; and *w′*, weighted mean excluding China and India.
a. 1984 data. b. 1969 data. c. 1974 data. d. 1979 data. e. 1986 data. f. 1976 data. g. 1981 data.
Source: Unesco data.

Table A-11. Repeaters in Primary Schools, Selected Years, 1965–85

Economy	*Repeaters as a percentage of total enrollment*					*Female repeaters as a percentage of female enrollment*				
	1965	*1970*	*1975*	*1980*	*1985*	*1965*	*1970*	*1975*	*1980*	*1985*
Low-income economies	**21.7 *m***	**20.9 *m***	**17.0 *m***	**15.6 *m***	**16.3 *m***	**23.1 *m***	**20.4 *m***	**19.6 *m***	**16.9 *m***	**16.5 *m***
	21.7 *u*	**19.4 *u***	**16.8 *u***	**16.2 *u***	**16.5 *u***	**23.5 *u***	**19.9 *u***	**18.5 *u***	**16.9 *u***	**17.1 *u***
	20.1 *w*	**20.1 *w***	**15.3 *w***	**13.5 *w***	**11.1 *w***	**21.1 *w***	**20.7 *w***	**16.4 *w***	**13.8 *w***	**10.9 *w***
Excluding China and India	**24.4 *w′***	**17.3 *w′***	**15.3 *w′***	**13.5 *w′***	**11.1 *w′***	**26.2 *w′***	**18.5 *w′***	**16.4 *w′***	**13.8 *w′***	**10.9 *w′***
1 Ethiopia	—	—	—	12.2[a]	9.5[b]	—	—	—	14.0[a]	11.2[b]
2 Bhutan	—	—	11.0[c]	12.9[d]	—	—	—	—	12.0[d]	—
3 Burkina Faso	17.9	16.1	17.5	17.1	14.1[e]	19.9	17.1	18.7	18.1	14.7[e]
4 Nepal	—	—	—	—	—	—	—	—	—	—
5 Bangladesh	—	—	18.5[c]	17.8[a]	—	—	—	18.7[c]	18.0[a]	—
6 Malawi	—	—	16.5	17.4	17.8	—	—	16.9	17.1	17.4
7 Zaire	—	22.9	21.1	18.8	—	—	23.4	21.7	19.2	—
8 Mali	25.8	26.3	22.9	29.6	29.6	—	—	—	30.4	30.3[e]
9 Myanmar	—	20.9	—	—	—	—	23.7	—	—	—
10 Mozambique	—	—	—	28.7[a]	23.5	—	—	—	30.7[a]	24.4
11 Madagascar	28.8	29.0	24.4[f]	—	—	29.0	28.1	23.7[f]	—	—
12 Uganda	—	—	10.2	9.6[d]	13.8[b]	—	—	—	—	13.7[b]
13 Burundi	—	21.9	25.6	30.2	17.7	—	20.6	24.1	30.4	17.7
14 Tanzania	—	1.6[g]	0.5	1.2[a]	1.0	—	—	0.4	1.2[a]	1.0[e]
15 Togo	—	33.9	28.8	30.9[d]	34.8	—	34.7	30.0	32.0[d]	38.1[e]
16 Niger	23.2	19.2	13.1	14.3	15.0	24.2	19.8	13.3	14.0	15.0
17 Benin	21.9[h]	18.8	21.2	19.6	26.9	24.9[h]	19.7	23.3	21.6	27.9
18 Somalia	11.1	—	—	—	—	—	—	—	—	—
19 Central African Republic	28.6	28.1	35.3	35.1	29.4	—	29.6	34.3	36.6	28.7
20 India	19.9	20.8	—	—	—	20.9	21.4	—	—	—

— Not available.

Note: *m*, median; *u*, unweighted mean; *w*, weighted mean; and *w′*, weighted mean excluding China and India.

a. 1981 data. b. 1986 data. c. 1976 data. d. 1979 data. e. 1984 data. f. 1974 data. g. 1969 data. h. 1964 data.

Source: Unesco data.

(Table continues on the following page.)

Table A-11 (continued)

Economy	Repeaters as a percentage of total enrollment					Female repeaters as a percentage of female enrollment				
	1965	1970	1975	1980	1985	1965	1970	1975	1980	1985
21 Rwanda	21.5	29.6	21.1[f]	5.7	12.1	21.9	29.0	20.5[f]	5.7	11.7
22 China	—	—	—	—	—	—	—	—	—	—
23 Kenya	—	5.1	5.2	12.9	—	—	4.8	—	12.8	—
24 Zambia	—	—	2.1	1.9	1.8[e]	—	—	1.8	1.6	1.6[e]
25 Sierra Leone	—	—	—	—	—	—	—	—	—	—
26 Sudan	—	2.4	—	0.0	0.0	—	2.8[g]	—	0.0	0.0
27 Haiti	—	—	—	20.9[d]	9.5	—	—	—	20.7[d]	9.5
28 Pakistan	—	—	—	—	—	—	—	—	—	—
29 Lesotho	—	19.8[g]	6.0	20.7	22.6	—	—	6.1	19.7	21.1
30 Ghana	—	3.0	2.3	2.1	—	—	3.3	2.4	2.2	—
31 Sri Lanka	—	22.2	15.4	10.4	8.2	—	19.7	—	—	8.8
32 Mauritania	—	—	14.8	14.0	18.3	—	—	17.5	16.7	20.3
33 Senegal	18.1	19.8	—	15.6	16.3	—	20.2	—	15.9	16.5
34 Afghanistan	—	—	26.9[c]	14.8	5.8	—	—	25.9[f]	15.8	5.7
35 Chad	—	25.9	34.9	—	23.8[b]	—	—	33.6	—	26.5[b]
36 Guinea	—	—	8.8[c]	21.9	27.1	—	—	—	—	30.8
37 Democratic Kampuchea	—	—	—	—	—	—	—	—	—	—
38 Lao People's Democratic Republic	—	—	—	—	—	—	—	—	—	—
39 Viet Nam	—	—	—	—	—	—	—	—	—	—
Lower-middle-income economies	**19.0 *m***	**15.8 *m***	**11.4 *m***	**11.7 *m***	**10.6 *m***	**18.4 *m***	**12.2 *m***	**11.1 *m***	**9.6 m**	**8.7 *m***
	19.6 *u*	**15.6 *u***	**13.1 *u***	**13.3 *u***	**12.2 *u***	**19.0 *u***	**13.0 *u***	**13.4 *u***	**12.2 *u***	**11.7 *u***
	19.6 *w*	**14.5 *w***	**11.9 *w***	**9.9 *w***	**10.2 *w***	**18.0 *w***	**12.6 *w***	**12.1 *w***	**9.7 *w***	**8.7 *w***
40 Liberia	—	—	—	—	—	—	—	—	—	—
41 People's Democratic Republic of Yemen	—	—	—	—	—	—	—	—	—	—
42 Indonesia	—	—	10.7	8.3	10.9	—	—	—	—	—
43 Arab Republic of Yemen	—	—	—	22.9	—	—	—	—	2.3	—

44	Philippines	—	—	—	2.4	1.8	—	—	—	2.0	1.8
45	Morocco	27.0	29.8	28.1	29.5	19.8	—	—	29.0[f]	27.8	18.1
46	Bolivia	—	—	—	—	—	—	—	—	—	—
47	Zimbabwe	—	—	—	0.0[a]	0.8[e]	—	—	—	0.0[a]	0.8[e]
48	Nigeria	—	—	—	—	—	—	—	—	—	—
49	Dominican Republic	—	22.3	—	18.0	12.8	—	20.8	—	—	—
50	Papua New Guinea	—	—	—	—	—	—	—	—	—	—
51	Côte d'Ivoire	—	—	20.9	19.6	28.7[e]	—	—	20.6	19.5	27.9[e]
52	Honduras	—	—	—	16.2	15.5[e]	—	—	—	15.4	14.5[e]
53	Egypt	—	4.5	6.7	7.9	1.5	—	5.7	6.9	8.8	1.3
54	Nicaragua	—	12.8	13.7[f]	16.9	15.4	—	12.3	—	16.0	14.6
55	Thailand	—	—	10.3	8.3	—	—	—	9.4	7.3	—
56	El Salvador	—	16.0[g]	7.5	8.8[a]	8.4[e]	—	15.7[g]	7.0[f]	8.3[a]	8.1[e]
57	Botswana	—	0.3	2.9	2.9	5.7	—	0.4	—	—	5.7
58	Jamaica	—	—	3.6	3.9	3.9[b]	—	—	3.2	3.6	3.5[b]
59	Cameroon	—	—	25.1	30.0	29.3	—	—	24.5	29.5	28.5
60	Guatemala	18.9	15.8	14.8	15.0	13.1	18.4	—	14.0[c]	—	—
61	Congo	26.5	32.6	26.1	25.7	29.9	27.2	33.1	25.6	25.2	29.1
62	Paraguay	21.9	17.7	15.3	13.6	10.6	20.7	15.3	13.8	12.1	9.3
63	Peru	—	17.0	10.2	18.8	14.1	—	16.4	—	18.3	—
64	Turkey	—	—	—	—	8.3	—	—	—	—	8.2
65	Tunisia	26.9	29.2	19.3	20.6	20.4	26.9	—	18.1	19.0	19.0
66	Ecuador	12.4	12.4	11.4	9.7	8.6[e]	12.5	12.0	11.0	10.4[d]	—
67	Mauritius	—	—	—	—	5.9	—	—	—	—	5.6
68	Colombia	19.0	16.6	15.4[f]	—	17.0	18.4	16.0	14.2[f]	—	16.8
69	Chile	—	10.4	12.5	—	—	—	9.4	11.1	—	—
70	Costa Rica	13.9	10.3	6.5	7.9	10.6	—	9.1	5.3	7.1	9.2
71	Jordan	—	6.4[g]	4.1	3.2	4.7[e]	—	6.1[g]	4.3	3.3	4.5[e]

— Not available.

Note: *m*, median; *u*, unweighted mean; *w*, weighted mean; and *w′*, weighted mean excluding China and India.

a. 1981 data. b. 1986 data. c. 1976 data. d. 1979 data. e. 1984 data. f. 1974 data. g. 1969 data. h. 1964 data.

Source: Unesco data.

(*Table continues on the following page.*)

Table A-11 (continued)

Economy	Repeaters as a percentage of total enrollment					Female repeaters as a percentage of female enrollment				
	1965	1970	1975	1980	1985	1965	1970	1975	1980	1985
72 Syrian Arab Republic	9.5	10.9	10.1	8.1	7.5	9.2	9.8	9.9	7.2	6.7
73 Lebanon	—	—	—	—	—	—	—	—	—	—
Upper-middle-income economies	**10.3 m**	**11.4 m**	**9.0 m**	**6.5 m**	**7.5 m**	**8.3 m**	**13.2 m**	**9.8 m**	**5.5 m**	**5.5 m**
	11.0 u	**12.2 u**	**9.3 u**	**8.3 u**	**8.8 u**	**10.1 u**	**14.0 u**	**9.7 u**	**7.3 u**	**6.3 u**
	9.5 w	**10.5 w**	**9.8 w**	**11.2 w**	**11.5 w**	**6.2 w**	**11.1 w**	**10.0 w**	**4.4 w**	**6.3 w**
74 Brazil	—	16.0	15.2	20.2	19.7	—	14.4[g]	14.1[f]	—	—
75 Malaysia	—	—	0.0[f]	0.0	0.0	—	—	0.0[f]	0.0	0.0
76 South Africa	—	—	—	—	—	—	—	—	—	—
77 Mexico	14.7	—	11.0	9.8	9.9	—	—	10.5	—	—
78 Uruguay	—	14.4	14.0[f]	12.4	11.3	—	—	11.8[f]	10.6	9.3
79 Hungary	4.2	3.8	2.7	2.1	3.0	—	—	2.1	1.6	2.1
80 Poland	—	5.2	2.5	2.2	2.9	—	—	—	—	—
81 Portugal	23.4	25.7	11.3	—	17.5[e]	22.4	24.3	9.8	—	15.3[e]
82 Yugoslavia	8.3	7.7	3.7	1.7	1.7	—	—	—	—	—
83 Panama	17.9	15.4	12.6	12.7	13.1	15.7	13.7	10.8	10.6	10.8
84 Argentina	12.3	11.4	8.7	—	—	10.9	10.0	7.7	—	—
85 Republic of Korea	0.5	0.1	0.0	0.0	0.0	0.5	0.1	0.0	0.0	0.0
86 Algeria	—	12.5[g]	12.5	11.7	7.5	—	12.7[g]	12.0	10.7	6.0
87 Venezuela	15.5	2.2	2.7	9.8	9.4	—	—	—	—	—
88 Gabon	—	33.0[g]	34.0	34.8	32.4	—	32.2[g]	34.4	34.4	—
89 Greece	6.6	5.2	2.9	1.1	0.2	5.6	4.3	2.4	0.9	0.1
90 Spain	—	—	—	6.4	4.9	—	—	—	5.5	—
91 Oman	—	—	9.3	—	11.7	—	—	9.1	—	9.5
92 Ireland	—	—	—	—	—	—	—	—	—	—
93 Trinidad and Tobago	—	—	—	3.9[a]	4.8[e]	—	—	—	3.9[a]	5.5[e]
94 Israel	—	—	—	—	—	—	—	—	—	—
95 Hong Kong	—	—	—	3.6	2.0[e]	—	—	—	2.7	1.6[e]
96 Singapore	—	—	7.8	6.6	1.1[e]	—	—	5.6	5.6	1.3[e]

97 Islamic Republic of Iran	—	9.1	—	—	10.2	—	6.5	—	—	9.4[b]
98 Iraq	—	20.6	15.7	10.2[d]	20.8	—	21.6	15.2	9.0[d]	18.9
99 Romania	6.1	—	—	—	—	5.2	—	—	—	—
High-income oil exporters	**20.1 *m***	**20.6 *m***	**14.4 *m***	**8.4 *m***	**5.7 *m***	**24.4 *m***	**16.7 *m***	**12.6 *m***	**7.7 *m***	**5.2 *m***
	20.1 *u*	**19.4 *u***	**14.0 *u***	**9.7 *u***	**7.8 *u***	**24.4 *u***	**17.9 *u***	**13.0 *u***	**7.9 *u***	**6.4 *u***
	22.7 *w*	**21.9 *w***	**15.3 *w***	**12.1 *w***	**11.0 *w***	**24.4 *w***	**19.6 *w***	**13.6 *w***	**8.7 *w***	**8.0 *w***
100 Saudi Arabia	—	20.6[g]	15.2	15.7	12.4	—	16.7[g]	11.7	9.9	8.8
101 Kuwait	15.4	11.9	10.8	6.2	5.2	—	12.2	11.0	6.3	5.2
102 United Arab Emirates	—	—	13.5	7.5	5.7	—	—	13.4	7.3	5.2
103 Libya	24.7	25.7	16.4	9.2	—	24.4	24.9	15.9	8.0	—
High-income industrial market economies	**0.0 *m***	**1.6 *m***	**2.5 *m***	**2.0 *m***	**1.5 *m***	**0.0 *m***	**1.3 *m***	**2.0 *m***	**1.4 *m***	**1.1 *m***
	2.5 *u*	**3.3 *u***	**4.0 *u***	**3.6 *u***	**3.1 *u***	**2.0 *u***	**2.8 *u***	**3.4 *u***	**2.6 *u***	**2.4 *u***
	3.2 *w*	**4.5 *w***	**3.4 *w***	**2.7 *w***	**2.2 *w***	**2.8 *w***	**4.0 *w***	**2.0 *w***	**1.2 *w***	**1.0 *w***
104 New Zealand	0.0	0.0	0.0	3.5	3.1	0.0	0.0	0.0	3.2	2.9
105 Italy	11.0	7.2	3.0	1.2	1.0[e]	9.8	6.2	2.4	1.0	0.8[e]
106 United Kingdom	—	—	—	—	—	—	—	—	—	—
107 Belgium	—	—	23.9	19.4	16.5	—	—	21.6	16.3	15.0
108 Austria	6.5	5.8	3.6[c]	—	—	—	4.5	—	—	—
109 Netherlands	—	3.1	2.5	2.5	2.2[e]	—	2.6	2.0	2.0	1.8[e]
110 France	—	13.2	9.2	9.2[d]	7.8	—	12.0	—	—	—
111 Australia	—	—	—	—	—	—	—	—	—	
112 Federal Republic of Germany	—	3.5	3.2	2.2	1.5	—	3.0	2.8	1.8	1.3
113 Finland	—	—	—	—	0.4[b]	—	—	—	—	0.2[b]
114 Denmark	0.0	0.0	0.0	0.0	0.0	0.0	0.0	—	0.0	0.0
115 Japan	0.0	0.0	0.0	0.0	0.0	0.0	0.0	0.0	0.0	0.0
116 Sweden	0.0	0.0	0.0	0.0	—	0.0	0.0	0.0	0.0	—

— Not available.

Note: *m*, median; *u*, unweighted mean; *w*, weighted mean; and *w′*, weighted mean excluding China and India.

a. 1981 data. b. 1986 data. c. 1976 data. d. 1979 data. e. 1984 data. f. 1974 data. g. 1969 data. h. 1964 data.

Source: Unesco data.

(Table continues on the following page.)

Table A-11 (continued)

Economy	Repeaters as a percentage of total enrollment					Female repeaters as a percentage of female enrollment				
	1965	1970	1975	1980	1985	1965	1970	1975	1980	1985
117 Canada	—	—	—	—	—	—	—	—	—	—
118 Norway	0.0	0.0	0.0	0.0	0.0	—	0.0	0.0	0.0	0.0
119 United States	—	—	—	—	—	—	—	—	—	—
120 Switzerland	—	—	2.5[c]	2.0	2.0	—	—	2.1[c]	1.7	1.7
Other economies										
121 Angola	—	—	—	29.2	—	—	—	—	—	—
122 Albania	—	—	—	—	—	—	—	—	—	—
123 Bulgaria	6.5	6.0	2.1	1.7	1.8	—	—	1.4	1.0	0.9
124 Czechoslovakia	4.3	3.9	0.8	0.9	1.3	—	—	0.6	0.7	0.4
125 Cuba	—	21.6	8.1	5.7	3.2	—	19.2	—	—	—
126 German Democratic Republic	—	—	—	—	—	—	—	—	—	—
127 Mongolia	—	—	—	1.8[a]	—	—	—	—	—	—
128 Democratic People's Republic of Korea	—	—	—	—	—	—	—	—	—	—
129 Union of Soviet Socialist Republics	2.7	2.7	0.8	0.5	0.8	—	—	—	—	—

— Not available.

Note: *m*, median; *u*, unweighted mean; *w*, weighted mean; and *w′*, weighted mean excluding China and India.

a. 1981 data. b. 1986 data. c. 1976 data. d. 1979 data. e. 1984 data. f. 1974 data. g. 1969 data. h. 1964 data.

Source: Unesco data.

Table A-12. Time Needed to Produce a Primary School Graduate, Selected Years, 1970–85

Economy	*Number of grades in the primary education system*	*Student-years needed per graduate*				*Difference between the student-years needed per graduate and years in the primary school system*			
		1970	*1975*	*1980*	*1985*	*1970*	*1975*	*1980*	*1985*
Low-income economies						**4.8 *m***	**4.8 *m***	**3.8 *m***	**4.0 *m***
						8.6 *u*	**4.5 *u***	**4.5 *u***	**5.2 *u***
1 Ethiopia	6	—	—	9.0	8.4	—	—	3.0	2.4
2 Bhutan	—	—	—	—	—	—	—	—	—
3 Burkina Faso	6	10.8	9.6	8.4	9.0	4.8	3.6	2.4	3.0
4 Nepal	—	—	—	—	—	—	—	—	
5 Bangladesh[a]	4, 5	—	13.5[b]	*14.5*	—	—	8.5	9.5	—
6 Malawi	8	—	16.0	15.2	14.4	—	8.0	7.2	6.4
7 Zaire	—	—	—	—	—	—	—	—	—
8 Mali	6	13.8	10.8	15.0[c]	15.0	7.8	4.8	9.0	9.0
9 Myanmar	—	—	—	—	—	—	—	—	—
10 Mozambique	—	—	—	—	—	—	—	—	—
11 Madagascar	—	—	—	—	—	—	—	—	—
12 Uganda	7	—	9.1	9.1[d]	—	—	2.1	2.1	—
13 Burundi[a]	7, 6	38.5[e]	*12.6*	9.6	8.4	31.5	6.6	3.6	2.4
14 Tanzania	7	9.8	7.7	7.7	7.7	2.8	0.7	0.7	0.7
15 Togo	6	10.8	9.6	15.0	13.2	4.8	3.6	9.0	7.2
16 Niger	—	—	—	—	—	—	—	—	—
17 Benin	—	—	—	—	—	—	—	—	—
18 Somalia	—	—	—	—	—	—	—	—	—
19 Central African Republic	6	15.6	12.0	13.2	25.8	9.6	6.0	7.2	19.8
20 India	—	—	—	—	—	—	—	—	—

— Not available.

Note: *m*, median; *u*, unweighted mean; *w*, weighted mean; and *w′*, weighted mean excluding China and India.

a. Length of the primary education system changed; italicized number flags the beginning of the new system. b. 1976 data. c. 1979 data. d. 1981 data. e. 1971 data. f. 1984 data.

Source: Unesco data.

(*Table continues on the following page.*)

Table A-12 (continued)

Economy	Number of grades in the primary education system	Student-years needed per graduate				Difference between the student-years needed per graduate and years in the primary school system			
		1970	1975	1980	1985	1970	1975	1980	1985
21 Rwanda[a]	6, 8	16.2	11.4	12.0	12.0	10.2	5.4	4.0	4.0
22 China	—	—	—	—	—	—	—	—	—
23 Kenya	—	—	—	—	—	—	—	—	—
24 Zambia	7	8.4[e]	8.4	7.7	7.7	1.4	1.4	0.7	0.7
25 Sierra Leone	—	—	—	—	—	—	—	—	—
26 Sudan	—	—	—	—	—	—	—	—	—
27 Haiti	6	—	—	10.8[d]	10.2[f]	—	—	4.8	4.2
28 Pakistan	—	—	—	—	—	—	—	—	—
29 Lesotho[a]	8, 7	24.8	14.0	12.6	12.6[f]	16.8	7.0	5.6	5.6
30 Ghana[a]	10, 6	12.0	7.2	6.6[b]	—	2.0	1.2	0.6	—
31 Sri Lanka	—	—	—	—	—	—	—	—	—
32 Mauritania	—	—	—	—	—	—	—	—	—
33 Senegal	6	9.0	—	7.8	7.8[f]	3.0	—	1.8	1.8
34 Afghanistan	—	—	—	—	—	—	—	—	—
35 Chad	—	—	—	—	—	—	—	—	—
36 Guinea	—	—	—	—	—	—	—	—	—
37 Democratic Kampuchea	—	—	—	—	—	—	—	—	—
38 Lao People's Democratic Republic	—	—	—	—	—	—	—	—	—
39 Viet Nam	—	—	—	—	—	—	—	—	—
Lower-middle-income economies						**3.8 *m*** **3.9 *u***	**2.4 *m*** **3.2 *u***	**2.7 *m*** **3.3 *u***	**1.8 *m*** **2.8 *u***
40 Liberia	—	—	—	—	—	—	—	—	—
41 People's Democratic Republic of Yemen	—	—	—	—	—	—	—	—	—
42 Indonesia	—	—	—	—	—	—	—	—	—
43 Arab Republic of Yemen	6	—	—	19.8	—	—	—	13.8	—

44 Philippines	—	—	—	—	—	—	—	—	—
45 Morocco	5	9.0	8.0	8.0	7.5	4.0	3.0	3.0	2.5
46 Bolivia	—	—	—	—	—	—	—	—	—
47 Zimbabwe	—	—	—	—	—	—	—	—	—
48 Nigeria	—	—	—	—	—	—	—	—	—
49 Dominican Republic	—	—	—	—	—	—	—	—	—
50 Papua New Guinea	—	—	—	—	—	—	—	—	—
51 Côte d'Ivoire	6	9.0[e]	7.8	7.8[c]	—	3.0	1.8	1.8	—
52 Honduras	6	—	—	11.4[c]	—	—	—	5.4	—
53 Egypt	6	7.2	7.2	6.6	6.6	1.2	1.2	0.6	0.6
54 Nicaragua	6	11.4	12.6[b]	11.4[c]	17.4	5.4	6.6	5.4	11.4
55 Thailand[a]	7, 6	—	14.0	7.8[c]	—	—	7.0	1.8	—
56 El Salvador	9	18.0[e]	17.1	11.7[d]	—	9.0	8.1	2.7	—
57 Botswana	7	10.5	8.4	9.1	7.7	3.5	1.4	2.1	0.7
58 Jamaica	—	—	7.8	—	—	1.8	—	—	—
59 Cameroon	—	—	—	—	—	—	—	—	—
60 Guatemala	6	12.6	12.0	10.8	10.2	6.6	6.0	4.8	4.2
61 Congo	6	10.2	9.6	9.6	10.8	4.2	3.6	3.6	4.8
62 Paraguay	—	—	—	—	—	—	—	—	—
63 Peru	6	—	7.8	9.0	—	—	1.8	3.0	—
64 Turkey	5	—	—	—	5.5	—	—	—	0.5
65 Tunisia	6	12.0	9.0	9.0	9.6	6.0	3.0	3.0	3.6
66 Ecuador	6	10.2	8.4	7.8	—	4.2	2.4	1.8	—
67 Mauritius	6	—	—	—	6.6	—	—	—	0.6
68 Colombia	—	—	—	—	—	—	—	—	—
69 Chile	8	9.6	11.2	—	—	1.6	3.2	—	—
70 Costa Rica	6	8.4	7.2	7.2	7.8	2.4	1.2	1.2	1.8

— Not available.

Note: m, median; u, unweighted mean; w, weighted mean; and w', weighted mean excluding China and India.

a. Length of the primary education system changed; italicized number flags the beginning of the new system. b. 1976 data. c. 1979 data. d. 1981 data. e. 1971 data. f. 1984 data.

Source: Unesco data.

(Table continues on the following page.)

Table A-12 (continued)

Economy	Number of grades in the primary education system	Student-years needed per graduate				Difference between the student-years needed per graduate and years in the primary school system			
		1970	1975	1980	1985	1970	1975	1980	1985
71 Jordan	6	7.8	6.6	6.6	—	1.8	0.6	0.6	—
72 Syrian Arab Republic	6	7.2	7.8	7.2	6.6	1.2	1.8	1.2	0.6
73 Lebanon	—	—	—	—	—	—	—	—	—
Upper-middle-income economies						**1.2 m** **1.5 u**	**1.8 m** **1.3 u**	**1.2 m** **1.0 u**	**1.2 m** **0.9 u**
74 Brazil	—	—	—	—	—	—	—	—	—
75 Malaysia	—	—	—	—	—	—	—	—	—
76 South Africa	—	—	—	—	—	—	—	—	—
77 Mexico	6	—	7.8	7.8	7.8	—	1.8	1.8	1.8
78 Uruguay	6	—	7.8[b]	7.8[c]	7.2	—	1.8	1.8	1.2
79 Hungary	—	—	—	—	—	—	—	—	—
80 Poland	—	—	—	—	—	—	—	—	—
81 Portugal	—	—	—	—	—	—	—	—	—
82 Yugoslavia	4	4.4	4.4	4.0	—	0.4	0.4	0.0	—
83 Panama	6	7.8	7.8	7.8	7.2	1.8	1.8	1.8	1.2
84 Argentina	7	9.8	9.1	—	—	2.8	2.1	—	—
85 Republic of Korea	6	6.0	6.0	6.0	6.0	0.0	0.0	0.0	0.0
86 Algeria	6	9.0[e]	8.4	7.8	6.6	3.0	2.4	1.8	0.6
87 Venezuela	6	8.4	7.8	7.8	7.8	2.4	1.8	1.8	1.8
88 Gabon	—	—	—	—	—	—	—	—	—
89 Greece	6	6.6	6.0	6.0	6.0[d]	0.6	0.0	0.0	0.0
90 Spain	—	—	—	—	—	—	—	—	—
91 Oman	—	—	—	—	—	—	—	—	—
92 Ireland	—	—	—	—	—	—	—	—	—
93 Trinidad and Tobago	—	—	—	—	—	—	—	—	—
94 Israel	—	—	—	—	—	—	—	—	—
95 Hong Kong	6	—	—	6.6	—	—	—	0.6	—

96 Singapore	6	6.6[e]	6.6	6.6	—	0.6	0.6	0.6	—
97 Islamic Republic of Iran	—	—	—	—	—	—	—	—	—
98 Iraq	—	—	—	—	—	—	—	—	—
99 Romania	—	—	—	—	—	—	—	—	—
High-income oil exporters									
100 Saudi Arabia	—	—	—	—	—	—	—	—	—
101 Kuwait	—	—	—	—	—	—	—	—	—
102 United Arab Emirates	—	—	—	—	—	—	—	—	—
103 Libya	—	—	—	—	—	—	—	—	—
High-income industrial market economies									
104 New Zealand	—	—	—	—	—	—	—	—	—
105 Italy	—	—	—	—	—	—	—	—	—
106 United Kingdom	—	—	—	—	—	—	—	—	—
107 Belgium	—	—	—	—	—	—	—	—	—
108 Austria	—	—	—	—	—	—	—	—	—
109 Netherlands	—	—	—	—	—	—	—	—	—
110 France	—	—	—	—	—	—	—	—	—
111 Australia	—	—	—	—	—	—	—	—	—
112 Federal Republic of Germany	—	—	—	—	—	—	—	—	—
113 Finland	—	—	—	—	—	—	—	—	—
114 Denmark	—	—	—	—	—	—	—	—	—
115 Japan	—	—	—	—	—	—	—	—	—
116 Sweden	—	—	—	—	—	—	—	—	—
117 Canada	—	—	—	—	—	—	—	—	—

— Not available.

Note: m, median; u, unweighted mean; w, weighted mean; and w', weighted mean excluding China and India.

a. Length of the primary education system changed; italicized number flags the beginning of the new system. b. 1976 data. c. 1979 data. d. 1981 data. e. 1971 data. f. 1984 data.

Source: Unesco data.

(Table continues on the following page.)

Table A-12 (continued)

Economy	Number of grades in the primary education system	Student-years needed per graduate				Difference between the student-years needed per graduate and years in the primary school system			
		1970	1975	1980	1985	1970	1975	1980	1985
118 Norway	—	—	—	—	—	—	—	—	—
119 United States	—	—	—	—	—	—	—	—	—
120 Switzerland	—	—	—	—	—	—	—	—	—
Other economies									
121 Angola	—	—	—	—	—	—	—	—	—
122 Albania	—	—	—	—	—	—	—	—	—
123 Bulgaria	—	—	—	—	—	—	—	—	—
124 Czechoslovakia	—	—	—	—	—	—	—	—	—
125 Cuba	—	—	—	—	—	—	—	—	—
126 German Democratic Republic	—	—	—	—	—	—	—	—	—
127 Mongolia	—	—	—	—	—	—	—	—	—
128 Democratic People's Republic of Korea	—	—	—	—	—	—	—	—	—
129 Union of Soviet Socialist Republics	—	—	—	—	—	—	—	—	—

— Not available.

Note: *m*, median; *u*, unweighted mean; *w*, weighted mean; and *w′*, weighted mean excluding China and India.

a. Length of the primary education system changed; italicized number flags the beginning of the new system. b. 1976 data. c. 1979 data. d. 1981 data. e. 1971 data. f. 1984 data.

Source: Unesco data.

Table A-13. Percentage of Primary School Cohort Reaching the Last Grade, Selected Years, 1970–86

Economy	1970	1975	1980	1985	Average, 1980–86
Low-income economies	**49.0 *m***	**57.8 *m***	**57.7 *m***	**59.2 *m***	**56.0 *m***
	48.1 *u*	**57.4 *u***	**59.8 *u***	**58.6 *u***	**57.5 *u***
1 Ethiopia	—	—	49.8	49.6	50.0
2 Bhutan	—	—	—	—	—
3 Burkina Faso	49.0	61.7	74.9	73.8	67.0
4 Nepal	—	—	—	—	—
5 Bangladesh	—	21.2[a]	20.4	—	—
6 Malawi	—	29.6	34.0	46.3	31.0
7 Zaire	—	—	—	—	60.0
8 Mali	44.2	57.8	36.4[b]	39.6	40.0
9 Myanmar	—	—	—	—	—
10 Mozambique	—	—	—	—	34.0
11 Madagascar	—	—	—	—	—
12 Uganda	—	77.6	75.7[c]	—	—
13 Burundi	8.6[d]	45.6	94.3	86.7	87.0
14 Tanzania	58.6	83.8	81.8	80.8	71.0
15 Togo	68.9	75.9	42.9	59.2	52.0
16 Niger	—	—	—	—	75.0
17 Benin	—	—	—	—	36.0
18 Somalia	—	—	—	—	—
19 Central African Republic	29.7	54.3	52.6	16.9	17.0
20 India	—	—	—	—	—
21 Rwanda	30.0	48.3	62.8	61.7	46.0
22 China	—	—	—	—	—
23 Kenya	—	—	—	—	—

— Not available.

Note: *m*, median; *u*, unweighted mean.

a. 1976 data. b. 1979 data. c. 1981 data. d. 1971 data. e. 1984 data.

Sources: Averages from World Conference on Education for All 1990; other data from Unesco.

(Table continues on the following page.)

Table A-13 (*continued*)

Economy	1970	1975	1980	1985	Average, 1980–86
24 Zambia	75.5[d]	80.3	83.5	80.1	80.0
25 Sierra Leone	—	—	—	—	36.0
26 Sudan	—	—	—	—	—
27 Haiti	—	—	31.2[c]	32.1[e]	32.0
28 Pakistan	—	—	—	—	—
29 Lesotho	21.9	34.1	48.4	51.7[e]	52.0
30 Ghana	71.6	76.0	81.5[b]	—	—
31 Sri Lanka	—	—	—	—	94.0
32 Mauritania	—	—	—	—	78.0
33 Senegal	71.1	—	85.9	82.9[e]	85.0
34 Afghanistan	—	—	—	—	78.0
35 Chad	—	—	—	—	78.0
36 Guinea	—	—	—	—	43.0
37 Democratic Kampuchea	—	—	—	—	—
38 Lao People's Democratic Republic	—	—	—	—	—
39 Viet Nam	—	—	—	—	—
Lower-middle-income economies	**69.1 *m*** **60.2 *u***	**74.5 *m*** **66.6 *u***	**74.2 *m*** **67.9 *u***	**80.5 *m*** **74.6 *u***	**73.5 *m*** **70.5 *u***
40 Liberia	—	—	—	—	—
41 People's Democratic Republic of Yemen	—	—	—	—	94.0
42 Indonesia	—	—	—	—	80.0
43 Arab Republic of Yemen	—	—	28.4	—	—
44 Philippines	—	—	—	—	75.0
45 Morocco	67.3	79.4	79.9	68.9	67.0
46 Bolivia	—	—	—	—	—
47 Zimbabwe	—	—	—	—	74.0
48 Nigeria	—	—	—	—	—

49 Dominican Republic	—	—	—	—	35.0
50 Papua New Guinea	—	—	—	—	—
51 Côte d'Ivoire	86.5[d]	96.2	89.5[b]	—	73.0
52 Honduras	—	—	36.7[b]	—	43.0
53 Egypt	77.9	83.9	85.1	93.4	95.0
54 Nicaragua	30.0	28.6[a]	37.5[b]	19.7	35.0
55 Thailand	—	43.0	76.7[b]	—	—
56 El Salvador	23.9[d]	28.9	59.7[c]	—	31.0
57 Botswana	49.6	81.8	73.3	89.3	89.0
58 Jamaica	—	74.5	—	—	82.0
59 Cameroon	—	—	—	—	70.0
60 Guatemala	30.8	32.2	41.3	35.5	36.0
61 Congo	73.4	70.5	74.2	70.7	71.0
62 Paraguay	—	—	—	—	50.0
63 Peru	—	65.8	70.2	—	—
64 Turkey	—	—	—	95.9	97.0
65 Tunisia	55.5	80.7	78.0	77.1	72.0
66 Ecuador	46.2	57.0	65.1	—	63.0
67 Mauritius	—	—	—	95.9	98.0
68 Colombia	—	—	—	—	57.0
69 Chile	77.3	66.7	—	—	85.0
70 Costa Rica	70.8	77.1	74.9	80.5	76.0
71 Jordan	71.4	86.7	97.0	—	96.0
72 Syrian Arab Republic	82.7	79.0	86.5	93.4	89.0
73 Lebanon	—	—	—	—	—
Upper-middle-income economies	**89.5 *m***	**82.0 *m***	**89.1 *m***	**85.9 *m***	**92.0 *m***
	81.9 *u*	**80.7 *u***	**84.9 *u***	**85.7 *u***	**84.0 *u***
74 Brazil	—	—	—	—	22.0

— Not available.

Note: *m*, median; *u*, unweighted mean.

a. 1976 data. b. 1979 data. c. 1981 data. d. 1971 data. e. 1984 data.

Sources: Averages from World Conference on Education for All 1990; other data from Unesco.

(*Table continues on the following page.*)

Table A-13 (continued)

Economy	1970	1975	1980	1985	Average, 1980–86
75 Malaysia	—	—	—	—	99.0
76 South Africa	—	—	—	—	—
77 Mexico	—	64.8	65.6	71.4	69.0
78 Uruguay	—	88.9[a]	88.1[b]	85.9	86.0
79 Hungary	—	—	—	—	92.0
80 Poland	—	—	—	—	93.0
81 Portugal	—	—	—	—	—
82 Yugoslavia	95.6	94.6	98.4	—	98.0
83 Panama	85.5	75.1	72.8	81.6	82.0
84 Argentina	63.7	65.7	—	—	—
85 Republic of Korea	95.2	92.4	93.9	99.0	99.0
86 Algeria	68.7[d]	71.2	76.5	90.2	91.0
87 Venezuela	59.5	61.2	68.5	73.1	73.0
88 Gabon	—	—	—	—	44.0
89 Greece	93.5	97.7	97.9	98.5[e]	99.0
90 Spain	—	—	—	—	98.0
91 Oman	—	—	—	—	92.0
92 Ireland	—	—	—	—	—
93 Trinidad and Tobago	—	—	—	—	—
94 Israel	—	—	—	—	93.0
95 Hong Kong	—	—	97.7	—	98.0
96 Singapore	93.7[d]	94.9	90.0	—	95.0
97 Islamic Republic of Iran	—	—	—	—	87.0
98 Iraq	—	—	—	—	73.0
99 Romania	—	—	—	—	—
High-income oil exporters	—	—	—	—	**88.0 *m***
	—	—	—	—	**69.0 *u***
100 Saudi Arabia	—	—	—	—	90.0

101 Kuwait	—	—	—	—	29.0
102 United Arab Emirates	—	—	—	—	88.0
103 Libya	—	—	—	—	—
High-income industrial market economies	—	—	—	—	**97.0 *m***
	—	—	—	—	**92.0 *u***
104 New Zealand	—	—	—	—	—
105 Italy	—	—	—	—	100.0
106 United Kingdom	—	—	—	—	—
107 Belgium	—	—	—	—	78.0
108 Austria	—	—	—	—	—
109 Netherlands	—	—	—	—	94.0
110 France	—	—	—	—	94.0
111 Australia	—	—	—	—	—
112 Federal Republic of Germany	—	—	—	—	97.0
113 Finland	—	—	—	—	97.0
114 Denmark	—	—	—	—	100.0
115 Japan	—	—	—	—	100.0
116 Sweden	—	—	—	—	100.0
117 Canada	—	—	—	—	—
118 Norway	—	—	—	—	100.0
119 United States	—	—	—	—	—
120 Switzerland	—	—	—	—	55.0
Other economies					
121 Angola	—	—	—	—	—
122 Albania	—	—	—	—	—
123 Bulgaria	—	—	—	—	86.0

— Not available.
Note: *m*, median; *u*, unweighted mean.
a. 1976 data. b. 1979 data. c. 1981 data. d. 1971 data. e. 1984 data.
Sources: Averages from World Conference on Education for All 1990; other data from Unesco.

(Table continues on the following page.)

Table A-13 (*continued*)

Economy	1970	1975	1980	1985	*Average, 1980–86*
124 Czechoslovakia	—	—	—	—	93.0
125 Cuba	—	—	—	—	91.0
126 German Democratic Republic	—	—	—	—	—
127 Mongolia	—	—	—	—	—
128 Democratic People's Republic of Korea	—	—	—	—	—
129 Union of Soviet Socialist Republics	—	—	—	—	—

— Not available.

Note: *m*, median; *u*, unweighted mean.

a. 1976 data. b. 1979 data. c. 1981 data. d. 1971 data. e. 1984 data.

Sources: Averages from World Conference on Education for All 1990; other data from Unesco.

Table A-14. Public Expenditure on Education, Selected Years, 1965–85

Economy	Total expenditure (millions of constant 1985 dollars)					Annual growth (percent)		
	1965	1970	1975	1980	1985	1965–75	1975–80	1980–85
Low-income economies	—	—	—	—	—	**6.1 *m***	**6.3 *m***	**2.9 *m***
	—	—	—	—	—	**6.1 *u***	**4.6 *u***	**8.0 *u***
	—	—	—	—	—	**5.1 *w***	**6.2 *w***	**9.3 *w***
Excluding China and India	—	—	—	—	—	**5.3 *w′***	**2.8 *w′***	**2.8 *w′***
1 Ethiopia	39.6	74.4	134.7	157.9	202.7	13.0	3.2	5.1
2 Bhutan	—	—	—	—	—	—	—	—
3 Burkina Faso	—	—	20.5[a]	25.7	28.7	—	5.8	2.2
4 Nepal	8.9	9.3	25.5	35.5	69.8	11.1	6.8	14.5
5 Bangladesh	119.1	—	113.7	202.2	299.6	-0.5	12.2	8.2
6 Malawi	17.3	25.0	19.8	31.4	37.4	1.4	9.7	3.6
7 Zaire	—	—	—	—	—	—	—	—
8 Mali	—	—	—	37.5	38.2	—	—	0.4
9 Myanmar	85.3	105.3	68.2	—	—	-2.2	—	—
10 Mozambique	—	—	—	—	—	—	—	—
11 Madagascar	110.0	—	73.4	134.7	78.8	-4.0	12.9	-10.2
12 Uganda	117.8	158.7	103.6	39.3	—	-1.3	-17.6	—
13 Burundi	—	—	—	28.0[b]	28.7	—	—	0.4
14 Tanzania	114.9	211.5	320.9	339.7	266.3	10.8	1.1	-4.8
15 Togo	7.1	10.8	21.0	41.3	35.3	11.5	14.5	-3.1
16 Niger	9.1	15.7	28.6	51.3	—	12.1	12.4	—
17 Benin	—	—	—	—	—	—	—	—
18 Somalia	3.9	6.1	14.0	8.3	—	13.6	-9.9	—
19 Central African Republic	—	—	29.7	24.9[b]	18.1	—	-4.3	-5.2
20 India	2,395.2	3,306.3	3,915.7	4,564.9	7,130.6	5.0	3.1	9.3

— Not available.

Note: *m*, median; *u*, unweighted mean; *w*, weighted mean; and *w′*, weighted mean excluding China and India.

a. 1976 data. b. 1979 data. c. 1984 data. d. 1974 data. e. 1981 data. f. 1986 data.

Sources: Unesco data; National Center for Education Statistics 1988.

(Table continues on the following page.)

Table A-14 (continued)

Economy	Total expenditure (millions of constant 1985 dollars)					Annual growth (percent)		
	1965	1970	1975	1980	1985	1965–75	1975–80	1980–85
21 Rwanda	—	20.1	22.0	38.6	51.4[c]	—	11.9	7.4
22 China	—	1,283.5	2,377.9	4,390.4	7,761.1	—	13.0	12.1
23 Kenya	84.7	119.8	238.8	352.7	375.6	10.9	8.1	1.3
24 Zambia	98.1	97.3	154.4	105.0	128.1[c]	4.6	–7.4	5.1
25 Sierra Leone	29.6	33.0	40.4	48.4	—	3.2	3.7	—
26 Sudan	132.0	201.3	302.3[d]	355.7	—	9.6	2.7	—
27 Haiti	—	—	16.6[a]	31.0	23.5	—	16.9	–5.4
28 Pakistan	186.4	238.4	373.8	448.6	693.0	7.2	3.7	9.1
29 Lesotho	4.3	5.1	12.1	21.0	19.0[c]	10.9	11.7	–2.5
30 Ghana	172.1	187.8	256.8	141.6	113.9	4.1	–11.2	–4.3
31 Sri Lanka	98.0	121.4	99.5	146.0	178.9	0.2	8.0	4.1
32 Mauritania	—	18.4	22.9[a]	—	—	—	—	—
33 Senegal	—	69.7	—	—	—	—	—	—
34 Afghanistan	—	—	—	—	—	—	—	—
35 Chad	12.3	—	—	—	—	—	—	—
36 Guinea	—	—	—	—	58.5[c]	—	—	—
37 Democratic Kampuchea	—	—	—	—	—	—	—	—
38 Lao People's Democratic Republic	—	—	—	2.0	4.9	—	—	19.6
39 Viet Nam	—	—	—	—	—	—	—	—
Lower-middle-income economies	—	—	—	—	—	**7.2 *m***	**7.2 *m***	**0.8 *m***
	—	—	—	—	—	**7.6 *u***	**7.7 *u***	**–0.8 *u***
	—	—	—	—	—	**7.3 *w***	**3.9 *w***	**–2.3 *w***
40 Liberia	15.8	19.4	20.7	66.6	—	2.7	26.3	—
41 People's Democratic Republic of Yemen	—	—	23.3	68.6	—	—	24.1	—
42 Indonesia	—	884.2	1,322.5	1,172.6	—	—	–2.4	—

43	Arab Republic of Yemen	—	8.3	—	243.3[e]	248.2	—	—	0.5
44	Philippines	343.7	474.1	467.9	534.5	402.4	3.1	2.7	–5.5
45	Morocco	158.4	220.8	392.4	640.4	—	9.5	10.3	—
46	Bolivia	47.0	84.5	117.8	161.0	12.1[c]	9.6	6.4	–47.6
47	Zimbabwe	—	98.6	133.0	253.0	385.4	—	13.7	8.8
48	Nigeria	937.1	—	2,490.5[d]	3,171.4[b]	911.3	11.5	5.0	–18.8
49	Dominican Republic	—	65.1	72.4[a]	93.1	75.3	—	6.5	–4.2
50	Papua New Guinea	53.9	70.6	141.4[a]	96.5[b]	—	9.2	–12.0	—
51	Côte d'Ivoire	120.7	195.7	313.8	547.8[b]	—	10.0	14.9	—
52	Honduras	44.9	58.8	83.0	100.9	144.9	6.3	4.0	7.5
53	Egypt	461.5	546.9	688.1	843.9[b]	1,714.9	4.1	5.2	12.5
54	Nicaragua	40.9	60.9	82.1	89.0	160.2	7.2	1.6	12.5
55	Thailand	230.3	473.8	677.4	969.4	1,449.5	11.4	7.4	8.4
56	El Salvador	69.1	90.2	137.3	163.7	110.2[c]	7.1	3.6	–9.4
57	Botswana	3.9	6.4	21.4	33.6	53.0	18.6	9.4	9.5
58	Jamaica	51.5	74.7	140.0	127.8	98.9	10.5	–1.8	–5.0
59	Cameroon	73.5	90.5	121.2	165.4	296.2	5.1	6.4	12.4
60	Guatemala	89.5	114.9	120.1	187.6[b]	168.7[c]	3.0	11.8	–2.1
61	Congo	—	42.4	84.4	87.7	101.4[c]	—	0.8	3.7
62	Paraguay	—	26.6	26.8	44.4	46.9	—	10.6	1.1
63	Peru	438.9	412.1	533.1	545.4	459.4	2.0	0.5	–3.4
64	Turkey	677.9	745.3	—	1,137.6	1,201.4	—	—	1.1
65	Tunisia	144.4	225.4	249.4	353.0	465.9	5.6	7.2	5.7
66	Ecuador	110.4	192.0	250.4	579.8	415.4	8.5	18.3	–6.5
67	Mauritius	14.6	16.1	25.5	44.9	38.7	5.7	12.0	–2.9
68	Colombia	301.6	344.2	520.5	578.2	0.0	5.6	2.1	—
69	Chile	342.8	628.0	448.4	727.5	630.1	2.7	10.2	–2.8
70	Costa Rica	69.4	110.1	192.6	278.1	162.1	10.7	7.6	–10.2
71	Jordan	—	69.6	88.1	205.2	267.9	—	18.4	5.5

— Not available.

Note: *m*, median; *u*, unweighted mean; *w*, weighted mean; and *w′*, weighted mean excluding China and India.

a. 1976 data. b. 1979 data. c. 1984 data. d. 1974 data. e. 1981 data. f. 1986 data.

Sources: Unesco data; National Center for Education Statistics 1988.

(*Table continues on the following page.*)

Table A-14 (continued)

Economy	Total expenditure (millions of constant 1985 dollars)					Annual growth (percent)		
	1965	1970	1975	1980	1985	1965–75	1975–80	1980–85
72 Syrian Arab Republic	137.4	229.3	444.5	674.1	1,020.3	12.5	8.7	8.6
73 Lebanon	—	—	—	—	—	—	—	—
Upper-middle-income economies	—	—	—	—	—	**9.4 m** **8.6 u** **8.7 w**	**7.2 m** **8.0 u** **8.6 w**	**3.6 m** **4.2 u** **0.8 w**
74 Brazil	1,602.4	2,610.1	4,480.1	7,168.3	7,987.3	10.8	9.9	2.2
75 Malaysia	356.7	464.5	944.7	1,418.2	1,917.0	10.2	8.5	6.2
76 South Africa	—	—	—	—	—	—	—	—
77 Mexico	1,381.7	1,998.6	4,246.1	4,701.9	4,424.4	11.9	2.1	−1.2
78 Uruguay	132.7	164.9	—	133.8	123.9	—	—	−1.5
79 Hungary	370.8	436.6	645.0	867.8	1,078.7	5.7	6.1	4.4
80 Poland	—	—	—	—	—	—	—	—
81 Portugal	130.9	208.5	613.1	848.5	897.3	16.7	6.7	1.1
82 Yugoslavia	829.2	1,254.3	1,853.5	2,150.4	1,553.9	8.4	3.0	−6.3
83 Panama	68.9	134.4	176.1	198.9	237.0	9.8	2.5	3.6
84 Argentina	1,773.0	1,798.7	1,651.0	2,643.0	1,200.7[f]	−0.7	9.9	−12.3
85 Republic of Korea	307.5	929.4	894.3	2,182.4	4,057.6	11.3	19.5	13.2
86 Algeria	808.6	1,825.7	2,164.5	3,515.5	3,441.3	10.3	10.2	−0.4
87 Venezuela	1,285.8	1,989.6	2,612.1	2,910.6	3,215.1	7.3	2.2	2.0
88 Gabon	35.0	45.3	67.6	79.9	154.7	6.8	3.4	14.1
89 Greece	321.5	412.1	509.5	683.1[b]	963.6	4.7	7.6	5.9
90 Spain	1,019.0	2,210.4	2,464.0	3,492.6	5,393.3	9.2	7.2	9.1
91 Oman	—	30.3	49.4	91.3	357.5	—	13.1	31.4
92 Ireland	374.1	548.4	820.5	1,068.4	1,118.3	8.2	5.4	0.9
93 Trinidad and Tobago	125.7	167.9	183.5	340.9	425.4	3.9	13.2	4.5
94 Israel	411.4	683.9	1,186.6	1,679.0	2,313.0[c]	11.2	7.2	8.3
95 Hong Kong	154.6	266.1	383.6	642.1	943.3[c]	9.5	10.9	10.1

96	Singapore	132.7	175.8	253.6	358.5	—	6.7	7.2	—
97	Islamic Republic of Iran	1,955.6	3,171.6	4,748.6[d]	9,611.1	6,559.8	10.4	12.5	–7.4
98	Iraq	—	—	—	—	—	—	—	—
99	Romania	—	—	—	—	—	—	—	—
High-income oil exporters		—	—	—	—	—	**11.0 *m***	**–1.5 m**	**8.2 *m***
							11.6 *u*	**5.2 *u***	**7.5 *u***
							14.7 *w*	**–0.9 *w***	**8.0 *w***
100	Saudi Arabia	1,132.8	1,364.0	6,032.6	5,589.9	8,143.5	18.2	–1.5	7.8
101	Kuwait	458.7	954.4	781.9	728.3	1,094.4	5.5	–1.4	8.5
102	United Arab Emirates	—	—	125.6	394.3	473.5	—	25.7	3.7
103	Libya	474.7	1,148.8	1,350.3	1,207.7	1,940.7	11.0	–2.2	10.0
High-income industrial market economies		—	—	—	—	—	**7.3 *m***	**2.0 *m***	**1.2 *m***
							7.2 *u*	**2.5 *u***	**1.2 *u***
							5.9 *w*	**4.1 *w***	**1.8 *w***
104	New Zealand	508.0	768.6	1,163.6	1,122.5	1,051.8	8.6	–0.7	–1.3
105	Italy	8,631.9	9,837.9	12,621.3	16,347.7[b]	—	3.9	6.7	—
106	United Kingdom	15,260.4	18,115.5	25,052.4	23,003.2	22,458.8	5.1	–1.7	–0.5
107	Belgium	1,895.4	—	4,200.6	4,645.1	4,840.0	8.3	2.0	0.8
108	Austria	1,166.9	1,958.2	2,933.4	3,366.9	3,800.8	9.7	2.8	2.5
109	Netherlands	4,122.5	6,477.3	8,674.2	9,464.4	8,519.9	7.7	1.8	–2.1
110	France	11,006.9	16,373.9	21,651.2	24,572.0	30,506.1[c]	7.0	2.6	5.6
111	Australia	2,847.0	4,533.0	8,318.0	8,643.0	9,914.0	11.3	0.8	2.8
112	Federal Republic of Germany	10,853.0	15,638.0	25,641.0	27,823.0	28,428.0	9.0	1.6	0.4
113	Finland	1,517.0	1,940.0	2,517.0	2,538.0	3,060.0	5.2	0.2	3.8
114	Denmark	1,967.0	2,823.0	3,524.0	3,437.0	3,526.0	6.0	–0.5	0.5
115	Japan	17,794.0	27,355.0	47,246.0	63,377.0	67,673.0	10.3	6.1	1.3

— Not available.

Note: *m*, median; *u*, unweighted mean; *w*, weighted mean; and *w′*, weighted mean excluding China and India.

a. 1976 data. b. 1979 data. c. 1984 data. d. 1974 data. e. 1981 data. f. 1986 data.

Sources: Unesco data; National Center for Education Statistics 1988.

(Table continues on the following page.)

Table A-14 (continued)

Economy	Total expenditure (millions of constant 1985 dollars)					Annual growth (percent)		
	1965	1970	1975	1980	1985	1965–75	1975–80	1980–85
116 Sweden	3,613.0	5,759.0	6,049.0	8,251.0	7,555.0	5.3	6.4	-1.7
117 Canada	8,941.0	16,623.0	18,826.0	21,833.0	23,749.0	7.7	3.0	1.7
118 Norway	1,363.0	1,859.0	2,746.0	3,423.0	3,685.0	7.3	4.5	1.5
119 United States	122,056.0	170,526.0	188,358.0	241,434.0	269,485.0	4.4	5.1	2.2
120 Switzerland	2,583.0	3,149.0	4,208.0	4,518.0	4,760.0	5.0	1.4	1.0
Other economies								
121 Angola	—	—	—	411.0[e]	321.0	—	—	-6.0
122 Albania	—	—	—	—	—	—	—	—
123 Bulgaria	—	—	—	—	—	—	—	—
124 Czechoslovakia	—	—	—	—	—	—	—	—
125 Cuba	—	—	—	—	—	—	—	—
126 German Democratic Republic	—	—	—	—	—	—	—	—
127 Mongolia	—	—	—	—	—	—	—	—
128 Democratc People's Republic of Korea	—	—	—	—	—	—	—	—
129 Union of Soviet Socialist Republics	—	—	—	—	—	—	—	—

— Not available.
Note: *m*, median; *u*, unweighted mean; *w*, weighted mean; and *w′*, weighted mean excluding China and India.
a. 1976 data. b. 1979 data. c. 1984 data. d. 1974 data. e. 1981 data. f. 1986 data.
Sources: Unesco data; National Center for Education Statistics 1988.

Table A-15. Expenditure on Education as a Percentage of GNP and of Government Expenditure, Selected Years, 1965–85

Economy	Percentage of GNP					Percentage of government expenditure				
	1965	1970	1975	1980	1985	1965	1970	1975	1980	1985
Low-income economies	**2.7 *m***	**3.2 *m***	**2.8 *m***	**3.1 *m***	**3.2 *m***	**15.1 *m***	**16.2 *m***	**15.0 *m***	**12.7 *m***	**15.3 *m***
	2.9 *u*	**3.1 *u***	**3.2 *u***	**3.3 *u***	**3.3 *u***	**15.5 *u***	**15.2 *u***	**15.0 *u***	**13.6 *u***	**15.0 *u***
	2.6 *w*	**2.3 *w***	**2.5 *w***	**2.7 *w***	**3.0 *w***	**9.9 *w***	**7.0 *w***	**7.2 *w***	**8.0 *w***	**9.9 *w***
Excluding China and India	**2.7 *w′***	**3.1 *w′***	**3.2 *w′***	**3.1 *w′***	**2.9 *w′***	**13.1 *w′***	**11.4 *w′***	**12.1 *w′***	**9.4 *w′***	**12.5 *w′***
1 Ethiopia	1.3	2.1	3.3	3.3	4.2	8.8	14.1	14.5	10.4	9.5
2 Bhutan	—	—	—	—	—	—	—	—	—	—
3 Burkina Faso	—	—	2.5[a]	2.6	2.6	—	—	19.0[a]	19.8	21.0
4 Nepal	0.6	0.6	1.5	1.8	2.8	8.4	6.7	11.5	12.4	10.8
5 Bangladesh	1.3	—	1.1	1.5	1.9	14.4	—	13.6	8.2	10.5[b]
6 Malawi	4.0	4.6	2.4	3.3	3.5	15.4	13.2	9.6	12.9[c]	8.5[d]
7 Zaire	—	—	—	—	—	—	—	—	—	—
8 Mali	—	—	—	3.7	3.7	28.7	—	—	30.8	30.2[d]
9 Myanmar	2.7	3.1	1.7	—	—	13.8	17.9	15.3	—	—
10 Mozambique	—	—	—	—	—	—	—	—	—	—
11 Madagascar	6.3	—	3.2	5.4	3.5	27.0	—	18.5	—	—
12 Uganda	3.8	3.9	2.5	1.2	—	18.6	17.8	17.0	11.3	—
13 Burundi	—	—	—	3.0[c]	2.6	—	—	—	17.5[c]	15.5
14 Tanzania	3.3	4.5	5.4	5.1	4.2	23.7	16.0	17.8	14.3	19.0
15 Togo	2.0	2.2	3.5	5.6	5.3	15.1	19.0	15.1	19.4	19.4
16 Niger	0.6	1.2	2.4	3.1	—	12.0	17.7	18.7	—	22.9
17 Benin	—	—	—	—	—	—	—	—	—	—
18 Somalia	0.8	1.0	2.1	1.0	—	6.9	7.6	12.5	8.7	—
19 Central African Republic	—	—	4.9	3.8[c]	2.6	—	—	20.1	20.9[c]	—
20 India	2.5	2.8	2.8	2.8	3.3	8.7	10.7	8.6	10.0	9.4

— Not available.

Note: *m*, median; *u*, unweighted mean; *w*, weighted mean; and *w′*, weighted mean excluding China and India.

a. 1976 data. b. 1986 data. c. 1979 data. d. 1984 data. e. 1974 data. f. 1981 data. g. 1971 data.

Sources: Unesco data; World Bank data.

(Table continues on the following page.)

Table A-15 (continued)

Economy	Percentage of GNP					Percentage of government expenditure				
	1965	1970	1975	1980	1985	1965	1970	1975	1980	1985
21 Rwanda	—	2.3	2.3	2.7	3.2[d]	—	26.6	25.3	21.6	25.1[d]
22 China	—	1.3	1.8	2.5	2.7	—	2.9	4.2	6.1	—
23 Kenya	4.6	5.0	6.3	6.9	6.5	22.7	17.6	19.4	18.1	14.8[d]
24 Zambia	5.1	4.5	6.7	4.5	5.4[d]	17.6	10.9	11.9	7.6	16.3[d]
25 Sierra Leone	3.6	3.2	3.4	3.8	—	—	17.5	19.6[e]	—	—
26 Sudan	2.6	3.9	5.5[e]	4.8	—	15.8	12.6	14.8[e]	9.1	—
27 Haiti	—	—	1.0[a]	1.5	1.2	—	—	7.9[a]	14.9	16.5
28 Pakistan	1.8	1.7	2.2	2.0	2.2	5.3	4.2	5.2	5.0	—
29 Lesotho	3.8	3.0	4.5	5.0	3.6[d]	15.1	16.2	23.5	14.8	—
30 Ghana	4.5	4.3	5.9	3.1	2.6	17.7	19.6	21.5	—	—
31 Sri Lanka	4.3	4.0	2.8	3.1	3.0	14.7	13.6	10.1	8.8	8.0
32 Mauritania	—	3.3	3.8[a]	—	—	—	21.9	14.3[a]	—	—
33 Senegal	—	3.8	—	—	—	—	21.3	—	—	—
34 Afghanistan	1.1	1.1	1.3[e]	2.0	—	11.2	—	—	12.7	—
35 Chad	2.3	—	—	—	—	17.1	—	—	—	—
36 Guinea	—	—	—	—	3.3[d]	—	—	—	—	15.3[d]
37 Democratic Kampuchea	3.7	5.8	—	—	—	18.5	23.5	—	—	—
38 Lao People's Democratic Republic	—	—	—	—	1.1	—	—	—	1.3	4.5
39 Viet Nam	—	—	—	—	—	—	—	—	—	—
Lower-middle-income economies	**3.0 *m***	**3.4 *m***	**3.6 *m***	**4.6 *m***	**3.9 *m***	**16.9 *m***	**17.4 *m***	**16.1 *m***	**16.0 *m***	**13.9 *m***
	3.3 *u*	**3.6 *u***	**4.1 *u***	**4.6 *u***	**4.0 *u***	**17.5 *u***	**18.1 *u***	**16.9 *u***	**16.4 *u***	**14.4 *u***
	2.8 *w*	**3.3 *w***	**3.2 *w***	**3.3 *w***	**2.8 *w***	**17.9 *w***	**17.1 *w***	**15.1 *w***	**13.5 *w***	**13.3 *w***
40 Liberia	2.2	2.0	1.9	5.7	—	8.7	9.5	11.6	24.3	—
41 People's Democratic Republic of Yemen	—	—	4.0	6.2	—	14.1	12.4	14.7	16.9	—
42 Indonesia	—	2.6	2.7	1.7	—	—	—	13.1	8.9	—

43	Arab Republic of Yemen	—	0.7	—	6.2[f]	5.3	5.1	—	—	—	—
44	Philippines	2.4	2.6	1.9	1.6	1.3	25.0	24.4	11.4	10.3	7.0[d]
45	Morocco	3.3	3.6	5.1	6.4	—	—	16.8	14.3	18.5	—
46	Bolivia	2.2	3.4	3.5	4.4	0.4[d]	24.7	28.4	33.6[e]	25.3	—
47	Zimbabwe	—	3.4	3.6	6.6	8.5	—	—	—	13.7	16.0
48	Nigeria	1.7	—	2.9[e]	3.3[c]	1.0	20.7	—	16.5[a]	16.2[c]	8.7
49	Dominican Republic	—	3.1	2.1[a]	2.3	1.8	—	15.9	14.3[a]	16.0	14.0
50	Papua New Guinea	4.6	4.5	7.7[a]	4.7[c]	—	14.4	13.2	—	14.2[c]	—
51	Côte d'Ivoire	5.2	5.5	6.6	9.2[c]	—	19.8	19.3	19.0	29.8[c]	—
52	Honduras	2.9	3.1	3.7	3.2	4.4	24.4	18.4	20.3	14.2	13.8
53	Egypt	4.8	4.8	5.1	4.5[c]	6.3	12.9	15.8	—	9.4[f]	11.5
54	Nicaragua	1.9	2.3	2.4	3.2	6.1	17.4	18.1	13.1	10.4	10.2
55	Thailand	2.4	3.2	3.5	3.4	3.9	20.1	17.3	21.0	20.6	18.5
56	El Salvador	2.8	2.9	3.4	3.9	3.0[d]	21.9	27.6	22.2	17.1	12.5[d]
57	Botswana	5.1	5.2	8.5	7.1	7.0	—	—	18.8	16.0	13.9
58	Jamaica	3.1	3.6	5.9	6.9	5.7	16.4	—	16.0	13.1	12.1
59	Cameroon	3.1	3.4	3.9	3.2	3.7	18.0	19.6	21.3	20.3	18.9
60	Guatemala	2.0	2.0	1.6	1.9[c]	1.8[d]	15.9	17.5	15.7	16.6[c]	12.4[d]
61	Congo	—	5.9	8.1	6.9	5.1[d]	—	23.7	18.2	23.6	9.8[d]
62	Paraguay	—	2.2	1.6	1.5	1.5	—	15.3	14.0	16.4	16.7
63	Peru	4.3	3.3	3.3	3.1	2.7	18.1	18.8	16.6	15.2	15.7
64	Turkey	3.7	3.0	—	2.8	2.3	19.4	13.7	—	10.5	—
65	Tunisia	5.7	7.0	5.2	5.4	5.9	29.7	23.2	16.4	16.4	—
66	Ecuador	3.0	4.2	3.2	5.6	3.7	—	23.2	25.9	33.3	20.6
67	Mauritius	2.8	3.1	3.6	5.3	3.8	11.9	11.5	9.6	11.6	9.8
68	Colombia	2.3	1.9	2.2	1.9	0.0	13.9	13.6	16.4	14.3	24.8[d]
69	Chile	3.5	5.1	4.1	4.6	4.4	15.1	22.0	12.0	11.9	15.3
70	Costa Rica	4.6	5.2	6.8	7.8	4.5	29.0	31.8	31.1	22.2	22.7

— Not available.

Note: *m*, median; *u*, unweighted mean; *w*, weighted mean; and *w′*, weighted mean excluding China and India.

a. 1976 data. b. 1986 data. c. 1979 data. d. 1984 data. e. 1974 data. f. 1981 data. g. 1971 data.

Sources: Unesco data; World Bank data.

(*Table continues on the following page.*)

Table A-15 (continued)

Economy	Percentage of GNP					Percentage of government expenditure				
	1965	1970	1975	1980	1985	1965	1970	1975	1980	1985
71 Jordan	—	4.7	5.1	6.5	6.9	9.2	9.3	8.1	11.3	13.0
72 Syrian Arab Republic	2.7	3.8	3.9	4.6	6.1	12.4	9.4	7.8	8.1	11.8
73 Lebanon	2.2	2.5	—	—	—	15.5	16.8	16.1	13.2	16.8
Upper-middle-income economies	**3.2 *m***	**3.4 *m***	**3.5 *m***	**3.7 *m***	**4.3 *m***	**12.2 *m***	**12.9 *m***	**11.9 *m***	**14.7 *m***	**14.9 *m***
	3.3 *u*	**3.6 *u***	**3.8 *u***	**4.2 *u***	**4.5 *u***	**13.5 *u***	**14.6 *u***	**12.8 *u***	**13.9 *u***	**14.1 *u***
	2.7 *w*	**3.2 *w***	**3.3 w**	**4.1 *w***	**3.9 *w***	**12.0 *w***	**12.9 *w***	**12.2 *w***	**14.6 *w***	**14.9 *w***
74 Brazil	2.6	2.9	3.0	3.5	3.7	11.9	10.6	18.6[e]	—	19.1
75 Malaysia	4.3	4.2	6.0	6.0	6.6	18.5	17.7	19.3	14.7	16.3
76 South Africa	—	—	—	—	—	—	—	—	—	—
77 Mexico	2.3	2.3	3.6	2.9	2.5	8.2	8.5	11.9	16.7	16.2
78 Uruguay	3.8	3.7	—	2.2	2.6	—	26.1	—	10.0	9.3
79 Hungary	—	—	4.1	4.7	5.4	10.1	6.9	4.2	5.2	6.4
80 Poland	—	—	—	—	—	9.4	8.7[g]	—	—	—
81 Portugal	1.4	1.7	4.0	4.4	4.6	6.3	6.6	11.2	—	—
82 Yugoslavia	4.3	4.9	5.4	4.7	3.4	—	23.3	24.4	32.5	—
83 Panama	4.1	5.5	5.7	5.0	5.2	—	22.1	21.3	19.0	18.7
84 Argentina	3.0	3.5[g]	2.5	3.6	1.8[b]	23.5	16.0[g]	9.5	15.1	7.5[b]
85 Republic of Korea	1.8	3.4	2.2	3.7	4.8	15.4	21.4	13.9	23.7	28.2
86 Algeria	4.9	7.8	6.7	7.8	6.1	18.5	31.6	23.0	24.3	15.6
87 Venezuela	3.3	4.1	4.5	4.4	5.4	18.0	22.9	15.9[a]	14.7	21.3
88 Gabon	3.2	3.1	2.1	2.8	5.0	14.0	16.2	13.2[e]	—	9.4
89 Greece	2.3	2.0	2.0	2.2[c]	2.9	12.2	9.6	8.0	8.4[d]	7.5
90 Spain	1.3	2.0	1.8	2.3	3.3	12.0	15.2	13.6	14.7	14.1
91 Oman	—	1.2[g]	1.6	2.1	4.0	—	2.8[g]	1.9	4.1	10.8[b]
92 Ireland	4.2	5.0	6.1	6.6	6.9	—	10.8	10.8	11.2[c]	8.9
93 Trinidad and Tobago	3.1	3.4	3.1	4.0	5.8	14.1	14.0	14.7	11.5	—
94 Israel	4.8	5.5	6.7	8.0	10.2[d]	9.8	8.1	7.6	7.3	9.2[d]

95 Hong Kong	2.0	2.6	2.7	2.5	2.8[d]	14.8	22.8	20.7	14.6	18.7[d]
96 Singapore	4.3	3.1	2.9	2.8	—	—	11.7	8.6	7.3	—
97 Islamic Republic of Iran	3.2	2.9	3.0[e]	7.2	3.8	8.2	9.6	9.2[e]	15.7	17.2
98 Iraq	5.3	—	4.3[a]	3.2[c]	—	23.1	—	6.9[a]	—	—
99 Romania	—	—	3.5	3.3	2.1[d]	7.8	8.0	6.4	6.7	—
High-income oil exporters	**4.0 *m***	**4.5 *m***	**4.5 *m***	**2.9 *m***	**6.4 *m***	**13.0 *m***	**11.2 *m***	**11.7 *m***	**8.1 *m***	**10.8 *m***
	4.3 *u*	**4.5 *u***	**5.0 *u***	**3.1 *u***	**5.9 *u***	**11.9 *u***	**12.8 *u***	**12.1 *u***	**7.1 *u***	**12.7 *u***
	4.4 *w*	**4.5 *w***	**6.9 *w***	**4.0 *w***	**7.1 *w***	**11.3 *w***	**11.9 *w***	**11.9 *w***	**8.3 *w***	**11.8 *w***
100 Saudi Arabia	6.2	4.8	10.3	5.4	9.2	13.0	9.8	11.7	8.7	11.2
101 Kuwait	2.8	4.2	3.0	2.4	4.5	6.9	11.2	10.0	8.1	9.5
102 United Arab Emirates	—	—	0.9	1.3	1.7	—	—	—	4.6[f]	10.4
103 Libya	4.0	4.5	5.9	3.4	8.2	15.7	17.4	14.5	—	19.8
High-income industrial market economies	**4.3 *m***	**5.1 *m***	**6.3 *m***	**5.9 *m***	**6.0 *m***	**18.5 *m***	**16.9 *m***	**15.2 *m***	**14.3 *m***	**13.2 *m***
	4.7 *u*	**5.5 *u***	**6.2 *u***	**6.2 *u***	**5.9 *u***	**18.3 *u***	**17.4 *u***	**16.7 *u***	**14.8 *u***	**13.8 *u***
	4.8 *w*	**5.5 *w***	**6.0 *w***	**6.2 *w***	**6.1 *w***	**17.8 *w***	**17.8 *w***	**17.0 *w***	**17.3 *w***	**15.6 *w***
104 New Zealand	3.7	4.9	6.1	6.0	4.9	11.1	—	17.1	14.5	18.4
105 Italy	4.1	3.4	3.9	4.4[c]	—	14.9	11.9	9.4	11.1[c]	—
106 United Kingdom	4.9	5.2	6.6	5.6	4.9	13.4	14.1	14.3	13.9	11.3[d]
107 Belgium	4.2	—	6.2	5.9	6.0	16.9	—	22.2	16.3	15.2
108 Austria	3.6	4.6	5.7	5.6	5.8	6.4	8.1	8.5	8.0	7.9
109 Netherlands	6.0	7.4	8.2	7.9	6.8	26.3	29.4	23.7	23.1	16.4
110 France	4.2	4.8	5.2	5.0	6.0[d]	22.4	24.9	29.5[e]	—	—
111 Australia	3.6	4.3	6.5	5.9	5.9	11.2	13.3	14.8	14.8	12.8
112 Federal Republic of Germany	3.0	3.5	5.1	4.7	4.5	—	9.2	10.7	10.1	9.2

— Not available.

Note: *m*, median; *u*, unweighted mean; *w*, weighted mean; and *w′*, weighted mean excluding China and India.

a. 1976 data. b. 1986 data. c. 1979 data. d. 1984 data. e. 1974 data. f. 1981 data. g. 1971 data.

Sources: Unesco data; World Bank data.

(Table continues on the following page.)

Table A-15 (continued)

Economy	Percentage of GNP					Percentage of government expenditure				
	1965	1970	1975	1980	1985	1965	1970	1975	1980	1985
113 Finland	5.8	5.9	6.3	5.5	5.7	22.5	—	13.0	11.2[c]	12.9
114 Denmark	5.7	6.9	7.8	6.9	6.3	22.8	16.9	15.2	9.5	13.4[b]
115 Japan	4.3	3.9	5.5	5.8	5.1	22.7	20.4	22.4	19.6	—
116 Sweden	5.8	7.6	7.0	9.1	7.8	—	—	13.4	14.1	12.6
117 Canada	5.8	8.6	7.6	7.3	7.0	18.5	24.1	17.8	17.3	12.7
118 Norway	5.3	6.0	7.1	7.2	6.4	26.8	15.5	14.7	13.8	13.6
119 United States	5.2	6.4	6.3	6.8	6.7	18.0	19.4	18.1	20.4	18.4
120 Switzerland	4.1	4.0	5.1	5.0	4.8	20.4	18.4	19.4	18.8	18.6
Other economies										
121 Angola	—	—	—	5.0[f]	3.4	—	—	—	—	10.8
122 Albania	—	—	—	—	—	10.1	11.6[g]	—	10.3	—
123 Bulgaria	—	—	—	—	—	9.2	9.1	8.5	—	—
124 Czechoslovakia	—	—	—	—	—	7.6	7.0	7.0	—	7.9
125 Cuba	—	—	—	—	—	11.1	18.4	30.1	—	14.7[b]
126 German Democratic Republic	—	—	—	—	—	—	—	—	—	—
127 Mongolia	—	—	—	—	—	16.6	15.6	—	—	—
128 Democratic People's Republic of Korea	—	—	—	—	—	—	—	—	—	—
129 Union of Soviet Socialist Republics	—	—	—	—	—	13.8	12.8	12.9	11.2	10.2[d]

— Not available.

Note: *m*, median; *u*, unweighted mean; *w*, weighted mean; and *w′*, weighted mean excluding China and India.

a. 1976 data. b. 1986 data. c. 1979 data. d. 1984 data. e. 1974 data. f. 1981 data. g. 1971 data.

Sources: Unesco data; World Bank data.

Table A-16. The Share of Recurrent Expenditure in Total Public Expenditure on Education, Selected Years, 1965–85

(percent)

Economy	*1965*	*1970*	*1975*	*1980*	*1985*
Low-income economies	**86.9 *m***	**93.0 *m***	**90.2 *m***	**87.0 *m***	**95.0 *m***
	83.4 *u*	**88.5 *u***	**86.8 *u***	**84.5 *u***	**91.3 *u***
	85.5 *w*	**92.5 *w***	**93.0 *w***	**90.4 *w***	**88.7 *w***
Excluding China and India	**78.9 *w′***	**84.6 *w′***	**83.3 *w′***	**82.0 *w′***	**87.6 *w′***
1 Ethiopia	83.9	81.3	82.2	79.5	84.3
2 Bhutan	67.2	56.7	55.2[a]	—	—
3 Burkina Faso	—	—	91.4[b]	93.0	95.3
4 Nepal	—	99.0[c]	—	—	—
5 Bangladesh	45.0	—	67.4	66.8	95.8
6 Malawi	78.9	73.7	91.9	75.6	72.2
7 Zaire	—	—	—	—	—
8 Mali	95.8	—	—	98.8	99.2
9 Myanmar	93.6	93.1	95.2	—	—
10 Mozambique	—	—	—	—	—
11 Madagascar	92.2	—	95.2	85.5	95.4
12 Uganda	86.4	82.4	93.0	88.3	—
13 Burundi	—	—	—	86.9[d]	92.7
14 Tanzania	87.4	82.8	80.4	84.3	—
15 Togo	97.2	88.8	96.3	96.4	94.6
16 Niger	94.9	93.8	87.0	47.0	—
17 Benin	—	—	—	—	—
18 Somalia	92.7	100.0	83.7	91.1	—
19 Central African Republic	—	—	86.9	97.2[d]	97.5[e]

— Not available.

Note: *m*, median; *u*, unweighted mean; *w*, weighted mean; and *w′*, weighted mean excluding China and India.

a. 1974 data. b. 1976 data. c. 1971 data. d. 1979 data. e. 1986 data. f. 1984 data.

Source: Unesco data.

(*Table continues on the following page.*)

Table A-16 (*continued*)

Economy	*1965*	*1970*	*1975*	*1980*	*1985*
20 India	89.5	94.6	99.1	98.8	97.6
21 Rwanda	—	98.8	99.9	84.7	97.8[f]
22 China	—	97.8	92.9	87.0	80.8
23 Kenya	92.0	93.9	95.4	92.1	93.8
24 Zambia	54.4	79.2	76.9	95.1	93.5[f]
25 Sierra Leone	78.6	89.0	90.2	95.3	—
26 Sudan	77.8	93.4	89.6[a]	92.2	—
27 Haiti	—	—	88.7[b]	80.1	99.8
28 Pakistan	65.1	60.8	69.6	73.1	77.8
29 Lesotho	81.1	91.3	76.0	79.9	83.7[f]
30 Ghana	85.8	87.8	77.9	55.7	97.1
31 Sri Lanka	94.6	93.0	93.6	83.8	76.7
32 Mauritania	—	95.0	98.7[b]	—	—
33 Senegal	—	97.8	—	—	—
34 Afghanistan	80.3	89.5	90.4[a]	90.0	—
35 Chad	92.9	—	—	—	—
36 Guinea	—	—	—	—	99.7[f]
37 Democratic Kampuchea	94.8	99.9[c]	—	—	—
38 Lao People's Democratic Republic	—	—	—	—	—
39 Viet Nam	—	—	—	—	—
Lower-middle-income economies	**88.6** ***m***	**91.6** ***m***	**85.0** ***m***	**89.4** ***m***	**92.6** ***m***
	86.6 ***u***	**88.6** ***u***	**82.9** ***u***	**86.1** ***u***	**88.5** ***u***
	82.7 ***w***	**86.0** ***w***	**76.0** ***w***	**80.2** ***w***	**86.6** ***w***
40 Liberia	88.3	—	85.0[b]	85.9	—
41 People's Democratic Republic of Yemen	88.1	97.5[c]	85.5[b]	98.9[d]	—
42 Indonesia	—	89.8	77.6	—	—
43 Arab Republic of Yemen	—	—	—	—	—
44 Philippines	98.8	98.8	81.6	96.0	93.4
45 Morocco	88.4	91.2	86.2	81.0	—

46	Bolivia	99.2	99.7	99.9	96.0	99.6[f]
47	Zimbabwe	—	96.1	91.8	97.4	88.6
48	Nigeria	73.8	—	54.2[a]	65.8[d]	85.9
49	Dominican Republic	—	93.1	82.4[b]	75.4	86.9
50	Papua New Guinea	81.0	82.0	90.2[b]	96.7[d]	—
51	Côte d'Ivoire	86.6	85.6	84.7	77.8[d]	—
52	Honduras	92.5	95.3	—	91.0	98.6
53	Egypt	92.2	92.7	86.4	82.4[d]	94.5
54	Nicaragua	94.9	98.7	79.8	87.5	96.7
55	Thailand	79.4	73.1	73.3	70.6	85.9
56	El Salvador	84.6	93.7	91.7	94.1	87.3[f]
57	Botswana	85.9	84.6	54.5	75.8	84.1
58	Jamaica	90.3	71.5	78.8	99.6	93.7
59	Cameroon	95.3	92.2	83.5	81.3	46.8
60	Guatemala	89.7	88.3	91.7[b]	89.4[d]	97.8[f]
61	Congo	—	96.6	82.6	93.8	92.3[f]
62	Paraguay	—	88.8	77.3	71.3	81.4
63	Peru	93.7	96.7	96.4	94.4	96.3
64	Turkey	76.3	71.0	—	83.7	83.4
65	Tunisia	70.1	84.4	89.2	87.6	90.2
66	Ecuador	86.3	78.4	—	94.0	93.7
67	Mauritius	88.7	92.0	87.4	89.9	92.8
68	Colombia	51.8	78.6	85.4	93.3	93.0[f]
69	Chile	90.2	—	94.4	94.9	99.6
70	Costa Rica	91.0	96.1	93.5	91.3	95.2
71	Jordan	89.0	83.1	83.2	79.2	87.1
72	Syrian Arab Republic	91.3	74.2	56.9	54.2	55.3
73	Lebanon	86.8	93.1	—	—	—

— Not available.

Note: *m*, median; *u*, unweighted mean; *w*, weighted mean; and *w′*, weighted mean excluding China and India.

a. 1974 data. b. 1976 data. c. 1971 data. d. 1979 data. e. 1986 data. f. 1984 data.

Source: Unesco data.

(Table continues on the following page.)

Table A-16 (*continued*)

	Economy	*1965*	*1970*	*1975*	*1980*	*1985*
	Upper-middle-income economies	**89.5 *m***	**87.2 *m***	**86.4 *m***	**86.3 *m***	**89.3 *m***
		88.1 *u*	**82.3 *u***	**84.6 *u***	**86.3 *u***	**87.4 *u***
		87.8 *w*	**78.6 *w***	**82.2 *w***	**86.2 *w***	**87.4 *w***
74	Brazil	—	76.3	—	—	—
75	Malaysia	83.3	91.5	84.9	83.0	85.4
76	South Africa	—	—	—	—	—
77	Mexico	76.5	66.2	65.0	91.7	90.1
78	Uruguay	90.9	—	—	94.7	96.0
79	Hungary	87.0	87.2	86.5	83.1	89.0
80	Poland	82.0	91.7[c]	—	—	—
81	Portugal	85.2	90.8	93.8	85.4	88.7
82	Yugoslavia	86.9	91.2	88.3	85.1	93.2
83	Panama	89.1	92.3	93.6	93.7	97.7
84	Argentina	92.8	—	93.8	84.5	—
85	Republic of Korea	90.2	77.1	74.4	84.3	79.7
86	Algeria	79.6	61.4	72.3	66.9	72.1
87	Venezuela	94.3	94.3	95.8	95.1	95.9[f]
88	Gabon	93.3	95.6	85.0	72.3	68.3
89	Greece	—	81.6	91.0	94.3[d]	95.2
90	Spain	—	66.6	86.3[a]	86.3[d]	89.5
91	Oman	—	38.5[c]	70.3	81.3	62.7
92	Ireland	81.0	83.9	86.6	86.6	91.0
93	Trinidad and Tobago	89.8	88.3	88.1	76.4	87.5
94	Israel	—	—	84.2	92.3	93.0[f]
95	Hong Kong	92.6	86.4	91.8	88.1	88.5[f]
96	Singapore	86.6	94.1	86.9	85.6	—
97	Islamic Republic of Iran	92.2	84.7	81.3[a]	88.4	88.6
98	Iraq	99.1	—	76.2[b]	97.2[d]	—
99	Romania	90.5	88.6	84.8	88.8	96.7

High-income oil exporters	**75.8 *m***	**95.0 *m***	**66.4 *m***	**71.3 *m***	**86.9 *m***
	79.1 *u*	**85.1 *u***	**64.0 *u***	**74.7 *u***	**85.3 *u***
	77.9 *w*	**85.1 *w***	**48.7 *w***	**66.9 *w***	**74.9 *w***
100 Saudi Arabia	75.8	95.0	42.7	63.5	69.6
101 Kuwait	99.2	96.6	80.4	93.1	97.6
102 United Arab Emirates	—	—	78.3	79.0	94.2
103 Libya	62.4	63.8	54.5	63.0	79.6
High-income industrial market economies	**78.7 *m***	**79.6 *m***	**86.1 *m***	**89.5 *m***	**91.7 *m***
	77.5 *u*	**80.1 *u***	**84.6 *u***	**88.5 *u***	**92.0 *u***
	78.0 *w*	**83.0 *w***	**86.3 *w***	**88.6 *w***	**93.2 *w***
104 New Zealand	78.7	79.4	79.7	89.9	91.2
105 Italy	83.0	97.2	89.2	86.5[d]	—
106 United Kingdom	77.6	85.2	89.6	94.1	96.0[f]
107 Belgium	91.5	—	91.7	98.9	94.9
108 Austria	79.1	77.8	78.5	85.3	90.1
109 Netherlands	78.9	79.7	82.2	86.7	87.6
110 France	62.6	64.1	87.6	92.5	94.5[f]
111 Australia	77.9	81.1	83.1	90.9	91.7
112 Federal Republic of Germany	67.0	72.1	80.3	86.4	90.3
113 Finland	81.3	89.8	86.1	91.7	92.8
114 Denmark	75.8	77.4	86.9	88.7	95.4[e]
115 Japan	76.0	72.7	71.4	67.7	94.7
116 Sweden	79.2	82.0	90.8	86.4	88.8
117 Canada	75.0	84.9	90.1	92.5	93.1
118 Norway	74.1	77.5	80.6	83.4	88.3
119 United States	80.3	86.7	90.2	93.2[d]	—
120 Switzerland	80.2	74.2	80.9	89.5	91.0

— Not available.

Note: *m*, median; *u*, unweighted mean; *w*, weighted mean; and *w′*, weighted mean excluding China and India.

a. 1974 data. b. 1976 data. c. 1971 data. d. 1979 data. e. 1986 data. f. 1984 data.

Source: Unesco data.

(*Table continues on the following page.*)

Table A-16 (*continued*)

Economy	1965	1970	1975	1980	1985
Other economies					
121 Angola	—	—	—	—	97.7
122 Albania	—	—	—	—	—
123 Bulgaria	90.5	86.0	92.1	95.9	89.6
124 Czechoslovakia	90.5	88.8	90.5	94.1	95.6
125 Cuba	73.9	—	74.8[a]	89.5	93.9
126 German Democratic Republic	—	—	—	—	—
127 Mongolia	91.9	92.7	—	—	—
128 Democratic People's Republic of Korea	—	—	—	—	—
129 Union of Soviet Socialist Republics	83.1	83.3	84.4	85.1	83.6

— Not available.

Note: *m*, median; *u*, unweighted mean; *w*, weighted mean; and *w′*, weighted mean excluding China and India.

a. 1974 data. b. 1976 data. c. 1971 data. d. 1979 data. e. 1986 data. f. 1984 data.

Source: Unesco data.

Table A-17. Public Recurrent Expenditure on Primary Education, Its Growth, and Its Share in Total Public Recurrent Expenditure on Education, Selected Years, 1965–85

Economy	Total (millions of 1985 dollars)					Annual growth (percent)			As a percentage of total public recurrent expenditure on education				
	1965	1970	1975	1980	1985	1965–75	1975–80	1980–85	1965	1970	1975	1980	1985
Low-income economies	—	—	—	—	—	**5.0 *m***	**4.8 *m***	**3.5 *m***	**45.9 *m***	**44.2 m**	**44.1 *m***	**41.7 *m***	**41.8 *m***
						3.7 *u*	**3.3 *u***	**3.8 *u***	**49.5 *u***	**46.3 *u***	**43.9 *u***	**41.9 *u***	**41.9 *u***
						9.0 *w*	**2.2 *w***	**7.6 *w***	**31.2 *w***	**28.8 *w***	**41.1 *w***	**39.5 *w***	**35.0 *w***
Excluding China and India	—	—	—	—	—	**4.4 *w′***	**3.5 *w′***	**4.6 *w′***	**45.9 *w′***	**42.9 *w′***	**43.3 *w′***	**44.0 *w′***	**46.6 *w′***
1 Ethiopia	13.9	27.3	40.5	52.8	86.5	11.3	5.4	10.4	41.7	45.1	36.6	42.0	50.6
2 Bhutan	—	—	—	—	—	—	—	—	—	22.0	13.6[a]	—	—
3 Burkina Faso	4.4	—	7.5	7.7	10.4	5.5	0.5	6.2	36.4	—	43.3	32.3	38.1
4 Nepal	—	2.4[b]	12.4	20.9	24.9	—	11.0	3.6	—	23.0[b]	—	—	—
5 Bangladesh	24.4	—	43.7	61.1	146.4	6.0	6.9	19.1	45.6	—	57.0	45.3	51.0
6 Malawi	7.9	7.8	8.1	9.2	11.1	0.3	2.6	3.8	58.3	42.3	44.6	38.9	41.3
7 Zaire	41.2	—	58.7	65.1	—	3.6	2.1	—	69.6	—	44.1	47.1	—
8 Mali	—	—	10.7	14.4	16.9	—	6.1	3.3	—	—	39.7	38.8	44.4
9 Myanmar	33.2	32.4[b]	—	—	—	—	—	—	41.6	28.6[b]	—	—	—
10 Mozambique	—	—	—	—	—	—	—	—	—	—	—	—	—
11 Madagascar	31.5	38.3	30.0	47.7	31.8	-0.5	9.7	-7.8	31.1	48.5	42.9	41.4	42.3
12 Uganda	—	52.4	39.6	15.0[c]	—	—	-14.9	—	—	40.0	41.1	16.2[c]	—
13 Burundi	7.7	—	7.7	10.4[d]	12.0	0.0	7.8	2.4	72.7	—	45.0	42.7[d]	45.0
14 Tanzania	53.7	72.6	96.3	142.5[d]	146.1	6.0	10.3	0.4	53.4	41.5	37.3	45.0[d]	—
15 Togo	—	6.5	—	11.7	14.2[e]	—	—	5.0	—	67.5	—	29.5	37.2[e]
16 Niger	5.0	8.9	—	17.4[c]	—	—	—	—	58.1	60.5	—	36.8[c]	—
17 Benin	—	12.2	14.4	—	—	—	—	—	—	49.8	44.9	—	—
18 Somalia	1.6	3.8	—	—	—	—	—	—	44.9	62.5	—	—	—

— Not available.

Note: *m*, median; *u*, unweighted mean; *w*, weighted mean; and *w′*, weighted mean excluding China and India.

a. 1974 data. b. 1971 data. c. 1981 data. d. 1979 data. e. 1984 data. f. 1986 data. g. 1976 data.

Sources: Unesco data; National Center for Education Statistics 1988.

(Table continues on the following page.)

Table A-17 (*continued*)

Economy	Total (millions of 1985 dollars)					Annual growth (percent)			As a percentage of total public recurrent expenditure on education				
	1965	1970	1975	1980	1985	1965–75	1975–80	1980–85	1965	1970	1975	1980	1985
19 Central African Republic	9.0	—	14.7	15.0	9.9[f]	5.0	0.4	—	59.7	—	56.9	63.1	49.4[f]
20 India	495.6	693.5	1,554.3	1,665.6	2,583.3	12.1	1.4	9.2	23.1	22.2	40.0	36.9	37.1
21 Rwanda	12.7	13.2	15.1	21.9	—	1.7	7.7	—	78.0	66.4	69.0	67.1	—
22 China	—	—	—	—	1,791.1	—	—	—	—	—	—	—	28.6
23 Kenya	42.4	44.8	143.9	193.5	210.6	13.0	6.1	1.7	54.4	39.8	63.1	59.6	59.8
24 Zambia	—	34.1	53.8	45.2	53.0[e]	—	-3.4	4.1	—	44.2	45.3	45.3	44.2[e]
25 Sierra Leone	8.6	—	9.6	—	—	1.1	—	—	36.9	—	26.5	—	—
26 Sudan	31.7	75.4	—	157.4	160.2	—	—	0.4	30.9	40.1	—	48.0	55.5
27 Haiti	11.8	11.5	9.5	14.7	12.0	-2.1	9.1	-4.0	63.7	65.1	63.0	59.3	51.0
28 Pakistan	53.1	57.5	106.8	129.4	216.8	7.2	3.9	10.9	43.7	39.7	41.1	39.4	40.2
29 Lesotho	2.3	2.8	4.4	6.5	6.2[e]	6.7	8.1	-1.2	64.1	59.3	47.7	38.6	39.1[e]
30 Ghana	53.0	64.6	49.0	23.1	29.4	-0.8	-14.0	4.9	35.9	39.2	24.5	29.3	26.6
31 Sri Lanka	58.2	68.7	19.8[a]	—	—	-11.3	—	—	62.8	60.9	—	—	—
32 Mauritania	—	—	9.4	11.5	12.9	—	4.1	2.3	—	—	45.3	35.4	25.1
33 Senegal	20.7	—	38.7[g]	42.3	—	5.9	2.2	—	38.5	—	47.8[g]	43.8	—
34 Afghanistan	—	—	—	—	—	—	—	—	31.6	28.6	38.3[a]	41.9	—
35 Chad	4.3	—	—	—	—	—	—	—	37.6	—	—	—	—
36 Guinea	13.9	—	—	17.5[d]	18.0[e]	—	—	0.6	46.2	—	—	24.7[d]	30.8[e]
37 Democratic Kampuchea	—	—	—	—	—	—	—	—	65.8	59.4[b]	—	—	—
38 Lao People's Democratic Republic	—	—	—	—	—	—	—	—	60.0	62.5	—	—	—
39 Viet Nam	—	—	—	—	—	—	—	—	—	—	—	—	—
Lower-middle-income economies	—	—	—	—	—	**6.1 *m***	**5.7 *m***	**3.9 *m***	**54.6 *m***	**51.2 *m***	**42.3 *m***	**43.7 *m***	**45.1 *m***
						6.0 *u*	**7.5 *u***	**3.4 *u***	**53.1 *u***	**51.8 *u***	**47.0 *u***	**43.3 *u***	**45.7 *u***
						5.8 *w*	**8.7 *w***	**4.7 *w***	**48.1 *w***	**49.8 *w***	**42.2 *w***	**35.2 *w***	**49.4 *w***
40 Liberia	6.5	—	5.7	8.7[d]	—	-1.3	11.2	—	46.3	—	—	17.6[d]	—

41 People's Democratic Republic of Yemen	—	—	—	—	—	—	—	—	—	—	—	—	—
42 Indonesia	—	261.4[b]	—	—	—	—	—	—	—	45.5[b]	—	—	—
43 Arab Republic of Yemen	—	—	—	—	145.9	—	—	—	—	—	—	—	—
44 Philippines	—	388.3	296.3[g]	315.2	240.2	—	1.6	-5.3	—	82.9	82.4[g]	61.4	63.9
45 Morocco	86.2	90.2	96.0[a]	183.5	—	1.2	11.4	—	61.5	44.8	39.5[g]	35.4	—
46 Bolivia	26.3	50.8	71.1	91.1	—	10.5	5.1	—	56.4	60.2	60.4	58.9	—
47 Zimbabwe	—	45.1	60.3	163.8	225.3	—	22.1	6.6	—	47.6	49.4	66.5	66.0
48 Nigeria	—	365.9	307.0[a]	697.5[c]	—	—	12.4	—	—	40.4	22.7[a]	17.2[c]	—
49 Dominican Republic	—	24.9	22.4[g]	25.8	30.5	—	3.6	3.4	—	41.1	37.5[g]	36.8	46.6
50 Papua New Guinea	—	21.3[b]	46.3	—	—	—	—	—	—	28.3[b]	39.6	—	—
51 Côte d'Ivoire	37.9	57.3	109.8	166.3[d]	—	11.2	10.9	—	36.2	34.2	41.3	39.0[d]	—
52 Honduras	29.5	36.0	—	56.8	70.1	—	—	4.3	71.1	64.2	—	61.9	49.1
53 Egypt	—	—	—	—	—	—	—	—	—	—	—	—	—
54 Nicaragua	23.7	34.8	36.1	34.8	67.0	4.3	-0.7	14.0	60.9	57.9	55.1	44.7	43.3
55 Thailand	120.1	185.3	310.3	395.4	727.4	10.0	5.0	13.0	65.7	53.5	62.5	57.8	58.4
56 El Salvador	38.9	49.0	72.4	95.3	—	6.4	5.7	—	66.6	57.9	57.5	61.9	—
57 Botswana	1.7	3.1	5.4	12.4[d]	20.4[a]	12.3	23.1	10.5	49.7	57.6	46.5	52.1[d]	43.2[e]
58 Jamaica	17.0	23.9	36.9	42.9	27.7	8.1	3.1	-8.4	36.5	44.7	33.5	33.7	29.9
59 Cameroon	—	—	—	—	—	—	—	—	—	—	—	—	—
60 Guatemala	—	56.0	66.0[g]	56.0[d]	—	—	-5.3	—	—	55.2	51.3[g]	33.4[d]	—
61 Congo	11.2	20.0	23.9	29.5	28.1[e]	7.9	4.3	-1.2	56.7	48.7	34.3	35.8	30.0[e]
62 Paraguay	—	15.3	—	—	14.0	—	—	—	—	64.8	—	—	36.6
63 Peru	171.9	158.8	209.3	232.1	157.3	2.0	2.1	-7.5	41.8	39.8	40.7	45.1	35.6
64 Turkey	—	—	—	416.3	459.9	—	—	2.0	—	—	—	43.7	45.9
65 Tunisia	53.4	—	94.0	127.3	184.8	5.8	6.3	7.7	52.8	—	42.3	41.2	44.0
66 Ecuador	39.4	69.2	—	112.2	177.0	—	—	9.5	41.3	45.9	—	20.6	45.5
67 Mauritius	9.5	10.2	10.7[a]	17.8	16.2	1.3	8.9	-1.9	73.3	68.8	66.9[a]	44.1	45.1

— Not available.

Note: *m*, median; *u*, unweighted mean; *w*, weighted mean; and *w′*, weighted mean excluding China and India.
a. 1974 data. b. 1971 data. c. 1981 data. d. 1979 data. e. 1984 data. f. 1986 data. g. 1976 data.
Sources: Unesco data; National Center for Education Statistics 1988.

(*Table continues on the following page.*)

Table A-17 (continued)

Economy	Total (millions of 1985 dollars)					Annual growth (percent)			As a percentage of total public recurrent expenditure on education				
	1965	1970	1975	1980	1985	1965–75	1975–80	1980–85	1965	1970	1975	1980	1985
68 Colombia	61.9	—	—	239.4	376.6	—	—	9.5	39.6	—	—	44.4	41.9
69 Chile	110.4	—	147.7	294.9	320.0	3.0	14.8	1.6	35.7	—	34.9	42.7	51.0
70 Costa Rica	38.2	54.1	67.0	71.2	54.2	5.8	1.2	–5.3	60.4	51.2	37.2	28.0	35.1
71 Jordan	—	—	—	—	—	—	—	—	54.6	—	—	—	—
72 Syrian Arab Republic	64.2	93.8	130.7	210.7	326.4	7.4	10.0	9.1	51.2	55.1	51.7	57.6	57.8
73 Lebanon	—	—	—	—	—	—	—	—	56.5	—	—	—	—
Upper-middle-income economies	—	—	—	—	—	**4.9 *m***	**6.0 *m***	**4.2 *m***	**51.0 *m***	**44.9 *m***	**39.1 *m***	**39.4 *m***	**37.7 *m***
						5.6 *u*	**7.4 *u***	**3.1 *u***	**50.1 *u***	**45.5 *u***	**45.2 *u***	**38.9 *u***	**37.3 u**
						3.6 *w*	**10.0 *w***	**0.2 *w***	**49.4 *w***	**45.6 *w***	**39.9 *w***	**38.6 *w***	**35.7 *w***
74 Brazil	—	—	2,180.1[g]	3,211.8	3,298.2	—	10.2	0.5	—	—	—	—	—
75 Malaysia	—	215.5[b]	—	412.1	618.6	—	—	8.5	—	44.9[b]	—	35.0	37.8
76 South Africa	—	—	—	—	—	—	—	—	—	—	—	—	—
77 Mexico	426.2	631.7	1,184.3	1,711.5	1,085.4	10.8	7.6	–8.7	40.3	47.7	42.9	39.7	27.2
78 Uruguay	54.1	74.4	—	61.4	44.9	—	—	–6.1	44.9	—	—	48.4	37.7
79 Hungary	131.2	147.3	212.2	282.1	354.1	4.9	5.9	4.7	40.7	38.7	38.0	39.1	36.9
80 Poland	—	—	—	649.6	928.5	—	—	7.4	29.0	34.1	27.6	28.7	33.7
81 Portugal	44.1	—	316.6	382.5	396.1	21.8	3.9	0.7	39.5	—	55.1	52.8	49.8
82 Yugoslavia	—	—	—	—	—	—	—	—	—	—	—	—	—
83 Panama	31.7	48.2	64.5	86.3	88.6	7.4	6.0	0.5	51.6	38.9	39.1	46.3	38.3
84 Argentina	819.2	260.3	418.6	895.3	887.9[e]	–6.5	16.4	–0.2	49.8	29.0	27.0	40.1	37.7[e]
85 Republic of Korea	184.0	461.0	414.8	917.4	1,509.9	8.5	17.2	10.5	66.4	64.3	62.4	49.9	46.7
86 Algeria	388.6	—	492.8[a]	668.6	—	2.7	5.2	—	60.3	—	39.5[a]	28.4	—
87 Venezuela	523.3	719.4	553.5	427.1	614.1[e]	0.6	–5.1	9.5	43.2	38.3	22.1	15.4	20.9[e]
88 Gabon	—	—	—	—	—	—	—	—	—	—	—	—	—
89 Greece	—	164.2	170.3	237.6[d]	289.6	—	8.7	3.4	—	48.8	36.7	36.9[d]	31.6
90 Spain	—	770.8	1,536.6[a]	2,003.6[d]	—	—	5.5	—	—	52.3	78.7[a]	58.9[d]	—
91 Oman	—	—	18.6[a]	—	167.5[f]	—	—	—	—	—	88.0[a]	—	50.9[f]

92 Ireland	171.8	197.1	254.9	241.6	296.8	4.0	-1.1	4.2	56.7	42.8	35.9	26.1	29.1
93 Trinidad and Tobago	61.3	—	—	122.3	177.0	—	—	7.7	54.2	—	—	46.9	47.5
94 Israel	179.9	240.4	326.4	523.2	767.5[e]	6.1	9.9	10.1	37.2	32.2	32.7	33.7	35.7[e]
95 Hong Kong	82.1	126.2	171.7	190.7	262.1[e]	7.7	2.1	8.3	57.4	54.9	48.7	33.7	31.4[e]
96 Singapore	66.4	73.2	83.8	109.8	—	2.4	5.6	—	57.7	44.2	38.1	35.8	—
97 Islamic Republic of Iran	991.0	1,366.1	1,206.6[a]	3,539.2	2,380.1	2.2	19.6	-7.6	54.9	50.9	31.3[a]	41.7	40.9
98 Iraq	—	—	—	—	—	—	—	—	67.9	60.1	45.3[g]	40.5[d]	—
99 Romania	—	—	—	—	—	—	—	—	50.4	50.4	70.4[a]	—	—
High-income oil exporters	—	—	—	—	—	**5.9 *m***	**-8.3 *m***	**23.4 *m***	**37.9 *m***	**19.3 *m***	**62.4 *m***	**29.5 *m***	**53.7 *m***
						5.9 *u*	**-8.3 *u***	**23.4 *u***	**37.9 *u***	**19.3 u**	**62.4 *u***	**29.5 *u***	**53.7 *u***
						5.9 *w*	**-8.3 *w***	**23.4 *w***	**37.9 *w***	**19.3 *w***	**67.5 *w***	**29.5 *w***	**53.7 *w***
100 Saudi Arabia	—	—	1,063.9[a]	—	—	—	—	—	—	—	75.7[a]	—	—
101 Kuwait	172.7	177.9	307.8	199.9	573.0	5.9	-8.3	23.4	37.9	19.3	49.0	29.5	53.7
102 United Arab Emirates	—	—	—	—	—	—	—	—	—	—	—	—	—
103 Libya	—	—	—	—	—	—	—	—	—	—	—	—	—
High-income industrial market economies	—	—	—	—	—	**5.8 *m***	**1.6 *m***	**1.5 *m***	**44.1 *m***	**31.4 *m***	**28.2 *m***	**27.9 *m***	**24.7 *m***
						4.3 *u*	**1.1 *u***	**0.9 *u***	**43.0 *u***	**31.9 *u***	**30.2 *u***	**29.2 *u***	**26.9 *u***
						4.4 *w*	**1.1 *w***	**2.1 *w***	**44.5 *w***	**40.9 *w***	**38.2 *w***	**32.8 *w***	**31.5 *w***
104 New Zealand	176.3	230.2	330.5	357.7	353.3	6.5	1.6	-0.2	44.1	37.7	35.6	35.4	36.9
105 Italy	3,265.0	2,711.0	3,376.8	4,126.4[d]	—	0.3	5.1	—	45.6	28.4	30.0	29.2d	—
106 United Kingdom	—	4,119.0	6,398.7	5,763.3	4,727.5[e]	—	-2.1	-4.8	—	26.7	28.5	26.6	21.7[e]
107 Belgium	509.8	592.8	982.3	1,163.7	1,133.0	6.8	3.4	-0.5	29.4	24.7	25.5	25.3	24.7
108 Austria	518.7	450.6	530.0	513.7	589.3	0.2	-0.6	2.8	56.2	29.6	23.0	17.9	17.2
109 Netherlands	821.8	1,070.8	1,448.6	1,579.8	1,690.2	5.8	1.7	1.4	25.3	20.8	20.3	19.2	22.6
110 France	1,649.9	2,493.6	4,353.7	4,999.2	5,950.4[e]	10.2	2.8	4.5	23.9	23.8	23.0	22.0	20.6[e]
111 Australia	707.7	—	—	—	—	—	—	—	31.9	—	—	—	—

— Not available.

Note: *m*, median; *u*, unweighted mean; *w*, weighted mean; and *w′*, weighted mean excluding China and India.

a. 1974 data. b. 1971 data. c. 1981 data. d. 1979 data. e. 1984 data. f. 1986 data. g. 1976 data.

Sources: Unesco data; National Center for Education Statistics 1988.

(Table continues on the following page.)

Table A-17 (*continued*)

Economy	Total (millions of 1985 dollars)					Annual growth (percent)			As a percentage of total public recurrent expenditure on education				
	1965	1970	1975	1980	1985	1965–75	1975–80	1980–85	1965	1970	1975	1980	1985
112 Federal Republic of Germany	3,176.3	4,345.7	3,750.1	3,856.4	3,586.6	1.7	0.6	-1.4	43.7	38.5	18.2	16.0	14.0
113 Finland	558.6	622.5	999.5	740.0	874.4	6.0	-5.8	3.4	45.3	35.8	46.1	31.8	30.8
114 Denmark	901.0	—	—	1,631.8	1,801.4[f]	—	—	—	60.4	—	—	53.5	43.8[f]
115 Japan	5,122.7	7,485.2	13,193.2	16,412.6	17,777.3	9.9	4.5	1.6	37.9	37.6	39.1	38.2	27.8
116 Sweden	—	—	—	—	—	—	—	—	—	—	—	—	—
117 Canada	—	—	—	—	—	—	—	—	—	—	—	—	—
118 Norway	—	—	—	—	—	—	—	—	—	—	—	—	—
119 United States	56,861.9	77,808.6	83,928.9	86,490.3	97,915.0	4.0	0.6	2.5	46.6	45.6	44.6	35.8	36.3
120 Switzerland	1,420.7	776.3	945.7[a]	—	—	-4.4	—	—	68.6	33.2	27.9[a]	—	—
Other economies													
121 Angola	—	—	—	—	—	—	—	—	—	—	—	—	—
122 Albania	—	—	—	—	—	—	—	—	—	—	—	—	—
123 Bulgaria	—	—	—	—	—	—	—	—	—	—	—	—	—
124 Czechoslovakia	—	—	—	—	—	—	—	—	39.2	36.4	41.2	40.9	43.8
125 Cuba	—	—	—	—	—	—	—	—	40.4	—	—	24.4	20.7
126 German Democratic Republic	—	—	—	—	—	—	—	—	—	—	—	—	—
127 Mongolia	—	—	—	—	—	—	—	—	—	—	—	—	—
128 Democratic People's Republic of Korea	—	—	—	—	—	—	—	—	—	—	—	—	—
129 Union of Soviet Socialist Republics	—	—	—	—	—	—	—	—	47.5	43.7	38.2	33.9	35.8

— Not available.

Note: *m*, median; *u*, unweighted mean; *w*, weighted mean; and *w′*, weighted mean excluding China and India.

a. 1974 data. b. 1971 data. c. 1981 data. d. 1979 data. e. 1984 data. f. 1986 data. g. 1976 data.

Sources: Unesco data; National Center for Education Statistics 1988.

Table A-18. Public Recurrent Expenditure on Primary Education as a Percentage of GNP, Selected Years, 1965–85

Economy	1965	1970	1975	1980	1985
Low-income economies	**1.2 *m***	**1.4 *m***	**1.1 *m***	**1.1 *m***	**1.2 *m***
	1.2 *u*	**1.3 *u***	**1.3 *u***	**1.4 *u***	**1.4 *u***
	0.7 *w*	**0.8 *w***	**1.1 *w***	**1.1 *w***	**1.0 *w***
Excluding China and India	**1.0 *w′***	**1.1 *w′***	**1.1 *w′***	**1.2 *w′***	**1.3 *w′***
1 Ethiopia	0.5	0.8	1.0	1.1	1.8
2 Bhutan	—	—	—	—	—
3 Burkina Faso	0.8	—	1.0	0.8	0.9
4 Nepal	—	0.2[a]	0.7	1.1	1.0
5 Bangladesh	0.3	—	0.4	0.5	0.9
6 Malawi	1.8	1.4	1.0	1.0	1.0
7 Zaire	1.2	—	1.4	1.6	—
8 Mali	—	—	1.4	1.4	1.6
9 Myanmar	1.1	0.9[a]	—	—	—
10 Mozambique	—	—	—	—	—
11 Madagascar	1.8	1.8	1.3	1.9	1.4
12 Uganda	—	1.3	1.0	0.4[b]	—
13 Burundi	1.4	—	1.0	1.1[c]	1.1
14 Tanzania	1.6	1.5	1.6	2.2[c]	2.3
15 Togo	—	1.3	—	1.6	2.2[d]
16 Niger	0.4	0.7	—	1.0[b]	—
17 Benin	—	2.0	2.1	—	—
18 Somalia	0.3	0.7	—	—	—
19 Central African Republic	1.9	—	2.4	2.3	1.4[e]
20 India	0.5	0.6	1.1	1.0	1.2
21 Rwanda	2.1	1.5	1.6	1.5	—

— Not available.

Note: *m*, median; *u*, unweighted mean; *w*, weighted mean; and *w′*, weighted mean excluding China and India.
a. 1971 data. b. 1981 data. c. 1979 data. d. 1984 data. e. 1986 data. f. 1974 data. g. 1976 data.
Sources: Unesco data; World Bank data.

(*Table continues on the following page.*)

Table A-18 (*continued*)

Economy	1965	1970	1975	1980	1985
22 China	—	—	—	—	0.6
23 Kenya	2.3	1.9	3.8	3.8	3.7
24 Zambia	—	1.6	2.3	1.9	2.2[d]
25 Sierra Leone	1.1	—	0.8	—	—
26 Sudan	0.6	1.4	—	2.1	2.4
27 Haiti	0.9	0.8	0.6	0.7	0.6
28 Pakistan	0.5	0.4	0.6	0.6	0.7
29 Lesotho	2.0	1.6	1.6	1.5	1.2[d]
30 Ghana	1.4	1.5	1.1	0.5	0.7
31 Sri Lanka	2.5	2.3	0.6[f]	—	—
32 Mauritania	—	—	1.7	1.8	1.9
33 Senegal	1.2	—	1.7[g]	2.0	—
34 Afghanistan	0.3	0.3	0.4[f]	0.7	—
35 Chad	0.8	—	—	—	—
36 Guinea	1.3	—	—	1.0[c]	1.0[d]
37 Democratic Kampuchea	2.3	3.2	—	—	—
38 Lao People's Democratic Republic	—	—	—	—	—
39 Viet Nam	—	—	—	—	—
Lower-middle-income economies	**1.5 *m***	**1.5 *m***	**1.6 *m***	**1.8 *m***	**1.6 *m***
	1.5 *u*	**1.6 *u***	**1.5 *u***	**1.7 *u***	**1.8 *u***
	1.3 *w*	**1.1 *w***	**1.0 *w***	**1.2 *w***	**1.5 *w***
40 Liberia	0.9	—	0.5	0.7[b]	—
41 People's Democratic Republic of Yemen	—	—	—	—	—
42 Indonesia	—	0.7[a]	—	—	—
43 Arab Republic of Yemen	—	—	—	—	3.1
44 Philippines	—	2.1	1.1[g]	0.9	0.8
45 Morocco	1.8	1.5	1.3[f]	1.8	—
46 Bolivia	1.3	2.0	2.1	2.5	—

47	Zimbabwe	—	1.6	1.6	4.3	5.0
48	Nigeria	—	0.6	0.4[f]	0.8[b]	—
49	Dominican Republic	—	1.2	0.7[g]	0.6	0.7
50	Papua New Guinea	—	1.3[a]	2.5	—	—
51	Côte d'Ivoire	1.6	1.6	2.3	2.8[c]	—
52	Honduras	1.9	1.9	—	1.8	2.1
53	Egypt	—	—	—	—	—
54	Nicaragua	1.1	1.3	1.1	1.3	2.6
55	Thailand	1.3	1.3	1.6	1.4	2.0
56	El Salvador	1.6	1.6	1.8	2.3	—
57	Botswana	2.2	2.5	2.1	3.0[c]	2.8[d]
58	Jamaica	1.0	1.2	1.6	2.3	1.6
59	Cameroon	—	—	—	—	—
60	Guatemala	—	1.0	0.8[g]	0.6[c]	—
61	Congo	2.0	2.8	2.3	2.3	1.4[d]
62	Paraguay	—	1.2	—	—	0.4
63	Peru	1.7	1.3	1.3	1.3	0.9
64	Turkey	—	—	—	1.0	0.9
65	Tunisia	2.1	—	1.9	1.9	2.3
66	Ecuador	1.1	1.5	—	1.1	1.6
67	Mauritius	1.8	2.0	1.5[f]	2.1	1.6
68	Colombia	0.5	—	—	0.8	1.1
69	Chile	1.1	—	1.4	1.9	2.2
70	Costa Rica	2.5	2.5	2.4	2.0	1.5
71	Jordan	—	—	—	—	—
72	Syrian Arab Republic	1.3	1.6	1.2	1.4	2.0
73	Lebanon	1.1	—	—	—	—

— Not available.

Note: *m*, median; *u*, unweighted mean; *w*, weighted mean; and *w′*, weighted mean excluding China and India.

a. 1971 data. b. 1981 data. c. 1979 data. d. 1984 data. e. 1986 data. f. 1974 data. g. 1976 data.

Sources: Unesco data; World Bank data.

(Table continues on the following page.)

Table A-18 (*continued*)

Economy	1965	1970	1975	1980	1985
Upper-middle-income economies	**1.5 *m***	**1.5 *m***	**1.2 *m***	**1.4 m**	**1.7 *m***
	1.7 *u*	**1.5 *u***	**1.3 *u***	**1.4 *u***	**1.6 *u***
	1.3 *w*	**1.1 *w***	**1.1 *w***	**1.5 *w***	**1.4 *w***
74 Brazil	—	—	1.4[g]	1.6	1.5
75 Malaysia	—	1.8[a]	—	1.8	2.1
76 South Africa	—	—	—	—	—
77 Mexico	0.7	0.7	1.0	1.1	0.6
78 Uruguay	1.5	1.7	—	1.0	0.9
79 Hungary	—	—	1.3	1.5	1.8
80 Poland	—	—	—	1.0	1.3
81 Portugal	0.5	—	2.1	2.0	2.0
82 Yugoslavia	—	—	—	—	—
83 Panama	1.9	2.0	2.1	2.2	1.9
84 Argentina	1.4	0.5	0.6	1.2	1.4[d]
85 Republic of Korea	1.1	1.7	1.0	1.6	1.8
86 Algeria	2.3	—	1.6[f]	1.5	—
87 Venezuela	1.4	1.5	0.9	0.6	1.0[d]
88 Gabon	—	—	—	—	—
89 Greece	—	0.8	0.7	0.8[c]	0.9
90 Spain	—	0.7	1.1[f]	1.3[c]	—
91 Oman	—	—	0.8[f]	—	1.9[e]
92 Ireland	2.0	1.8	1.9	1.5	1.8
93 Trinidad and Tobago	1.5	—	—	1.4	2.4
94 Israel	2.1	1.9	1.9	2.5	3.4[d]
95 Hong Kong	1.1	1.2	1.2	0.7	0.8[d]
96 Singapore	2.1	1.3	1.0	0.9	—
97 Islamic Republic of Iran	1.6	1.3	0.8[f]	2.6	1.4
98 Iraq	3.6	3.2	1.5[g]	1.3[c]	—
99 Romania	—	—	—	—	—

High-income oil exporters		**1.1 *m* 1.1 *u* 1.1 *w***	**0.8 *m* 0.8 *u* 0.8 *w***	**1.6 *m* 1.6 *u* 1.7 *w***	**0.7 *m* 0.7 *u* 0.7 *w***	**2.4 *m* 2.4 *u* 2.4 *w***
100	Saudi Arabia	—	—	2.0[f]	—	—
101	Kuwait	1.1	0.8	1.2	0.7	2.4
102	United Arab Emirates	—	—	—	—	—
103	Libya	—	—	—	—	—
High-income industrial market economies		**1.3 *m* 1.5 *u* 1.9 *w***	**1.1 *m* 1.3 *u* 2.0 *w***	**1.4 *m* 1.5 *u* 2.1 *w***	**1.5 *m* 1.6 *u* 1.9 *w***	**1.3 *m* 1.5 *u* 1.9 *w***
104	New Zealand	1.3	1.5	1.7	1.9	1.7
105	Italy	1.5	0.9	1.1	1.1[c]	—
106	United Kingdom	—	1.2	1.7	1.4	1.1[d]
107	Belgium	1.1	1.0	1.4	1.5	1.4
108	Austria	1.6	1.1	1.0	0.9	0.9
109	Netherlands	1.2	1.2	1.4	1.3	1.3
110	France	0.6	0.7	1.0	1.0	1.2[d]
111	Australia	0.9	—	—	—	—
112	Federal Republic of Germany	0.9	1.0	0.8	0.7	0.6
113	Finland	2.1	1.9	2.5	1.6	1.6
114	Denmark	2.6	—	—	3.3	3.1[e]
115	Japan	1.2	1.1	1.5	1.5	1.3
116	Sweden	—	—	—	—	—
117	Canada	—	—	—	—	—
118	Norway	—	—	—	—	—
119	United States	2.4	2.9	2.8	2.4	2.4
120	Switzerland	2.2	1.0	1.1[f]	—	—

— Not available.

Note: *m*, median; *u*, unweighted mean; *w*, weighted mean; and *w′*, weighted mean excluding China and India.

a. 1971 data. b. 1981 data. c. 1979 data. d. 1984 data. e. 1986 data. f. 1974 data. g. 1976 data.

Sources: Unesco data; World Bank data.

(Table continues on the following page.)

Table A-18 (*continued*)

Economy	*1965*	*1970*	*1975*	*1980*	*1985*
Other economies					
121 Angola	—	—	—	—	—
122 Albania	—	—	—	—	—
123 Bulgaria	—	—	—	—	—
124 Czechoslovakia	—	—	—	—	—
125 Cuba	—	—	—	—	—
126 German Democratic Republic	—	—	—	—	—
127 Mongolia	—	—	—	—	—
128 Democratic People's Republic of Korea	—	—	—	—	—
129 Union of Soviet Socialist Republics	—	—	—	—	—

— Not available.

Note: *m*, median; *u*, unweighted mean; *w*, weighted mean; and *w′*, weighted mean excluding China and India.

a. 1971 data. b. 1981 data. c. 1979 data. d. 1984 data. e. 1986 data. f. 1974 data. g. 1976 data.

Sources: Unesco data; World Bank data.

Table A-19. Recurrent Expenditure on Primary Education, by Purpose, Latest Available Year, 1979–86

Economy	Year	Recurrent expenditure (percent)			
		For teacher emoluments	For teaching materials	For adminis-tration	Other[a]
Low-income economies		**95.3 *m***	**1.3 *m***	**0.7 *m***	**0.2 *m***
		87.3 *u*	**3.5 *u***	**3.9 *u***	**5.3 *u***
		85.3 *w*	**2.9 *w***	**0.6 *w***	**11.2 w**
Excluding China and India		**84.7 *w′***	**8.6 *w′***	**2.9 *w′***	**3.9 *w′***
1 Ethiopia	—	—	—	—	—
2 Bhutan	—	—	—	—	—
3 Burkina Faso	1985	95.8	0.0	2.8	1.4
4 Nepal	1980	76.8	7.2	9.6	6.4
5 Bangladesh	1982	54.0	0.0	0.0	46.0
6 Malawi	1985	91.5	0.0	0.0	8.5
7 Zaire	—	—	—	—	—
8 Mali	1985	98.3	1.7	0.0	0.0
9 Myanmar	—	—	—	—	—
10 Mozambique	—	—	—	—	—
11 Madagascar	1985	99.5	0.0	0.5	0.0
12 Uganda	—	—	—	—	—
13 Burundi	1985	98.6	0.5	0.8	0.0
14 Tanzania	1979	58.2	28.9	2.6	10.4
15 Togo	1984	98.2	1.8	0.0	0.0
16 Niger	—	—	—	—	—
17 Benin	—	—	—	—	—
18 Somalia	—	—	—	—	—
19 Central African Republic	1986	97.8	1.6	0.0	0.7

— Not available.

Note: *m*, median; *u*, unweighted mean; *w*, weighted mean; and *w′*, weighted mean excluding China and India.

a. Includes scholarships, welfare services, and undistributed expenditure.

Source: Unesco data.

(*Table continues on the following page.*)

Table A-19 (*continued*)

Economy	Year	Recurrent expenditure (percent): For teacher emoluments	For teaching materials	For adminis-tration	Other[a]
20 India	—	—	—	—	—
21 Rwanda	1981	94.7	3.5	1.7	0.0
22 China	—	—	—	—	—
23 Kenya	—	—	—	—	—
24 Zambia	1982	89.6	3.1	7.2	0.2
25 Sierra Leone	—	—	—	—	—
26 Sudan	—	—	—	—	—
27 Haiti	1986	69.8	1.0	29.0	0.2
28 Pakistan	—	—	—	—	—
29 Lesotho	1984	99.9	0.0	0.0	0.1
30 Ghana	—	—	—	—	—
31 Sri Lanka	—	—	—	—	—
32 Mauritania	—	—	—	—	—
33 Senegal	—	—	—	—	—
34 Afghanistan	—	—	—	—	—
35 Chad	—	—	—	—	—
36 Guinea	—	—	—	—	—
37 Democratic Kampuchea	—	—	—	—	—
38 Lao People's Democratic Republic	—	—	—	—	—
39 Viet Nam	—	—	—	—	—
Lower-middle-income economies		**91.0 *m***	**1.9 *m***	**2.7 *m***	**1.3 *m***
		89.7 *u*	**2.5 *u***	**4.9 *u***	**2.9 *u***
		89.0 *w*	**2.2 *w***	**4.2 *w***	**4.6 *w***
40 Liberia	1979	78.3	8.0	0.0	13.7
41 People's Democratic Republic of Yemen	—	—	—	—	—
42 Indonesia	—	—	—	—	—
43 Arab Republic of Yemen	—	—	—	—	—

44	Philippines	—	—	—	—	—
45	Morocco	—	—	—	—	—
46	Bolivia	1981	98.0	0.0	0.0	2.0
47	Zimbabwe	1984	89.6	9.2	0.1	1.2
48	Nigeria	—	—	—	—	—
49	Dominican Republic	1986	96.7	0.0	2.6	0.7
50	Papua New Guinea	—	—	—	—	—
51	Côte d'Ivoire	1979	85.3	2.1	2.7	9.9
52	Honduras	1982	92.7	1.8	1.7	3.8
53	Egypt	—	—	—	—	—
54	Nicaragua	1986	92.5	3.9	3.3	0.2
55	Thailand	1986	90.2	3.1	3.8	2.9
56	El Salvador	—	—	—	—	—
57	Botswana	—	—	—	—	—
58	Jamaica	1985	88.6	2.5	5.5	3.4
59	Cameroon	—	—	—	—	—
60	Guatemala	1979	97.1	1.5	0.2	1.3
61	Congo	—	—	—	—	—
62	Paraguay	—	—	—	—	—
63	Peru	1985	77.6	0.5	21.7	0.2
64	Turkey	—	—	—	—	—
65	Tunisia	—	—	—	—	—
66	Ecuador	1986	79.2	0.1	20.1	0.6
67	Mauritius	—	—	—	—	—
68	Colombia	—	—	—	—	—
69	Chile	—	—	—	—	—
70	Costa Rica	1985	98.4	0.0	1.6	0.0
71	Jordan	—	—	—	—	—

— Not available.

Note: m, median; u, unweighted mean; w, weighted mean; and w', weighted mean excluding China and India.

a. Includes scholarships, welfare services, and undistributed expenditure.

Source: Unesco data.

(*Table continues on the following page.*)

Table A-19 (*continued*)

Economy	Year	Recurrent expenditure (percent) For teacher emoluments	For teaching materials	For adminis-tration	Other[a]
72 Syrian Arab Republic	1981	91.7	1.9	5.8	0.5
73 Lebanon	—	—	—	—	—
Upper-middle-income economies		**87.9 *m***	**0.6 *m***	**1.1 *m***	**8.4 *m***
		80.1 *u*	**1.5 *u***	**4.6 *u***	**13.8 *u***
		84.1 *w*	**2.4 *w***	**5.8 *w***	**7.7 *w***
74 Brazil					
75 Malaysia	1984	89.5	5.0	3.7	1.9
76 South Africa	—	—	—	—	—
77 Mexico	1984	87.9	3.2	0.4	8.5
78 Uruguay	—	—	—	—	—
79 Hungary	1986	33.7	0.0	21.3	45.0
80 Poland	—	—	—	—	—
81 Portugal	—	—	—	—	—
82 Yugoslavia	—	—	—	—	—
83 Panama	—	—	—	—	—
84 Argentina	1986	46.9	0.0	2.2	50.8
85 Republic of Korea	1986	79.8	1.7	17.0	1.5
86 Algeria	1979	98.1	0.2	0.0	1.6
87 Venezuela	1984	86.9	0.6	4.1	8.4
88 Gabon	—	—	—	—	—
89 Greece	1982	93.9	2.5	1.1	2.4
90 Spain	—	—	—	—	—
91 Oman	—	—	—	—	—
92 Ireland	1986	88.7	0.2	0.0	11.0
93 Trinidad and Tobago	1985	79.1	3.1	0.1	17.6
94 Israel	—	—	—	—	—
95 Hong Kong	1984	96.1	0.3	0.8	2.7

Country	Year				
96 Singapore	—	—	—	—	—
97 Islamic Republic of Iran	—	—	—	—	—
98 Iraq	—	—	—	—	—
99 Romania	—	—	—	—	—
High-income oil exporters		**63.5 *m***	**1.2 *m***	**31.6 *m***	**3.7 *m***
		63.5 *u*	**1.2 *u***	**31.6 *u***	**3.7 *u***
		63.5 *w*	**1.2 w**	**31.6 *w***	**3.7 *w***
100 Saudi Arabia	—	—	—	—	—
101 Kuwait	1986	63.5	1.2	31.6	3.7
102 United Arab Emirates	—	—	—	—	—
103 Libya	—	—	—	—	—
High-income industrial market economies		**76.1 *m***	**0.8 *m***	**1.5 *m***	**18.1 *m***
		73.0 *u*	**2.6 *u***	**5.4 *u***	**19.1 *u***
		60.3 *w*	**4.4 *w***	**1.8 *w***	**33.5 *w***
104 New Zealand	1986	79.2	9.7	2.6	8.5
105 Italy	1983	90.7	0.1	3.3	5.9
106 United Kingdom	—	—	—	—	—
107 Belgium	1982	87.4	0.2	0.3	12.1
108 Austria	1986	72.2	0.0	26.8	1.0
109 Netherlands	1985	75.7	1.4	0.0	22.9
110 France	1982	76.5	0.0	0.0	23.5
111 Australia	—	—	—	—	—
112 Federal Republic of Germany	1986	86.7	0.0	0.0	13.3
113 Finland	1984	53.7	5.4	4.8	36.1
114 Denmark	1980	52.4	4.3	16.3	27.1
115 Japan	1982	55.1	4.8	0.0	40.2

— Not available.

Note: *m*, median; *u*, unweighted mean; *w*, weighted mean; and *w′*, weighted mean excluding China and India.

a. Includes scholarships, welfare services, and undistributed expenditure.

Source: Unesco data.

(Table continues on the following page.)

Table A-19 (*continued*)

Economy	Year	Recurrent expenditure (percent)			
		For teacher emoluments	For teaching materials	For administration	Other[a]
116 Sweden	—	—	—	—	—
117 Canada	—	—	—	—	—
118 Norway	—	—	—	—	—
119 United States	—	—	—	—	—
120 Switzerland	—	—	—	—	—
Other economies					
121 Angola	—	—	—	—	—
122 Albania	—	—	—	—	—
123 Bulgaria	—	—	—	—	—
124 Czechoslovakia	1983	43.8	22.5	0.0	33.7
125 Cuba	1984	57.2	1.6	0.0	41.2
126 German Democratic Republic	—	—	—	—	—
127 Mongolia	—	—	—	—	—
128 Democratic People's Republic of Korea	—	—	—	—	—
129 Union of Soviet Socialist Republics	—	—	—	—	—

— Not available.

Note: *m*, median; *u*, unweighted mean; *w*, weighted mean; and *w′*, weighted mean excluding China and India.

a. Includes scholarships, welfare services, and undistributed expenditure.

Source: Unesco data.

Table A-20. Public Recurrent Expenditure per Student in Primary School, Selected Years, 1965–85

Economy	1985 constant dollars					Percentage of GNP per capita				
	1965	1970	1975	1980	1985	1965	1970	1975	1980	1985
Low-income economies	**40.7 *m***	**40.8 *m***	**40.7 *m***	**29.4 *m***	**31.2 *m***	**20.0 *m***	**14.1 *m***	**13.0 *m***	**11.7 *m***	**11.9 *m***
	43.8 *u*	**46.0 *u***	**45.3 *u***	**42.8 *u***	**36.0 *u***	**19.4 *u***	**19.1 *u***	**17.4 *u***	**15.5 *u***	**15.9 *u***
	14.7 *w*	**18.1 *w***	**24.6 *w***	**24.9 *w***	**21.7 *w***	**7.4 *w***	**7.8 *w***	**11.0 *w***	**10.4 *w***	**8.1 *w***
Excluding China and India	**26.5 *w′***	**41.3 *w′***	**26.6 *w′***	**29.2 *w′***	**32.2 *w′***	**12.8 *w′***	**14.0 *w′***	**11.9 *w′***	**12.2 *w′***	**13.5 *w′***
1 Ethiopia	36.7	41.7	37.4	24.8	35.3	31.2	33.4	29.8	19.3	31.1
2 Bhutan	—	—	—	—	—	—	—	—	—	—
3 Burkina Faso	49.1	—	53.1	38.2	29.6	45.1	—	43.9	27.3	21.2
4 Nepal	—	6.1[a]	22.9	19.6	13.7	—	—	16.8	14.6	9.0
5 Bangladesh	5.7	—	5.2	7.4	16.4	3.7	—	3.8	4.9	10.3
6 Malawi	23.4	21.5	12.6	11.4	11.8	21.6	17.8	8.1	7.4	8.1
7 Zaire	19.9	—	16.6	15.5	—	10.4	—	8.8	10.0	—
8 Mali	—	—	42.4	49.5	57.8	—	—	32.7	32.4	41.1
9 Myanmar	14.8	10.1[a]	—	—	—	11.4	7.9[a]	—	—	—
10 Mozambique	—	—	—	—	—	—	—	—	—	—
11 Madagascar	46.7	40.8	24.8	27.7	19.9	16.1	12.6	8.1	9.7	10.4[b]
12 Uganda	—	72.8	40.7	10.7[c]	—	—	17.3	11.0	4.0[c]	—
13 Burundi	52.4	—	59.4	65.1[d]	31.1	30.5	—	28.8	27.8[d]	13.1
14 Tanzania	69.8	84.8	60.5	44.4[d]	46.1	23.5	24.2	16.3	12.4[d]	16.1
15 Togo	—	28.4	—	23.1	31.3[b]	—	11.6	—	8.0	14.4[b]
16 Niger	80.8	100.5	—	73.7[c]	—	21.5	30.6	—	25.2[c]	—
17 Benin	—	78.6	55.4	—	—	—	33.4	25.0	—	—
18 Somalia	55.4	116.5	—	—	—	36.9	72.5	—	—	—
19 Central African Republic	70.1	—	66.4	60.9	36.1[e]	26.1	—	22.0	21.7	—
20 India	9.8	12.2	23.7	22.5	29.9	5.1	5.6	10.5	9.4	10.7

— Not available.

Note: *m*, median; *u*, unweighted mean; *w*, weighted mean; and *w′*, weighted mean excluding China and India.

a. 1971 data. b. 1984 data. c. 1981 data. d. 1979 data. e. 1986 data. f. 1974 data. g. 1976 data.

Sources: Unesco data; World Bank data.

(Table continues on the following page.)

Table A-20 (*continued*)

Economy	1985 constant dollars					Percentage of GNP per capita				
	1965	1970	1975	1980	1985	1965	1970	1975	1980	1985
21 Rwanda	38.5	31.5	37.6	31.1	—	20.0	13.1	16.9	11.0	—
22 China	—	—	—	—	13.4	—	—	—	—	4.9
23 Kenya	40.7	31.4	49.9	49.3	44.8	21.6	15.1	18.0	16.1	15.8
24 Zambia	—	49.1	61.7	43.4	41.8[b]	—	9.5	13.0	10.4	11.4[b]
25 Sierra Leone	68.0	—	46.6	—	—	20.2	—	11.5	—	—
26 Sudan	74.2	91.3	—	107.5	92.2	17.9	24.3	—	26.9	30.1
27 Haiti	41.6	31.3	19.5	22.9	13.8	12.4	10.3	6.1	5.9	4.1
28 Pakistan	18.8	16.2	22.9	27.0	32.8	8.6	6.1	8.7	8.7	8.7
29 Lesotho	13.7	15.3	19.8	26.5	20.8[b]	11.6	9.7	8.7	8.5	6.0[b]
30 Ghana	37.5	45.5	41.5	16.0	18.4	7.7	9.0	9.6	3.9	5.6
31 Sri Lanka	26.3	41.1	13.5[f]	—	—	12.8	17.2	5.6[f]	—	—
32 Mauritania	—	—	186.3	127.0	91.6	—	—	45.5	30.3	24.3
33 Senegal	94.6	—	111.9[g]	100.8	—	22.3	—	25.3[e]	27.0	—
34 Afghanistan	—	—	—	—	—	—	—	—	—	—
35 Chad	26.2	—	—	—	—	16.1	—	—	—	—
36 Guinea	81.0	—	—	66.6[d]	62.5[b]	30.8	—	—	21.0[d]	21.4[b]
37 Democratic Kampuchea	—	—	—	—	—	—	—	—	—	—
38 Lao People's Democratic Republic	—	—	—	—	—	—	—	—	—	—
39 Viet Nam	—	—	—	—	—	—	—	—	—	—
Lower-middle-income economies	**72.7 *m***	**74.8 *m***	**74.3 *m***	**79.2 *m***	**99.5 *m***	**10.3 *m***	**11.0 *m***	**9.2 *m***	**9.5 *m***	**10.9 *m***
	75.0 *u*	**77.7 *u***	**87.4 *u***	**94.1 *u***	**95.6 *u***	**11.6 *u***	**11.3 *u***	**10.8 *u***	**10.8 *u***	**10.9 *u***
	57.8 *w*	**66.6 *w***	**63.0 *w***	**68.1 *w***	**80.9 *w***	**9.3 *w***	**8.6 *w***	**8.3 *w***	**7.5 *w***	**9.2 *w***
40 Liberia	77.8	—	54.8	64.6[d]	—	12.9	—	8.1	9.4[d]	—
41 People's Democratic Republic of Yemen	—	—	—	—	—	—	—	—	—	—
42 Indonesia	—	16.8[a]	—	—	—	—	5.6a	—	—	—
43 Arab Republic of Yemen	—	—	—	—	148.7	—	—	—	—	25.5

44	Philippines	—	55.7	37.9[g]	37.1	26.9	—	11.4	6.3[g]	5.7	4.7
45	Morocco	77.3	76.7	67.9[f]	84.5	—	21.6	19.1	16.1[f]	16.4	—
46	Bolivia	53.1	74.8	82.7	93.1	—	9.7	12.8	12.0	14.3	—
47	Zimbabwe	—	61.3	69.9	132.6	101.7	—	11.1	11.6	24.2	18.8
48	Nigeria	—	104.1	57.2[f]	48.7[c]	—	—	10.7	8.3[g]	4.6[c]	—
49	Dominican Republic	—	32.6	24.0[g]	23.3	25.0	—	6.9	3.7[g]	3.3	3.9
50	Papua New Guinea	—	108.8[a]	194.3	—	—	—	16.5[a]	28.1	—	—
51	Côte d'Ivoire	107.1	113.9	163.2	174.3[d]	—	21.0	17.6	23.3	23.3[d]	—
52	Honduras	104.0	94.3	—	94.5	91.5	15.6	13.1	—	11.1	12.2
53	Egypt	—	—	—	—	—	—	—	—	—	—
54	Nicaragua	114.9	122.0	105.7	73.7	119.3	9.3	9.6	7.5	7.4	14.9
55	Thailand	25.9	32.9	46.4	53.5	101.7	8.3	8.0	9.9	8.9	14.3
56	El Salvador	101.9	96.1	98.2	114.3	—	12.3	11.1	9.7	12.2	—
57	Botswana	25.7	37.3	46.4	79.2[d]	97.2[b]	18.6	18.8	13.9	16.9[d]	13.9[b]
58	Jamaica	64.7	67.4	99.2	119.3	80.3	5.6	5.7	8.5	13.9	—
59	Cameroon	—	—	—	—	—	—	—	—	—	—
60	Guatemala	—	110.7	103.2[g]	74.0[d]	—	—	10.0	7.8[g]	5.0[d]	—
61	Congo	60.0	83.0	74.9	75.5	61.3[b]	11.5	13.9	9.9	9.5	5.6[b]
62	Paraguay	—	36.1	—	—	24.5	—	6.9	—	—	2.9
63	Peru	90.4	67.8	73.7	73.4	42.4	10.2	7.1	6.9	7.2	4.8
64	Turkey	—	—	—	73.6	69.3	—	—	—	8.0	6.8
65	Tunisia	72.7	—	100.8	120.8	143.1	13.4	—	11.7	11.7	13.1
66	Ecuador	50.8	70.3	—	73.1	105.7	6.9	9.0	—	5.7	9.0[b]
67	Mauritius	70.6	67.8	70.2[f]	138.2	115.1	10.3	11.0	8.7[f]	15.6	11.4
68	Colombia	27.2	—	—	57.4	93.2	3.8	—	—	4.8	7.8
69	Chile	72.4	—	64.2	134.9	155.2	6.3	—	6.1	9.5	13.2
70	Costa Rica	137.9	154.8	185.4	204.2	149.4	13.6	12.5	13.0	12.7	10.4
71	Jordan	—	—	—	—	—	—	—	—	—	—
72	Syrian Arab Republic	90.9	101.4	102.6	135.4	160.8	9.7	10.6	6.8	8.1	10.1

— Not available.

Note: *m*, median; *u*, unweighted mean; *w*, weighted mean; and *w'*, weighted mean excluding China and India.

a. 1971 data. b. 1984 data. c. 1981 data. d. 1979 data. e. 1986 data. f. 1974 data. g. 1976 data.

Sources: Unesco data; World Bank data.

(Table continues on the following page.)

Table A-20 (continued)

Economy	1985 constant dollars 1965	1970	1975	1980	1985	Percentage of GNP per capita 1965	1970	1975	1980	1985
73 Lebanon	—	—	—	—	—	—	—	—	—	—
Upper-middle-income economies	**186.0 *m***	**188.8 *m***	**255.2 *m***	**249.2 *m***	**296.1 *m***	**10.6 *m***	**8.9 *m***	**8.1 *m***	**9.1 *m***	**12.5 *m***
	221.5 *u*	**228.8 *u***	**267.9 *u***	**341.0 *u***	**404.0 *u***	**10.1 *u***	**8.9 *u***	**9.3 *u***	**10.6 *u***	**11.9 *u***
	164.9 *w*	**164.4 *w***	**164.6 *w***	**228.1 *w***	**197.3 *w***	**9.3 *w***	**7.0 *w***	**7.0 *w***	**9.3 *w***	**8.5 *w***
74 Brazil	—	—	110.1[g]	142.1	133.2	—	—	7.6[g]	8.5	8.3
75 Malaysia	—	124.9[a]	—	205.1	281.3	—	—	—	12.0	15.2
76 South Africa	—	—	—	—	—	—	—	—	—	—
77 Mexico	61.6	68.3	103.3	116.7	71.8	4.5	4.2	5.5	5.0	3.2
78 Uruguay	161.4	210.1	—	185.4	126.1	12.3	13.2	—	8.9	7.6
79 Hungary	92.8	132.0	201.9	242.7	272.8	—	—	13.4	14.0	14.6
80 Poland	—	—	—	155.9	193.4	—	—	—	8.2	10.3
81 Portugal	49.4	—	262.8	308.4	320.6	5.0	—	16.3	15.9	16.7
82 Yugoslavia	—	—	—	—	—	—	—	—	—	—
83 Panama	155.8	188.8	192.8	255.7	260.5	12.3	11.9	10.9	12.5	12.5
84 Argentina	262.2	76.9	117.2	228.5	200.4[b]	9.8	3.8	4.7	8.9	9.4[b]
85 Republic of Korea	37.2	80.2	74.1	162.1	310.9	6.3	9.5	6.4	10.6	15.2
86 Algeria	286.2	—	195.1[f]	214.4	—	20.5	—	9.8[f]	8.9	—
87 Venezuela	360.1	406.5	262.5	168.8	226.6[b]	8.4	8.9	5.7	3.8	6.5[b]
88 Gabon	—	—	—	—	—	—	—	—	—	—
89 Greece	—	180.9	182.0	264.1[d]	326.2	—	7.9	6.3	8.0[d]	9.8
90 Spain	—	196.2	416.1[f]	555.2[d]	—	—	6.1	10.5[f]	13.7[d]	—
91 Oman	—	—	382.3[f]	—	847.1[e]	—	—	12.6[f]	—	—
92 Ireland	454.0	493.1	629.7	575.2	706.3	11.3	13.2	14.8	12.1	15.3
93 Trinidad and Tobago	288.9	—	—	732.2	1,051.6	6.4	—	—	9.3	16.9
94 Israel	404.0	501.9	609.7	841.3	1,104.3[b]	12.1	12.0	12.0	15.5	20.3[b]
95 Hong Kong	137.4	170.6	267.2	353.0	488.8[b]	6.7	6.5	8.1	7.0	7.8[b]
96 Singapore	186.0	201.4	255.2	376.5	—	11.3	7.5	6.6	7.2	—
97 Islamic Republic of Iran	384.9	400.0	292.9[f]	737.5	350.6	15.1	10.4	6.1[f]	21.1	8.9

98	Iraq	—	—	—	—	—	—	—	—	—	—
99	Romania	—	—	—	—	—	—	—	—	—	—
	High-income oil exporters	**3,205.8 *m***	**2,355.9 *m***	**2,226.4 *m***	**1,341.8 *m***	**3,312.6 *m***	**10.1 *m***	**7.7 *m***	**16.3 *m***	**6.1 *m***	**23.3 *m***
		3,205.8 *u*	**2,355.9 *u***	**2,226.4 *u***	**1,341.8 *u***	**3,312.6 *u***	**10.1 *u***	**7.7 *u***	**16.3 *u***	**6.1 *u***	**23.3 *u***
		3,205.8 *w*	**2,355.9 *w***	**1,859.7 *w***	**1,341.8 *w***	**3,312.6 *w***	**10.1 *w***	**7.7 *w***	**18.4 *w***	**6.1 *w***	**23.3 *w***
100	Saudi Arabia	—	—	1,700.1[f]	—	—	—	—	21.8[f]	—	—
101	Kuwait	3,205.8	2,355.9	2,752.6	1,341.8	3,312.6	10.1	7.7	10.7	6.1	23.3
102	United Arab Emirates	—	—	—	—	—	—	—	—	—	—
103	Libya	—	—	—	—	—	—	—	—	—	—
	High-income industrial market economies	**722.3 *m***	**754.7 *m***	**1,049.5 *m***	**1,332.5 *m***	**1,578.9 *m***	**11.7 *m***	**11.5 *m***	**14.4 *m***	**15.9 *m***	**17.6 *m***
		1,050.8 *u*	**998.6 *u***	**1,309.5 *u***	**1,635.2 *u***	**1,977.9 *u***	**14.3 *u***	**12.8 *u***	**15.4 *u***	**17.8 *u***	**17.7 *u***
		1,219.0 *w*	**1,524.3 *w***	**1,874.7 *w***	**2,132.5 *w***	**2,608.8 *w***	**16.5 *w***	**17.1 *w***	**19.3 *w***	**19.2 *w***	**20.4 *w***
104	New Zealand	479.5	574.9	844.4	938.2	1,072.8	9.1	10.4	13.7	15.7	16.3
105	Italy	722.3	558.2	698.6	915.6[d]	—	17.7	10.5	12.1	13.8[d]	—
106	United Kingdom	—	709.4	1,117.6	1,173.6	1,105.8[b]	—	11.3	16.6	16.0	14.2[b]
107	Belgium	530.0	606.4	1,042.8	1,381.9	1,551.4	10.9	9.8	15.0	17.4	18.9
108	Austria	1,100.4	847.1	1,056.1	1,283.0	1,714.0	24.4	14.9	15.6	16.1	19.9
109	Netherlands	583.2	732.2	996.6	1,184.8	1,150.8	10.5	10.9	12.9	14.0	13.3
110	France	298.7	504.8	946.1	1,084.3	1,442.0[b]	5.5	7.5	12.0	11.9	15.5[b]
111	Australia	454.7	—	—	—	—	6.6	—	—	—	—
112	Federal Republic of Germany	903.5	1,093.9	960.8	1,385.3	1,578.9	9.1	14.7	11.9	14.5	15.4
113	Finland	1,214.3	1,611.7	2,202.8	1,982.1	2,298.0	21.2	22.6	26.1	20.5	21.2
114	Denmark	2,049.8	—	—	3,754.4	4,596.6[e]	28.4	—	—	38.4	—
115	Japan	516.0	777.2	1,272.9	1,387.8	1,602.2	12.4	11.7	16.6	14.8	14.5
116	Sweden	—	—	—	—	—	—	—	—	—	—
117	Canada	—	—	—	—	—	—	—	—	—	—

— Not available.

Note: *m*, median; *u*, unweighted mean; *w*, weighted mean; and *w′*, weighted mean excluding China and India.

a. 1971 data. b. 1984 data. c. 1981 data. d. 1979 data. e. 1986 data. f. 1974 data. g. 1976 data.

Sources: Unesco data; World Bank data.

(Table continues on the following page.)

Table A-20 (*continued*)

Economy	1985 constant dollars					Percentage of GNP per capita				
	1965	1970	1975	1980	1985	1965	1970	1975	1980	1985
118 Norway	—	—	—	—	—	—	—	—	—	—
119 United States	1,751.0	2,291.9	2,756.6	3,150.9	3,644.0	16.3	17.5	19.9	20.3	21.7
120 Switzerland	3,057.1	1,675.0	1,818.4[f]	—	—	—	12.2	12.1[f]	—	—
Other economies										
121 Angola	—	—	—	—	—	—	—	—	—	—
122 Albania	—	—	—	—	—	—	—	—	—	—
123 Bulgaria	—	—	—	—	—	—	—	—	—	—
124 Czechoslovakia	—	—	—	—	—	—	—	—	—	—
125 Cuba	—	—	—	—	—	—	—	—	—	—
126 German Democratic Republic	—	—	—	—	—	—	—	—	—	—
127 Mongolia	—	—	—	—	—	—	—	—	—	—
128 Democratic People's Republic of Korea	—	—	—	—	—	—	—	—	—	—
129 Union of Soviet Socialist Republics	—	—	—	—	—	—	—	—	—	—

— Not available.
Note: *m*, median; *u*, unweighted mean; *w*, weighted mean; and *w′*, weighted mean excluding China and India.
a. 1971 data. b. 1984 data. c. 1981 data. d. 1979 data. e. 1986 data. f. 1974 data. g. 1976 data.
Sources: Unesco data; World Bank data.

Table A-21. Total Recurrent Expenditure on Primary Education per Inhabitant, Selected Years, 1965–85

(constant 1985 dollars)

Economy	1965	1970	1975	1980	1985
Low-income economies	**2.5 *m***	**3.2 *m***	**3.3 *m***	**2.9 *m***	**2.8 *m***
	2.8 *u*	**3.5 *u***	**3.9 *u***	**4.1 *u***	**3.8 *u***
	1.4 *w*	**1.7 *w***	**2.5 *w***	**2.6 *w***	**2.5 *w***
Excluding China and India	**2.0 *w′***	**3.1 *w′***	**2.5 *w′***	**2.9 *w′***	**3.1 *w′***
1 Ethiopia	0.5	0.9	1.2	1.4	2.0
2 Bhutan	—	—	—	—	—
3 Burkina Faso	0.9	—	1.2	1.1	1.3
4 Nepal	—	0.2[a]	1.0	1.4	1.5
5 Bangladesh	0.4	—	0.6	0.7	1.5
6 Malawi	2.0	1.7	1.5	1.5	1.5
7 Zaire	2.4	—	2.6	2.5	—
8 Mali	—	—	1.8	2.2	2.3
9 Myanmar	1.4	1.2[a]	—	—	—
10 Mozambique	—	—	—	—	—
11 Madagascar	5.2	5.7	3.9	5.5	3.1
12 Uganda	—	5.4	3.6	1.2[b]	—
13 Burundi	2.5	—	2.1	2.6[c]	2.6
14 Tanzania	4.6	5.4	6.0	7.9[c]	6.6
15 Togo	—	3.2	—	4.5	4.8[d]
16 Niger	1.3	2.1	—	3.1[b]	—
17 Benin	—	4.6	4.8	—	—
18 Somalia	0.5	1.1	—	—	—
19 Central African Republic	5.2	—	7.2	6.6	3.7[e]

— Not available.

Note: *m*, median; *u*, unweighted mean; *w*, weighted mean; and *w′*, weighted mean excluding China and India.

a. 1971 data. b. 1981 data. c. 1979 data. d. 1984 data. e. 1986 data. f. 1974 data. g. 1976 data.

Sources: Unesco data; World Bank data.

(*Table continues on the following page.*)

Table A-21 (*continued*)

Economy	1965	1970	1975	1980	1985
20 India	1.0	1.3	2.5	2.4	3.4
21 Rwanda	4.0	3.6	3.5	4.3	—
22 China	—	—	—	—	1.7
23 Kenya	4.3	3.9	10.5	11.6	10.3
24 Zambia	—	8.2	11.1	8.0	8.2[d]
25 Sierra Leone	3.5	—	3.3	—	—
26 Sudan	2.6	5.4	—	8.4	7.3
27 Haiti	2.9	2.6	1.9	2.7	2.0
28 Pakistan	1.0	0.9	1.5	1.6	2.3
29 Lesotho	2.4	2.6	3.7	4.8	4.1[d]
30 Ghana	6.8	7.5	5.0	2.2	2.3
31 Sri Lanka	5.2	5.5	1.5[f]	—	—
32 Mauritania	—	—	6.9	7.4	7.3
33 Senegal	5.3	—	7.6[g]	7.4	—
34 Afghanistan	—	—	—	—	—
35 Chad	1.3	—	—	—	—
36 Guinea	3.4	—	—	3.3[d]	3.0[d]
37 Democratic Kampuchea	—	—	—	—	—
38 Lao People's Democratic Republic	—	—	—	—	—
39 Viet Nam	—	—	—	—	—
Lower-middle-income economies	**10.5 *m***	**10.7 *m***	**14.1 *m***	**14.3 *m***	**16.0 *m***
	10.2 *u*	**11.0 *u***	**12.9 *u***	**15.1 *u***	**16.3 *u***
	7.9 *w*	**8.3 *w***	**8.5 *w***	**10.8 *w***	**12.8 w**
40 Liberia	5.5	—	3.6	4.8[c]	—
41 People's Democratic Republic of Yemen	—	—	—	—	—
42 Indonesia	—	2.2[a]	—	—	—
43 Arab Republic of Yemen	—	—	—	—	18.3
44 Philippines	—	10.3	6.7[g]	6.4	4.3
45 Morocco	6.5	5.9	5.7[f]	9.5	—

46	Bolivia	6.8	11.7	14.5	16.4	—
47	Zimbabwe	—	8.6	9.9	23.4	26.8
48	Nigeria	—	5.5	4.2[f]	8.0[b]	—
49	Dominican Republic	—	5.6	4.3[g]	4.5	4.8
50	Papua New Guinea	—	8.6[a]	17.2	—	—
51	Côte d'Ivoire	8.4	10.4	16.3	20.8[c]	—
52	Honduras	12.9	13.7	—	15.5	16.0
53	Egypt	—	—	—	—	—
54	Nicaragua	13.5	17.0	15.0	12.6	20.5
55	Thailand	3.9	5.2	7.5	8.5	14.1
56	El Salvador	12.9	13.7	17.7	21.1	—
57	Botswana	3.1	5.0	7.2	14.3[c]	19.7[d]
58	Jamaica	9.7	12.8	18.1	19.7	11.9
59	Cameroon	—	—	—	—	—
60	Guatemala	—	10.7	10.7[g]	8.3[c]	—
61	Congo	10.5	16.6	17.4	18.4	15.4[d]
62	Paraguay	—	6.5	—	—	3.8
63	Peru	15.0	12.0	13.8	13.4	8.1
64	Turkey	—	—	—	9.3	9.1
65	Tunisia	11.5	—	16.8	19.9	25.5
66	Ecuador	7.6	11.4	—	13.8	18.9
67	Mauritius	12.6	12.3	12.3[f]	18.4	15.9
68	Colombia	3.3	—	—	9.2	13.3
69	Chile	12.9	—	14.3	26.5	26.4
70	Costa Rica	25.6	31.3	34.0	32.1	21.8
71	Jordan	—	—	—	—	—
72	Syrian Arab Republic	12.1	15.0	17.6	23.9	31.2
73	Lebanon	—	—	—	—	—

— Not available.

Note: *m*, median; *u*, unweighted mean; *w*, weighted mean; and *w*′, weighted mean excluding China and India.

a. 1971 data. b. 1981 data. c. 1979 data. d. 1984 data. e. 1986 data. f. 1974 data. g. 1976 data.

Sources: Unesco data; World Bank data.

(Table continues on the following page.)

Table A-21 (*continued*)

Economy	1965	1970	1975	1980	1985
Upper-middle-income economies	**32.6 *m***	**26.5 *m***	**33.6 *m***	**33.8 *m***	**37.9 *m***
	33.5 *u*	**33.4 *u***	**35.8 *u***	**46.1 *u***	**56.3 *u***
	23.6 *w*	**24.4 *w***	**25.7 *w***	**35.7 *w***	**31.5 *w***
74 Brazil	—	—	19.7[g]	26.5	24.3
75 Malaysia	—	19.4[a]	—	29.9	39.4
76 South Africa	—	—	—	—	—
77 Mexico	9.5	12.0	19.1	24.3	13.8
78 Uruguay	20.1	26.5	—	21.3	15.3
79 Hungary	12.9	14.2	20.1	26.3	33.2
80 Poland	—	—	—	18.3	25.0
81 Portugal	4.8	—	33.6	38.6	39.0
82 Yugoslavia	—	—	—	—	—
83 Panama	23.9	31.5	36.9	44.1	40.6
84 Argentina	36.8	10.9	16.1	31.7	29.7[d]
85 Republic of Korea	6.4	14.4	11.8	24.1	36.8
86 Algeria	32.6	—	31.7[f]	35.8	—
87 Venezuela	58.3	67.8	43.7	28.4	36.4[d]
88 Gabon	—	—	—	—	—
89 Greece	—	18.7	18.8	24.9[c]	29.2
90 Spain	—	22.8	43.7[f]	54.0[c]	—
91 Oman	—	—	25.3[f]	—	129.4[e]
92 Ireland	59.7	66.8	80.2	71.0	84.3
93 Trinidad and Tobago	68.4	—	—	111.7	149.4
94 Israel	70.2	80.8	94.5	134.9	184.1[d]
95 Hong Kong	22.2	32.0	39.4	37.8	48.8[d]
96 Singapore	35.2	35.3	37.0	45.5	—
97 Islamic Republic of Iran	41.2	48.1	37.3[f]	92.3	53.8
98 Iraq	—	—	—	—	—
99 Romania	—	—	—	—	—

High-income oil exporters		**366.7 *m* 366.7 *u* 366.7 *w***	**239.1 *m* 239.1 *u* 239.1 *w***	**230.0 *m* 230.0 *u* 173.6 *w***	**145.7 *m* 145.7 *u* 145.7 *w***	**334.7 *m* 334.7 *u* 334.7 w**
100	Saudi Arabia	—	—	154.3[f]	—	—
101	Kuwait	366.7	239.1	305.7	145.7	334.7
102	United Arab Emirates	—	—	—	—	—
103	Libya	—	—	—	—	—
High-income industrial market economies		**66.8 *m* 105.5 *u* 148.2 *w***	**73.0 *m* 103.5 *u* 179.0 *w***	**106.6 *m* 130.6 *u* 200.6 *w***	**113.3 *m* 144.8 *u* 206.9 *w***	**114.9 *m* 159.6 *u* 236.1 *w***
104	New Zealand	67.1	81.6	107.1	114.9	108.8
105	Italy	62.8	50.5	60.9	73.3[c]	—
106	United Kingdom	—	74.2	113.7	102.3	83.7[d]
107	Belgium	54.0	61.5	100.3	118.2	114.9
108	Austria	71.5	60.7	70.3	68.0	78.0
109	Netherlands	66.8	82.1	106.0	111.6	116.6
110	France	33.8	49.1	82.6	93.1	108.3[d]
111	Australia	62.1	—	—	—	—
112	Federal Republic of Germany	54.2	71.7	60.7	62.6	58.8
113	Finland	122.4	135.1	212.2	154.8	178.2
114	Denmark	189.4	—	—	318.5	351.8[e]
115	Japan	51.8	71.7	117.9	140.5	147.2
116	Sweden	—	—	—	—	—
117	Canada	—	—	—	—	—
118	Norway	—	—	—	—	—
119	United States	292.6	379.5	388.6	379.8	409.2
120	Switzerland	242.6	123.9	146.8[f]	—	—

— Not available.

Note: *m*, median; *u*, unweighted mean; *w*, weighted mean; and *w′*, weighted mean excluding China and India.

a. 1971 data. b. 1981 data. c. 1979 data. d. 1984 data. e. 1986 data. f. 1974 data. g. 1976 data.

Sources: Unesco data; World Bank data.

(Table continues on the following page.)

Table A-21 (continued)

Economy	*1965*	*1970*	*1975*	*1980*	*1985*
Other economies					
121 Angola	—	—	—	—	—
122 Albania	—	—	—	—	—
123 Bulgaria	—	—	—	—	—
124 Czechoslovakia	—	—	—	—	—
125 Cuba	—	—	—	—	—
126 German Democratic Republic	—	—	—	—	—
127 Mongolia	—	—	—	—	—
128 Democratic People's Republic of Korea	—	—	—	—	—
129 Union of Soviet Socialist Republics	—	—	—	—	—

— Not available.

Note: *m*, median; *u*, unweighted mean; *w*, weighted mean; and *w'*, weighted mean excluding China and India.

a. 1971 data. b. 1981 data. c. 1979 data. d. 1984 data. e. 1986 data. f. 1974 data. g. 1976 data.

Sources: Unesco data; World Bank data.

Table A-22. Expenditure on Primary School Teachers and Teaching Materials, Latest Year Available, 1979–86

	Expenditure on teaching materials per student			Average teacher salary		
Economy	*Year*	*Constant 1985 dollars*	*As a percentage of GNP per capita*	*Year*	*Constant 1985 dollars*	*As a multiple of GNP per capita*
Low-income economies		**0.5 *m***	**0.23 *m***		**1,369.2 *m***	**6.1 *m***
		1.5 *u*	**0.53 *u***		**1,687.8 *u***	**6.6 *u***
		0.8 *w*	**0.28 *w***		**1,037.2 *w***	**3.9 *w***
Excluding China and India		**2.1 *w′***	**1.04 *w′***		**1,024.2 *w′***	**4.9 *w′***
1 Ethiopia	1981	0.7	0.55	1981	1,608.8	12.6
2 Bhutan	—	—	—	—	—	—
3 Burkina Faso	1984	0.3	0.23	1985	1,769.9	12.7
4 Nepal	1980	1.4	1.04	1980	575.4	4.3
5 Bangladesh	1986	0.0	0.00	1986	309.1	2.1
6 Malawi	—	—	—	1986	849.7	5.8
7 Zaire	—	—	—	—	—	—
8 Mali	1985	1.0	0.71	1985	1,931.8	13.7
9 Myanmar	—	—	—	—	—	—
10 Mozambique	—	—	—	—	—	—
11 Madagascar	1985	0.0	0.00	1985	750.5	3.5
12 Uganda	—	—	—	—	—	—
13 Burundi	1985	0.3	0.13	1985	1,704.0	7.2
14 Tanzania	1979	12.8	3.58	1979	1,072.0	3.0
15 Togo	1986	0.6	0.24	1984	1,369.2	6.3
16 Niger	—	—	—	—	—	—
17 Benin	—	—	—	—	—	—
18 Somalia	—	—	—	—	—	—

— Not available.

Note: *m*, median; *u*, unweighted mean; *w*, weighted mean; and *w′*, weighted mean excluding China and India.

Sources: Unesco data; World Bank data; National Center for Education Statistics 1988.

(*Table continues on the following page.*)

Table A-22 (continued)

Economy	Expenditure on teaching materials per student: Year	Constant 1985 dollars	As a percentage of GNP per capita	Average teacher salary: Year	Constant 1985 dollars	As a multiple of GNP per capita
19 Central African Republic	1986	0.4	0.14	1986	1,958.6	6.9
20 India	1986	0.4	0.14	1986	1,041.9	3.6
21 Rwanda	1983	2.2	0.73	1981	3,135.8	10.4
22 China	—	—	—	—	—	—
23 Kenya	—	—	—	1984	1,295.4	4.6
24 Zambia	1982	1.8	0.44	1982	2,483.9	6.1
25 Sierra Leone	—	—	—	—	—	—
26 Sudan	—	—	—	—	—	—
27 Haiti	1986	0.2	0.06	1986	491.7	1.4
28 Pakistan	—	—	—	—	—	—
29 Lesotho	1982	0.0	0.00	1984	1,097.7	3.2
30 Ghana	1984	0.1	0.03	1986	787.9	2.0
31 Sri Lanka	—	—	—	—	—	—
32 Mauritania	1983	2.5	0.61	1984	5,629.5	13.7
33 Senegal	1981	3.1	0.87	1981	3,476.3	9.8
34 Afghanistan	—	—	—	—	—	—
35 Chad	—	—	—	—	—	—
36 Guinea	—	—	—	1979	2,104.3	6.6
37 Democratic Kampuchea	—	—	—	—	—	—
38 Lao People's Democratic Republic	—	—	—	—	—	—
39 Viet Nam	—	—	—	—	—	—
Lower-middle-income economies		**1.4 *m***	**0.13 *m***		**2,569.4 *m***	**2.6 *m***
		1.9 *u*	**0.24 *u***		**2,711.7 *u***	**3.1 *u***
		1.8 *w*	**0.18 *w***		**2,114.7 *w***	**2.1 *w***

No.	Country	Year			Year		
40	Liberia	1979	5.2	0.75	1979	1,833.6	2.7
41	People's Democratic Republic of Yemen	—	—	—	—	—	—
42	Indonesia	—	—	—	—	—	—
43	Arab Republic of Yemen	—	—	—	—	—	—
44	Philippines	—	—	—	—	—	—
45	Morocco	1979	0.5	0.04	1979	2,900.3	5.7
46	Bolivia	—	—	—	1982	1,525.4	2.6
47	Zimbabwe	1986	7.1	1.20	1984	3,549.9	6.6
48	Nigeria	—	—	—	—	—	—
49	Dominican Republic	—	—	—	1986	898.5	1.3
50	Papua New Guinea	—	—	—	—	—	—
51	Côte d'Ivoire	1979	3.6	0.48	1979	5,805.8	7.8
52	Honduras	1982	1.6	0.21	1982	3,307.3	4.2
53	Egypt	—	—	—	—	—	—
54	Nicaragua	1986	4.7	0.61	1986	3,523.5	4.5
55	Thailand	1986	3.1	0.39	1986	1,861.6	2.3
56	El Salvador	—	—	—	—	—	—
57	Botswana	1986	0.0	0.00	1986	1,938.8	2.3
58	Jamaica	1985	2.0	0.27	1985	2,508.9	3.4
59	Cameroon	—	—	—	—	—	—
60	Guatemala	1979	1.1	0.07	1979	2,489.6	1.7
61	Congo	1980	0.0	0.00	1980	4,077.4	5.1
62	Paraguay	—	—	—	—	—	—
63	Peru	1985	0.2	0.02	1985	1,145.4	1.3
64	Turkey	1986	0.1	0.01	1985	205.4	0.2
65	Tunisia	1984	2.2	0.21	1984	4,225.6	3.9
66	Ecuador	1986	0.1	0.01	1986	2,241.2	2.0
67	Mauritius	1984	0.0	0.00	1986	1,161.5	0.9

— Not available.

Note: *m*, median; *u*, unweighted mean; *w*, weighted mean; and *w′*, weighted mean excluding China and India.

Sources: Unesco data; World Bank data; National Center for Education Statistics 1988.

(Table continues on the following page.)

Table A-22 (continued)

Economy	Expenditure on teaching materials per student: Year	Constant 1985 dollars	As a percentage of GNP per capita	Average teacher salary: Year	Constant 1985 dollars	As a multiple of GNP per capita
68 Colombia	—	—	—	1981	2,630.0	2.2
69 Chile	1981	0.7	0.05	1981	2,845.7	2.0
70 Costa Rica	1983	1.7	0.13	1986	4760.3	3.2
71 Jordan	—	—	—	—	—	—
72 Syrian Arab Republic	1984	—	0.19	1984	4,220.9	2.6
73 Lebanon	—	—	—	—	—	—
Upper-middle-income economies		**4.4 *m***	**0.22 *m***		**7,219.6 *m***	**2.2 *m***
		13.6 *u*	**0.46 *u***		**8,394.7 *u***	**2.3 *u***
		5.5 *w*	**0.29 *w***		**5,465.5 *w***	**2.1 *w***
74 Brazil	—	—	—	—	—	—
75 Malaysia	1984	13.6	0.71	1984	6,191.5	3.2
76 South Africa	—	—	—	—	—	—
77 Mexico	1986	1.8	0.09	1986	1,733.3	0.9
78 Uruguay	1980	10.9	0.52	1986	1,955.3	1.0
79 Hungary	—	—	—	1986	1,345.8	0.7
80 Poland	—	—	—	—	—	—
81 Portugal	1985	0.0	0.00	1985	5,055.7	2.6
82 Yugoslavia	—	—	2.41	—	—	—
83 Panama	1985	4.4	0.21	1986	5,522.5	2.4
84 Argentina	—	—	—	1984	3,839.5	1.8
85 Republic of Korea	1986	5.6	0.24	1979	9,841.6	4.2
86 Algeria	1980	2.3	0.10	1980	7,408.4	3.1
87 Venezuela	1984	1.3	0.04	1984	5,125.1	1.5
88 Gabon	—	—	—	—	—	—
89 Greece	1982	7.2	0.22	1985	7,219.6	2.2

90	Spain	—	—	—	1979	16,657.7	4.1
91	Oman	1986	32.4	0.65	—	—	—
92	Ireland	1986	1.4	0.03	1986	16,647.4	3.3
93	Trinidad and Tobago	1985	32.7	0.52	1985	18,369.0	2.9
94	Israel	1984	88.3	1.63	1984	11,890.7	2.2
95	Hong Kong	1984	1.7	0.03	1984	12,843.4	2.1
96	Singapore	1982	0.0	0.00	1982	11,063.6	1.8
97	Islamic Republic of Iran	—	—	—	—	—	—
98	Iraq	—	—	—	—	—	—
99	Romania	—	—	—	—	—	—
High-income oil exporters			**40.8** *m*	**0.29** *m*		**40,227.0** *m*	**2.9** *m*
			40.8 *u*	**0.29** *u*		**40,227.0** *u*	**2.9** *u*
			40.8 *w*	**0.29** *w*		**40,227.0** *w*	**2.9** *w*
100	Saudi Arabia	—	—	—	—	—	—
101	Kuwait	1986	40.8	0.29	1986	40,227.0	2.9
102	United Arab Emirates	—	—	—	—	—	—
103	Libya	—	—	—	—	—	—
High-income industrial market economies			**68.2** *m*	**0.76** *m*		**19,849.1** *m*	**1.9** *m*
			63.7 *u*	**0.71** *u*		**18,859.3** *u*	**1.9** *u*
			52.4 *w*	**0.62** *w*		**21,149.3** w	**1.7** *w*
104	New Zealand	1986	100.5	1.26	1986	16,449.3	2.1
105	Italy	1983	1.0	0.01	1983	12,151.7	1.7
106	United Kingdom	1984	34.0	—	—	—	—
107	Belgium	1985	1.6	0.02	1986	20,305.7	2.2
108	Austria	—	—	—	1986	13,329.3	1.3
109	Netherlands	1985	16.4	0.19	1985	14,561.5	1.7

— Not available.
Note: *m*, median; *u*, unweighted mean; *w*, weighted mean; and *w*′, weighted mean excluding China and India.
Sources: Unesco data; World Bank data; National Center for Education Statistics 1988.

(Table continues on the following page.)

Table A-22 (*continued*)

Economy	Expenditure on teaching materials per student			Average teacher salary		
	Year	Constant 1985 dollars	As a percentage of GNP per capita	Year	Constant 1985 dollars	As a multiple of GNP per capita
110 France	1984	77.9	0.84	1982	19,887.5	2.1
111 Australia	—	—	—	—	—	—
112 Federal Republic of Germany	—	—	—	1986	23,300.9	1.9
113 Finland	1984	112.7	1.04	1986	19,849.1	1.6
114 Denmark	1980	161.3	1.65	1980	23,452.3	2.4
115 Japan	1982	68.2	0.69	1982	19,498.1	2.0
116 Sweden	—	—	—	—	—	—
117 Canada	—	—	—	—	—	—
118 Norway	—	—	—	—	—	—
119 United States	—	—	—	1985	24,667.0	1.5
120 Switzerland	—	—	—	—	—	—
Other economies						
121 Angola	—	—	—	—	—	—
122 Albania	—	—	—	—	—	—
123 Bulgaria	—	—	—	—	—	—
124 Czechoslovakia	—	—	—	—	—	—
125 Cuba	—	—	—	—	—	—
126 German Democratic Republic	—	—	—	—	—	—
127 Mongolia	—	—	—	—	—	—
128 Democratic People's Republic of Korea	—	—	—	—	—	—
129 Union of Soviet Socialist Republics	—	—	—	—	—	—

— Not available.

Note: m, median; u, unweighted mean; w, weighted mean; and w', weighted mean excluding China and India.

Sources: Unesco data; World Bank data; National Center for Education Statistics 1988.

Table A-23. Preschool Enrollment Rate, Selected Years, 1965–80

(percent)

	Economy	1965	1970	1975	1980
	Low-income economies	**0.0 *m***	**0.0 *m***	**0.0 *m***	**1.0 *m***
		0.2 *u*	**0.9 *u***	**1.8 *u***	**3.2 *u***
1	Ethiopia	—	—	—	—
2	Bhutan	—	—	—	—
3	Burkina Faso	0.0	0.0	0.0	0.0
4	Nepal	—	5.0	0.0	0.0
5	Bangladesh	—	—	—	—
6	Malawi	—	—	—	—
7	Zaire	0.0	0.0	—	—
8	Mali	—	—	—	—
9	Myanmar	—	—	—	—
10	Mozambique	0.0	0.0	—	—
11	Madagascar	—	—	—	—
12	Uganda	—	—	—	—
13	Burundi	0.0	0.0	0.0	0.0
14	Tanzania	—	—	—	—
15	Togo	1.0	2.0	2.0	2.0
16	Niger	0.0	—	0.0	0.0
17	Benin	—	—	—	—
18	Somalia	0.0	0.0	0.0	0.0
19	Central African Republic	1.0	3.0	3.0	—
20	India	0.0	0.0	—	1.0
21	Rwanda	—	—	—	—
22	China	—	—	—	12.0

— Not available.
Note: *m*, median; *u*, unweighted mean.
Source: O'Connor 1988.

(*Table continues on the following page.*)

Table A-23 (*continued*)

Economy	1965	1970	1975	1980
23 Kenya	—	—	—	—
24 Zambia	—	—	—	1.0
25 Sierra Leone	—	0.0	—	—
26 Sudan	0.0	1.0	—	5.0
27 Haiti	—	—	—	—
28 Pakistan	—	—	—	—
29 Lesotho	—	—	—	—
30 Ghana	1.0	1.0	1.0	—
31 Sri Lanka	—	—	—	—
32 Mauritania	—	—	—	—
33 Senegal	—	—	—	1.0
34 Afghanistan	0.0	0.0	0.0	0.0
35 Chad	0.0	0.0	—	—
36 Guinea	—	—	—	—
37 Democratic Kampuchea	—	—	—	—
38 Lao People's Democratic Republic	—	—	—	—
39 Viet Nam	—	—	12.0	19.0
Lower-middle-income economies	**1.0 *m***	**2.0 *m***	**3.0 *m***	**4.5 *m***
	2.9 *u*	**3.7 *u***	**5.4 *u***	**7.3 *u***
40 Liberia	—	—	—	—
41 People's Democratic Republic of Yemen	—	—	0.0	—
42 Indonesia	1.0	2.0	3.0	4.0
43 Arab Republic of Yemen	—	—	—	—
44 Philippines	1.0	1.0	—	1.0
45 Morocco	0.0	0.0	12.0	16.0
46 Bolivia	9.0	8.0	5.0	10.0
47 Zimbabwe	—	—	—	
48 Nigeria	—	—	—	—
49 Dominican Republic	1.0	1.0	1.0	3.0

	Economy				
50	Papua New Guinea	—	—	0.0	0.0
51	Côte d'Ivoire	0.0	0.0	0.0	0.0
52	Honduras	1.0	2.0	3.0	4.0
53	Egypt	—	—	—	—
54	Nicaragua	3.0	3.0	2.0	6.0
55	Thailand	1.0	2.0	3.0	5.0
56	El Salvador	3.0	4.0	6.0	9.0
57	Botswana	—	—	—	—
58	Jamaica	3.0	3.0	42.0	47.0
59	Cameroon	2.0	1.0	—	3.0
60	Guatemala	2.0	2.0	3.0	4.0
61	Congo	0.0	0.0	—	1.0
62	Paraguay	1.0	1.0	2.0	2.0
63	Peru	2.0	3.0	7.0	7.0
64	Turkey	0.0	0.0	0.0	0.0
65	Tunisia	1.0	—	—	—
66	Ecuador	1.0	1.0	2.0	3.0
67	Mauritius	10.0	9.0	8.0	10.0
68	Colombia	1.0	—	3.0	5.0
69	Chile	4.0	5.0	11.0	10.0
70	Costa Rica	2.0	3.0	6.0	8.0
71	Jordan	—	—	—	—
72	Syrian Arab Republic	—	—	—	—
73	Lebanon	21.0	34.0	—	28.0
Upper-middle-income economies		**5.0 *m***	**6.0 *m***	**14.0 *m***	**15.0 *m***
		9.2 *u*	**13.0 *u***	**17.5 *u***	**21.5 *u***
74	Brazil	—	2.0	3.0	6.0
75	Malaysia	—	—	—	—
76	South Africa	—	—	—	—

— Not available.
Note: *m*, median; *u*, unweighted mean.
Source: O'Connor 1988.

(Table continues on the following page.)

Table A-23 (continued)

Economy	*1965*	*1970*	*1975*	*1980*
77 Mexico	4.0	5.0	5.0	9.0
78 Uruguay	9.0	7.0	14.0	15.0
79 Hungary	29.0	32.0	42.0	57.0
80 Poland	21.0	29.0	39.0	42.0
81 Portugal	1.0	2.0	5.0	7.0
82 Yugoslavia	5.0	6.0	10.0	16.0
83 Panama	2.0	3.0	5.0	7.0
84 Argentina	6.0	9.0	14.0	19.0
85 Republic of Korea	0.0	1.0	1.0	1.0
86 Algeria	—	—	—	—
87 Venezuela	2.0	3.0	14.0	14.0
88 Gabon	—	—	—	—
89 Greece	7.0	11.0	16.0	20.0
90 Spain	19.0	26.0	28.0	37.0
91 Oman	—	—	—	—
92 Ireland	—	44.0	42.0	40.0
93 Trinidad and Tobago	—	—	—	—
94 Israel	—	—	—	—
95 Hong Kong	10.0	33.0	40.0	42.0
96 Singapore	2.0	2.0	2.0	9.0
97 Islamic Republic of Iran	0.0	0.0	3.0	3.0
98 Iraq	1.0	1.0	2.0	3.0
99 Romania	25.0	22.0	42.0	47.0
High-income oil exporters	**6.5 *m***	**1.0 *m***	**1.0 *m***	**1.5 *m***
	3.5 *u*	**5.0 *u***	**5.0 *u***	**3.5 *u***
100 Saudi Arabia	0.0	1.0	1.0	2.0
101 Kuwait	12.0	14.0	13.0	11.0
102 United Arab Emirates	1.0	0.0	1.0	0.0
103 Libya	1.0	—	—	1.0

High-income industrial market economies		**13.0 *m*** **19.0 *u***	**19.0 *m*** **23.5 *u***	**26.5 *m*** **33.9 *u***	**25.5 *m*** **36.3 *u***
104	New Zealand	6.0	8.0	11.0	15.0
105	Italy	29.0	35.0	47.0	52.0
106	United Kingdom	5.0	9.0	14.0	1.0
107	Belgium	57.0	64.0	67.0	61.0
108	Austria	—	—	—	—
109	Netherlands	38.0	41.0	50.0	47.0
110	France	42.0	53.0	63.0	66.0
111	Australia	13.0	7.0	15.0	14.0
112	Federal Republic of Germany	19.0	25.0	47.0	52.0
113	Finland	5.0	7.0	20.0	23.0
114	Denmark	1.0	5.0	13.0	19.0
115	Japan	14.0	19.0	23.0	28.0
116	Sweden	13.0	20.0	38.0	44.0
117	Canada	10.0	19.0	23.0	21.0
118	Norway	3.0	4.0	16.0	23.0
119	United States	13.0	16.0	34.0	32.0
120	Switzerland	—	—	30.0	35.0
Other economies					
121	Angola	0.0	0.0	—	—
122	Albania	8.0	15.0	—	23.0
123	Bulgaria	56.0	51.0	58.0	60.0
124	Czechoslovakia	30.0	35.0	37.0	52.0
125	Cuba	7.0	11.0	—	—
126	German Democratic Republic	—	—	—	—
127	Mongolia	—	—	—	—
128	Democratic People's Republic of Korea	—	—	—	—
129	Union of Soviet Socialist Republics	—	—	—	—

— Not available.

Note: *m*, median; *u*, unweighted mean.

Source: O'Connor 1988.

Technical Notes

General Information

Growth rates have been computed using the compound growth rate equation,

$$X_t = X_0(1 + r)^t$$

where X is the variable, t is time, and r is the growth rate.

Weighted and group unweighted averages and medians for growth rates and ratios (percentages) for each income group are provided. Totals for absolute quantities are provided and reflect sums of constant cases. Constant cases are those countries for which data are available for all years canvassed by the particular table.

Value indicators are expressed in constant 1985 dollars. Local currencies were converted by applying the GDP deflator and the 1985 official exchange rate.

In the notes that follow, terms in italics (other than publication titles) correspond to column headings in the tables.

Table A-1

Data on population age 6–11 for the years 1965 to 1985 were interpolated from the *UN Population Estimates and Projections: 1988 Revision* (United Nations 1988). The UN data come in the form of five-year cohorts; Sprague Multipliers were used to apportion five-year totals into single-year populations. This method is described in *Manual III, Methods for Population Projections by Sex and Age* (United Nations Department of Economic and Social Affairs 1956). World Bank estimates were interpolated from data provided by the Population, Health and Nutrition Division of the Bank's Population and Human Resources Department. Differences between World Bank and UN population data for 1985 are due to differences in their respective methodologies for calculating population.

Table A-2

Total enrollment includes students of all ages enrolled in both public and private primary schools, unless otherwise noted below. Most enrollment data come from the Unesco data base official enrollment series. When official series data are missing, data from the Unesco data base estimated series are used, unless otherwise indicated. *Annual growth* in enrollment is calculated using the compound growth equation described under "General Information." *Females as a percentage of total* is the number of females divided by total enrollment.

Algeria: Data for 1975 and 1980 include students enrolled in OUCFA schools.
Belgium: Data include special education students.
Czechoslovakia: In 1984 the length of primary education was shortened from nine to eight years.
El Salvador: Data for 1975 include students enrolled in evening schools.
Finland: Data include integrated special education students.
Federal Republic of Germany: Because of a new method of estimation, data for 1985 are not comparable with those of previous years.
Ghana: Data refer to enrollment in public schools only.
Haiti: In 1976 the length of primary education was shortened from seven to six years.
Jordan: Data for 1985 include students enrolled in United Nations Relief and Works Administration (UNRWA) schools.
Kenya: In 1985 the length of primary education was increased from seven to eight years.
Lao People's Democratic Republic: In 1976 the length of primary education was shortened from six to five years.
Liberia: Data include preschool enrollments.
Madagascar: In 1976 the length of primary education was shortened from six to five years.
Mauritania: In 1977 the length of primary education was shortened from seven to six years.
Mozambique: Data for 1980 include preschool enrollments.
Nepal: In 1981 the length of primary education was increased from three to five years.
Netherlands: In 1985 the length of primary education was increased from six to eight years.
Pakistan: Data include preschool enrollments.
Philippines: Data for 1980 refer to enrollment in public schools only.
Rwanda: In 1979 the length of primary education was increased from six to eight years.
Somalia: The length of primary education was increased from four to six years in 1976 and then to eight years in 1980.

Sri Lanka: In 1975 the length of primary education was increased from five to six years. Data for 1975 refer to enrollment in public schools only.

Switzerland: Data for 1975 refer to enrollment in public schools only and include special education students.

Tanzania: Data are for the mainland only.

Thailand: In 1978 the length of primary education was shortened from seven to six years.

Trinidad and Tobago: Data refer to enrollment in government-maintained and -aided schools only and include intermediate departments of secondary schools.

Uganda: Data for 1975 and 1980 refer to enrollment in government-maintained and -aided schools only.

Viet Nam: Gross enrollment data for 1965 and 1970 are extrapolated from the enrollment data for children age 6–11 reported in table A-3.

People's Democratic Republic of Yemen: In 1979 the length of primary education was increased from five to six years.

Table A-3

Total includes children age 6–11 enrolled in both public and private schools, unless otherwise indicated for table A-2; data are from the Unesco data base. *Annual growth* in enrollment is calculated using the compound growth rate equation described under "General Information." *Enrolled children age 6–11 as a percentage of population age 6–11* is calculated as the number of children age 6–11 who are enrolled in primary school divided by the population age 6–11 (UN estimates reported in table A-1) adjusted for each country's primary school age structure. Thus, for countries where the entrance age is 7, the number of children age 6–11 enrolled in primary school is divided by the population age 7–11. It is assumed that an insignificant number of 6-year-olds are counted as enrolled in primary school in such cases. For countries where the official primary school completion age is 10 years old or younger, no adjustment was made. For India, however, the population age 6–9 was used for the denominator for 1965–80, and the population age 6–10 was used for 1985, reflecting changes in the primary education system. Ratios of more than 100 percent are the result of inconsistencies in the enrollment data or errors due to the interpolation of the population data.

Table A-4

Total gross enrollment ratios are derived from the total enrollment of all ages, from table A-2, divided by the population of primary school age. For *female* gross enrollment ratios, the total female enrollment is divided by the female school-age population. Single-age enrollment data are obtained through the Unesco data base. Single-year population data are interpolated from the *UN Population Estimates and Projections: 1988 Revision* (United Nations 1988) using Sprague Multipliers, as in table A-1. Because of inconsistencies in either the enrollment or the population data, enrollment ratios for Congo, Gabon, Libya, and Viet Nam have been suppressed for certain years. For Switzerland, no enrollment ratios are shown because of the lack of a uniform education structure.

Table A-5

The *distribution* data are calculated from the total enrollment in each grade of primary school divided by total primary school enrollment.

Table A-6

The *distribution* data are calculated from the total enrollment in each grade divided by the total primary enrollments of grades 1–5.

Table A-7

The *ratio* of female to male students is calculated from the female enrollment divided by the male enrollment for each grade in the primary system.

Table A-8

Data in this table cover both public and private schools unless otherwise indicated below. The *student-school ratio* is calculated as the number of students reported in table A-2 divided by the number of schools.

El Salvador: Data for 1975 include evening schools.
Federal Republic of Germany: Because of a new method of estimation, data for 1985 are not comparable with those of previous years.
Ghana: Data refer to public schools only.
Jordan: Data for 1985 include UNRWA schools.
Liberia: Data for 1980 include schools offering the first stage of secondary education.
Mozambique: Data include preschools.
Netherlands: In 1985 the length of primary education was increased from six to eight years.
Nicaragua: Data for 1975 include preschools.
Pakistan: Data include preschools.
Philippines: Data for 1980 refer to public schools only.

Switzerland: Data for 1975 refer to public schools only.

Tanzania: Data are for the mainland only. Data for 1980 refer to government-maintained and -aided schools only.

Trinidad and Tobago: Data refer to government-maintained and -aided schools only and include intermediate departments of secondary schools.

Union of Soviet Socialist Republics: Data include general secondary schools.

Uganda: Data for 1975 and 1980 refer to government-maintained and -aided schools only.

Table A-9

Total primary teachers includes full- and part-time teachers in both public and private schools unless otherwise noted below. Instruction personnel without teaching functions are excluded. Most of the data on number of teachers come from the Unesco data base official series. When data from the official series are missing, data from the Unesco data base estimated series are generally used. However, for the following countries and years, the number of teachers is calculated from the number of students reported in table A-2, divided by student-teacher ratios reported for these countries and years in the *Statistical Yearbook 1988* (Unesco 1988): Canada (1985); Chad (1975); Denmark (1975); El Salvador (1975); France (1975, 1980); India (1980); Jamaica (1975, 1980); Liberia (1980); Morocco (1975, 1980); New Zealand (1975); Nicaragua (1975); Spain (1975); Sri Lanka (1975, 1985); United Arab Emirates (1975, 1980, 1985); Zimbabwe (1975).

Student-teacher ratios were computed from the total number of students enrolled (table A-2) divided by the total number of teachers in this table.

Afghanistan: Data for 1975 do not include teachers or students in primary classes attached to middle and secondary schools.

Belgium: Data include teachers and students in preschools.

Burundi: Data for 1980 refer to teachers and students in public education only.

China: There are two types of teachers in China—Gongban, paid from the central budget, and Minban, locally or privately paid by communities. Data refer primarily to Gongban but after 1975 also include those Minban who received small subsidies from the central government. Numbers for all years remain low. The student-teacher ratio has remained approximately 18:1 since 1965.

Egypt: Data do not include teachers and students in primary schools under the auspices of Al-Azhar University.

Finland: Data include teachers and students in integrated special education programs.

Federal Republic of Germany: Because of a new method of estimation, data for 1985 are not comparable with those of previous years.

Ghana: Data refer to teachers and students in public schools only.
Israel: Data do not include teachers or students in intermediate classes attached to primary schools.
Jordan: Data for 1985 include students enrolled in UNRWA schools.
Liberia: Data for 1985 include teachers and students at the first stage of secondary education.
Netherlands: Data for 1975 refer to full-time teachers only.
Pakistan: Data include preschool teachers and students.
Philippines: Data for 1980 refer to teachers and students in public schools only.
Switzerland: Data for 1975 refer to teachers and students in public schools only and include those in special education programs.
Tanzania: Data are for the mainland only. Data for 1980 refer to teachers and students in government-maintained and -aided schools only.
Trinidad and Tobago: Data refer to teachers and students in government-maintained and -aided schools only and include those in intermediate departments of secondary schools.
Uganda: Data for 1975 and 1980 refer to teachers and students in government-maintained and -aided schools only.
People's Democratic Republic of Yemen: Except for 1975, data refer to teachers and students in public schools only.

Table A-10

The data are calculated from the number of primary students enrolled in private schools divided by total primary enrollments (table A-2). Care should be taken when interpreting these data because some countries classify aided schools as private while others do not.

Table A-11

Repeaters as a percentage of total enrollment were calculated from the total number of repeaters in both public and private primary schools divided by total primary enrollment, unless otherwise indicated below. *Female repeaters as a percentage of female enrollment* were calculated from the number of female repeaters in both public and private primary schools divided by female primary enrollment, unless otherwise indicated.

Belgium: Data include special education students.
Bulgaria: Data for 1975 include repeaters enrolled in evening schools.
Chad: Repeaters enrolled in private Muslim schools are excluded.
El Salvador: Repeaters enrolled in evening schools are included.
Ethiopia: Only repeaters enrolled in public schools are included.

Greece: Repeaters enrolled in evening schools are included.
Japan: A policy of automatic promotion is practiced in primary education.
Republic of Korea: A policy of automatic promotion is practiced in primary education.
Kuwait: Data for 1975 and 1980 include only repeaters enrolled in public schools.
Malaysia: A policy of automatic promotion is practiced in primary education.
Tanzania: Data are for the mainland only.
Uganda: Data are for government-maintained and -aided schools only.
United Arab Emirates: Only repeaters enrolled in public schools are included.
Uruguay: Except for 1985, only repeaters enrolled in public schools are included.
People's Democratic Republic of Yemen: Only repeaters in public schools are included.

Table A-12

Student-years needed per graduate reflect the total number of student-years invested in each primary school completer given prevailing promotion, repetition, and dropout rates. These data were prepared by Unesco for the Education and Employment Division of the World Bank's Population and Human Resources Department.

When two numbers are reported under *number of grades in the primary education system,* the length of the primary cycle changed between 1970 and 1985, and the year of change is indicated by italicized data. The *difference between the student-years needed per graduate and years in the primary school system* reflects the "flow efficiency" of the education system: the smaller the difference, the greater the system's flow efficiency.

Table A-13

The data reflect the percentage of entering students who can be expected to reach the final year of primary school given prevailing dropout rates. Averages for 1980–86 come from World Conference on Education for All (1990); other data were calculated by Unesco for the Education and Employment Division of the World Bank's Population and Human Resources Department.

Table A-14

Total expenditure includes both capital and recurrent expenditures at every level of administration according to the constitution of the country (central or federal authorities, provincial or regional authorities, and local authorities, unless otherwise indicated below). Local currencies were converted to constant 1985 dollars using the method described under "General Information." Totals are not reported for the summary groups because data are missing for a significant number of countries. *Annual growth* rates were calculated with the compound growth rate equation using constant 1985 dollar figures.

Algeria: Data for 1985 refer to expenditure by the Ministry of Basic and Secondary Education only.

Argentina: Data for 1986 refer to expenditure by the Ministry of Education only.

Bangladesh: Except for 1975, data refer to expenditure by the Ministry of Education only.

Belgium: Data refer to expenditure by the Ministry of Education only.

Bolivia: Except for 1975 and 1980, expenditure on universities is not included.

Brazil: Data for 1985 do not include expenditure by the municipalities.

Central African Republic: Data for 1985 refer to expenditure by the Ministry of Education only.

Colombia: Data refer to expenditure by the Ministry of Education only.

Ecuador: Data for 1975 refer to expenditure by the Ministry of Education only.

Egypt: Data for 1985 do not include expenditure by Al-Azhar University.

France: Data are for metropolitan France.

Italy: Data for 1985 refer to expenditure by the Ministry of Education only.

Jamaica: Data for 1975 and 1980 refer to expenditure by the Ministry of Education only.

Japan: Data for 1985 refer to total public and private expenditure on education.

Jordan: Data for 1975 do not include expenditure on universities.

Kuwait: Data for 1985 do not include expenditure by the Public Authority for Applied Education and Training.

Lebanon: Data refer to expenditure by the Ministry of Education only.

Mexico: Data for 1985 refer to expenditure by the Ministry of Education only.

Nicaragua: Data for 1985 refer to expenditure by the Ministry of Education only.

Nigeria: From 1985, data refer to expenditure by the federal government only.

Somalia: Data do not include expenditure on tertiary education.

Sudan: Data for 1985 do not include expenditure on tertiary education.

Tanzania: Data for 1980 and 1985 refer to expenditure by the Ministry of Education only.

United States: Data for 1985 are from the *1988 Digest of Education Statistics* (National Center for Education Statistics 1988).

Table A-15

Expenditure on education as *percentage of* GNP is calculated from total public expenditure on education (as reported in table A-14) divided by the gross national product. Expenditure on education as a *percentage of government expenditure* is calculated from total public expenditure on education (as reported in table A-14) divided by government expenditure. All calculations use local currency in current values. GNP data are from the World Bank and total government expenditures are from the Unesco data base. Weighted averages are calculated using constant 1985 dollar figures for total public expenditure on education, GNP, and government expenditure.

Table A-16

Recurrent expenditure on education includes expenditure on administration, emoluments to teachers and supporting teaching staff, schoolbooks and other teaching materials, scholarships, welfare services, and maintenance of school buildings. Data on the share of recurrent expenditure in total public expenditure on education are calculated from recurrent expenditure divided by total public expenditure on education (as reported in table A-14). All calculations use local currency in current values. Weighted averages were calculated using constant 1985 dollars.

Dominican Republic: Data for 1980 and 1985 recurrent expenditure refer to expenditure by the Ministry of Education only.

Japan: Data for 1975 and 1980 recurrent expenditure do not include public subsidies for private education.

Paraguay: Except for 1985, data for recurrent expenditure refer to expenditure by the Ministry of Education only.

Philippines: Data for 1975 recurrent expenditure do not include expenditure on the state university and colleges.

Syria: Data for recurrent expenditure do not include expenditure on tertiary education.

Table A-17

Total expenditure statistics are not reported for the summary groups because data are missing for a significant number of countries. *Annual growth* rates were calculated with the compound growth rate equation described under "General Information." Public recurrent expenditure on primary education *as a percentage of total public recurrent expenditure on education* was calculated from recurrent expenditure on primary education divided by total public recurrent expenditure on education (as reported in table A-16). Calculations use local currency expressed in current values. Weighted averages were calculated using constant 1985 dollars.

For the following countries and years, data include recurrent expenditure for preschool education: Argentina (1975, 1980, 1984); Belgium (1975, 1980, 1985); Bolivia (1980); Brazil (1976, 1980, 1985); Chile (1975, 1985); Colombia (1985); Congo (1980, 1984); Costa Rica (1975, 1980, 1985); Denmark (1980); Honduras (1985); India (1975, 1980, 1985); Iraq (1976, 1979); Ireland (1975); Liberia (1979); Mali (1985); Pakistan (1985); Portugal (1980); Thailand (1985); Turkey (1985); United Kingdom (1975, 1980); Uruguay (1985); Venezuela (1975).

Colombia: Data for 1985 refer to capital and recurrent expenditures on primary education.

Côte d'Ivoire: Data for 1975 include foreign aid.

Dominican Republic: Data refer to recurrent expenditure on primary education by the Ministry of Education only.

India: Data for 1985 refer to recurrent expenditure on primary education by the Education Department of the central and state governments only.

Kenya: Data for 1975 and 1980 refer to expenditure on primary education by the Ministry of Education only.

Mexico: Data refer to recurrent expenditure on primary education by the Ministry of Education only.

Nepal: Data refer to "regular" and "development" expenditure for primary education.

Thailand: For 1975, data on total public recurrent expenditure on primary education include public subsidies to private education at the primary and secondary levels.

United States: Data are from the *1988 Digest of Education Statistics* (National Center for Education Statistics 1988).

Venezuela: Data for 1980 and 1984 refer to recurrent expenditure on primary education by the central government only.

Table A-18

Data are calculated from the total public recurrent expenditure on primary education (as reported in table A-17) divided by the gross national product. Calculations use local currencies in current values. Weighted averages use 1985 constant dollars. GNP figures are from the World Bank.

Table A-19

Data are calculated from the expenditure for each purpose divided by public recurrent expenditures on primary education (as reported in table A-17), using local currency expressed in current values. Weighted averages use 1985 constant dollars. Reported figures of 0.0 percent indicate expenditure of less than 0.05 percent.

Table A-20

Expenditure in *1985 constant dollars* is calculated from the total public recurrent expenditure on primary education (as reported in table A-17) divided by the total number of primary students (as reported in table A-2). Expenditure as a *percentage of* GNP *per capita* is calculated from the expenditure per student divided by gross national product per capita, which is derived by dividing the total GNP in constant 1985 dollars by the total mid-year population. Both GNP and population figures are from the World Bank.

Table A-21

Data are calculated from the total public recurrent expenditure on primary education (as reported in table A-17) divided by the total mid-year population. Population figures are from the World Bank.

Table A-22

Expenditure on teaching materials per student is calculated from the fraction of recurrent expenditure on teaching materials (reported as a percentage in table A-19) multiplied by the public recurrent expenditure on primary education (reported in 1985 dollars in table A-17) divided by the total number of primary students in 1985 (table A-2). Expenditure figures of 0.0 indicate that less than $0.05 is spent per student on teaching materials. *Average teacher salary* is calculated as the fraction of public expenditure on teacher salaries (reported as a percentage in table A-19) multiplied by the public recurrent expenditure on primary education (reported in 1985 dollars in table A-17) divided by the number of teachers in 1985 (table A-9). GNP per capita figures are calculated in the same manner as in table A-20.

United States: Average teacher salaries are from the *1988 Digest of Education Statistics* (National Center for Education Statistics 1988).

Bibliography

The word "processed" describes informally reproduced works that may not be commonly available through libraries.

Agarwal, D., S. Upadhyay, A. Tripathi, and K. Agarwal. 1987. *Nutritional Status, Physical Work Capacity and Mental Function in School Children*. New Delhi: Nutritional Foundation of India.

Al-Hariri, Rafedi. 1987. "Islam's Point of View on Women's Education in Saudi Arabia." *Comparative Education* 23(1):51–57.

Ali, Anthony, and Augustine Akubue. 1988. "Nigerian Primary Schools' Compliance with Nigerian National Policy on Education: An Evaluation of Continuous Assessment Practices." *Evaluation Review* 12(6):625–37.

Altbach, Philip G. 1983. "Key Issues of Textbook Provision in the Third World." *Prospects: Quarterly Review of Education* 13(3):315–25.

Amadeo, E., and J. M. Camargo. 1989. "The Political Economy of Budget Cuts: A Suggested Scheme of Analysis." International Labour Organisation, Training Policies Department, Geneva. Processed.

Anastasi, A. 1988. *Psychological Testing*. 6th ed. New York: Macmillan Co.

Anderson, J., and C. Herencia. 1983. "L'image de la femme et de l'homme dans les livres scolaires peruviens." Unesco Document ED-83/WS/93. Paris: Unesco (United Nations Educational, Scientific, and Cultural Organization).

Anderson, Lorin W., Doris W. Ryan, and B. J. Shapiro. 1989. *The IEA Classroom Environment Study*. Oxford: Pergamon Press.

Anderson, Mary B. 1988. *Improving Access to Schooling in the Third World: An Overview*. BRIDGES Research Report Series, Issue 1. Cambridge, Mass.: Harvard University.

Anzalone, Stephen J. 1991. "Educational Technology and the Improvement of General Education in Developing Countries." In Lockheed, Marlaine E., John Middleton, and Greta S. Nettleton, eds., "Educational Technology: Sustainable and Effective Use." Education and Employment Background Paper PHREE/91/32. World Bank, Population and Human Resources Department, Washington, D.C. Processed.

Anzalone, Stephen J., and Stephen D. McLaughlin. 1984. *Electronic Aids in a Developing Country: Improving Basic Skills in Lesotho.* Amherst, Mass.: Center for International Education, University of Massachusetts.

Anzalone, Stephen J., and M. Shyam. 1989. "Final External Evaluation: Radio Education Teacher Training II Project." Institute for International Research, Arlington, Va. Processed.

APEID (Asia and the Pacific Programme of Educational Innovation for Development). 1984a. *Towards Universalization of Primary Education in Asia and the Pacific.* Vol. 7: *Pakistan.* Bangkok: Unesco Regional Office for Education in Asia and the Pacific.

_________. 1984b. *Towards Universalization of Primary Education in Asia and the Pacific.* Vol. 9: *Philippines.* Bangkok: Unesco Regional Office for Education in Asia and the Pacific.

_________. 1984c. *Towards Universalization of Primary Education in Asia and the Pacific.* Vol. 10: *Republic of Korea.* Bangkok: Unesco Regional Office for Education in Asia and the Pacific.

_________. 1989. *Multigrade Teaching in Single Teacher Primary Schools.* Bangkok: Unesco Regional Office for Education in Asia and the Pacific.

Armer, Michael, and Robert Youtz. 1971. "Formal Education and Individual Modernity in an African Society." *American Journal of Sociology* 76(January):604–26.

Armitage, Jane, João Bautista-Neto, Ralph Harbison, Donald Holsinger, and Raimundo Leite. 1986. "School Quality and Achievement in Rural Brazil." EDT Discussion Paper 25. World Bank, Population and Human Resources Department, Washington, D.C. Processed.

Arriagada, Ana-María. 1981. "Determinants of Sixth Grade Student Achievement in Colombia." Paper prepared for Bank Research Project RPO 671-60 on "Textbook Availability and Educational Quality." World Bank, Population and Human Resources Department, Washington, D.C. Processed.

_________. 1983. "Determinants of Sixth Grade Student Achievement in Peru." Paper presented at the annual meeting of the Comparative and International Education Society, Atlanta, Ga.

Ashworth, A. 1982. "International Differences in Infant Mortality and the Impact of Malnutrition: A Review." *Human Nutrition: Clinical Nutrition* 36c(1):7–23.

Au, Kathryn H. 1977. "Cognitive Training and Reading Achievement." Paper presented at the meeting of the Association for the Advancement of Behavioral Therapy, Atlanta, Ga.

Auerhan, J., S. Ramakrishnan, R. Romain, G. Stoikov, L. Tiburcio, and P. Torres. 1985. "Institutional Development in Education and Training in Sub-Saharan African Countries." EDT Discussion Paper 22. World Bank, Population and Human Resources Department, Washington, D.C. Processed.

Austin, J. D. 1978. "Homework Research in Mathematics." *School Science and Mathematics* 79:115–22.

Avalos, Beatrice. 1986. "Teacher Effectiveness: An Old Theme with New Questions." Paper presented at a seminar on Quality of Teaching in Lesser Developed Countries, Alternative Models, International Movement toward Educational Change, Bali.

Avalos, Beatrice, and Wadi Haddad. 1981. *A Review of Teacher Effectiveness Research in Africa, India, Latin America, Middle East, Malaysia, Philippines, and Thailand: Synthesis of Results*. Ottawa, Canada: International Development Research Center.

Baker, Victoria J. 1988a. *Blackboard in the Jungle: Formal Education in Disadvantaged Rural Areas: A Sri Lankan Case*. Delft, The Netherlands: Eburon.

__________. 1988b. "Schooling and Disadvantage in Sri Lankan and Other Rural Situations." *Comparative Education* 24(3):377–88.

Bautista, Arturo, Patrick A. Barker, John T. Dunn, Max Sánchez, and Donald L. Kaiser. 1982. "The Effects of Oral Iodized Oil on Intelligence, Thyroid Status, and Somatic Growth in School-Age Children from an Area of Endemic Goiter." *American Journal of Clinical Nutrition* 35(1):127–34.

Beeby, Clarence E. 1966. *The Quality of Education in Developing Countries*. Cambridge, Mass.: Harvard University Press.

__________. 1979. *Assessment of Indonesian Education: A Guide in Planning*. Wellington: New Zealand Council for Educational Research.

Behrman, Jere R., and Nancy Birdsall. 1983. "The Quality of Schooling: The Standard Focus on Quantity Alone Is Misleading." *American Economic Review* 73(5):928–46.

Behrman, Jere R., and Anil B. Deolalikar. 1988. "Wages and Schooling in Indonesia: How Does Incorporation of Repetition and Dropout Rates Alter the Apparent Schooling Impact?" Preliminary draft, August.

Beishuizen, Jos, Judith Tobin, and Peter R. Weston. 1988. *The Use of Microcomputers in Teaching and Learning*. Berwyn, Penn.: Swets North America.

Benavot, Aaron. 1985. "Education and Economic Development in the Modern World." Ph.D. diss., Stanford University, Stanford, Calif.

__________. Forthcoming. *Education and Economic Development in the Modern World*. Oxford: Pergamon Press.

Benavot, Aaron, and David Kamens. 1989. "The Curricular Content of Primary Education in Developing Countries." Policy, Planning and Research Working Paper 237. World Bank, Population and Human Resources Department, Washington, D.C. Processed.

Bennett, Nicholas. 1991. "How Can Schooling Help Improve the Lives of the Poorest? The Need for Radical Reform." In Henry M. Levin and Marlaine E. Lockheed, eds., "Effective Schools in Developing Countries." Education and Employment Background Paper PHREE/91/38. World Bank, Population and Human Resources Department, Washington, D.C. Processed.

Benson, Charles. 1985. "Nigeria: Education Sector Expenditure Review." World Bank, Washington, D.C. Processed.

Bequele, Assefa, and Jo Boyden, eds. 1988. *Combating Child Labour*. Geneva: International Labour Office.

Berg, Alan D., and Susan Brems. 1986. "Micronutrient Deficiencies: Present Knowledge of Effects and Control." PHN Technical Note 86-32. World Bank, Population and Human Resources Department, Washington, D.C. Processed.

Berk, R. A., ed. 1984. *A Guide to Criterion-Referenced Test Construction*. Baltimore, Md.: Johns Hopkins University Press.

Berman, Paul, and Milbrey W. McLaughlin. 1977. *Federal Programs Supporting Educational Change.* Vol. 7: *Factors Affecting Implementation and Continuation.* Santa Monica, Calif.: Rand Corporation.

Bianchi, P., M. Carnoy, and M. Castells. 1988. *Economic Reform and Technology Transfer in China.* Stanford, Calif.: Stanford University, CERAS.

Birdsall, Nancy. 1989. "Pragmatism, Robin Hood, and Other Themes: Good Government and Social Well-Being in Developing Countries." Draft report written for Rockefeller Foundation, New York.

Birdsall, Nancy, and M. Louise Fox. 1985. "Why Males Earn More: Location and Training of Brazilian Schoolteachers." *Economic Development and Cultural Change* 33(3):533–56.

Bleichrodt, N., P. Drenth, and A. Querido. 1980. "Effects of Iodine Deficiency on Mental and Psychomotor Abilities." *American Journal of Physical Anthropology* 53(1):55–67.

Boissiere, M., John B. Knight, and Richard H. Sabot. 1985. "Earnings, Schooling, Ability, and Cognitive Skills." *American Economic Review* 75(5):1016–30.

Bolvin, John O., and others. 1978. "Analytical Case Study of the Korean Educational Development Institute: Third Interim Report." Washington, D.C.: American Association of Colleges for Teacher Education.

Botswana, Ministry of Education and Culture. 1986. *Education and Human Resources Sector Review: Botswana.* Gaborone, Botswana.

Botswana, Ministry of Finance and Development Planning. 1984. *Botswana Education and Human Resources Sector Assessment.* Washington, D.C.: United States Agency for International Development.

Bray, Mark. 1987a. *Are Small Schools the Answer? Cost-Effective Strategies for Rural School Provision.* London: The Commonwealth Secretariat.

_________. 1987b. *School Clusters in the Third World: Making Them Work.* Paris: Unesco-UNICEF Co-operative Program.

_________. 1989. *Multiple-Shift Schooling: Design and Operation for Cost-Effectiveness.* London: The Commonwealth Secretariat.

Bray, Mark, and Kevin Lillis, eds. 1988. *Community Financing of Education: Issues and Policy Implications in Less Developed Countries.* Oxford: Pergamon Press.

Brenner, M. 1982. "Student Attitudes and School Attendance in Liberia." Paper presented at the annual meeting of the American Anthropological Association, Washington, D.C.

Brown, Byron W., and Daniel H. Saks. 1987. "The Microeconomics of the Allocation of Teachers' Time and Student Learning." *Economics of Education Review* 6(4):319–32.

Buchanan, J. M. 1963. "The Economics of Earmarked Taxes." *Journal of Political Economy* 71(5):457–69.

Burden, Paul R. 1984. "Are Teacher Career Ladder Plans Feasible in Rural and Small Schools?" Paper presented at the annual Rural and Small Schools Conference, Manhattan, Kans.

Butterworth, Barbara, D. M. Karmacharya, and R. Martin. 1983. *Radio Education Teacher Training Program: Final Evaluation Report.* Kathmandu, Nepal: United

States Agency for International Development.

Cameron, John, Robert Cowan, Paul Hurst, Martin McLean, and Brian Holmes. 1983. *International Handbook of Educational Systems*. 3 vols. New York: J. Wiley and Sons.

Campbell, Patricia, and Susan Klein. 1982. "Equity Issues in Education." In H. Mitzel, ed., *Encyclopedia of Educational Research*, 5th ed., 581–87. London: Free Press.

Carraher, Terrezinha N., David W. Carraher, and Analucia D. Schliemann. 1987. "Written and Oral Mathematics." *Journal for Research in Mathematics Education* 18(2):83–97.

CERID/WEI (Research Center for Educational Innovation and Development). 1984. *Determinants of Educational Participation in Rural Nepal*. Kathmandu, Nepal: Tribhuvan University Press.

Chall, Jeanne S. 1983. *Stages of Reading Development*. New York: McGraw-Hill.

Chamie, Mary. 1983. *National, Institutional, and Household Factors Affecting Young Girls' School Attendance in Developing Societies*. Washington, D.C.: United States Agency for International Development, International Center for Research on Women and Office of Human Resources.

Chang, Patricia M., Karen Lussier, and Marc Ventresca. 1989. "Schooling the Nation: Passage of Compulsory Schooling Rules and Establishment of Central Education Agencies, 1810–1985." Paper presented at the annual meeting of the American Educational Research Association, San Francisco.

Chapman, David W. 1983. "Career Satisfaction of Teachers." *Educational Research Quarterly* 7(3):40–50.

_________. 1990. "Monitoring Implementation." In David W. Chapman and Carol A. Carrier, eds., *Improving Educational Quality: A Global Perspective*. Westport, Conn.: Greenwood Press.

Chapman, David W., and Carol A. Carrier, eds. 1990. *Improving Educational Quality: A Global Perspective*. Westport, Conn.: Greenwood Press.

Chen, Lincoln C., and Nevin S. Scrimshaw. 1983. *Diarrhea and Malnutrition: Interactions, Mechanisms, and Interventions*. New York: Plenum Press.

Chernichovsky, Dov, and Oey A. Meesook. 1985. *School Enrollment in Indonesia*. World Bank Staff Working Paper 746. Washington, D.C.

Chesterfield, Ray, and H. Ned Seelye. 1987. *Process Evaluation of the Programa Nacional de Educación Bilingüe (PRONEBI)*. Final report submitted to USAID/Guatemala and PRONEBI. Washington, D.C.: United States Agency for International Development.

Chivore, B. R. S. 1986. "Form IV Pupils' Perception of and Attitude towards the Teaching Profession in Zimbabwe." *Comparative Education* 22(3):232–53.

_________. 1988. "Factors Determining the Attractiveness of the Teaching Profession in Zimbabwe." *International Review of Education* 34(1):59–77.

Chung, Fay. 1989a. "Government and Community Partnership in the Financing of Education in Zimbabwe." Address presented at the 1989 conference of the *International Journal of Educational Development*, Oxford.

_________. 1989b. "Policies for Primary and Secondary Education in Zimbabwe: Alternatives to the World Bank Perspective." *Zimbabwe Journal of Educational Research* 1(1):22–42.

Cochrane, Susan H. 1979. *Fertility and Education: What Do We Really Know?* Baltimore, Md.: Johns Hopkins University Press.

_________. 1986. "The Effects of Education on Fertility and Mortality." EDT Discussion Paper 26. World Bank, Population and Human Resources Department, Washington, D.C. Processed.

Cohen, Elizabeth G. 1986. *Designing Groupwork.* New York: Teachers College Press.

Colbert de Arboleda, Clara Victoria. 1987. "Universalización de la Primaria en Colombia: El programa de Escuela Nueva." In Fundación para la Educación Superior (FES), eds., *La Educación Rural en Colombia: Situación, Experiencias y Perspectivas.* Bogotá: FES.

Colclough, Christopher, and Keith Lewin. 1990. "Education for All in Low Income and Adjusting Countries: The Challenge for the 1990s." Draft document prepared for the World Conference on Education for All, Jomtien, Thailand.

Cole, Nancy S. 1990. "Conceptions of Educational Achievement." *Educational Researcher* 19(3):2–7.

Coleman, Albert B., and Elmer J. Clark. 1983. "An Analysis of the Rural-Urban Balance for Education in Developing Countries: A Case Study of Liberia." ERIC Reproduction Service Document ED 234976. Iowa City: University of Iowa.

Colletta, Nat J., and Margaret Sutton. 1989. "Achieving and Sustaining Universal Primary Education: International Experience Relevant to India." Policy, Planning and Research Working Paper 166. World Bank, Population and Human Resources Department, Washington, D.C. Processed.

Commonwealth Secretariat. 1984a. *Country Papers.* Vol. 4(g): *Gambia.* London.

_________. 1984b. "Report on the Conference of Commonwealth Education Ministers and Commonwealth Secretariat." London.

Coombs, Philip H., with Roy Prosser and Manzoor Ahmed. 1973. *New Paths to Learning for Rural Children and Youth.* New York: UNICEF (United Nations Children's Fund), International Council for Educational Development.

Cope, Janet, Carmelle Denning, and Lucie Ribeiro. 1989. "Content Analysis of Reading and Mathematics Textbooks in Fifteen Developing Countries." Book Development Council, London. Processed.

Cornia, Giovanni A., Richard Jolly, and Frances Stewart, eds. 1987. *Adjustment with a Human Face.* Oxford: Clarendon Press.

Cotlear, Daniel. 1986. "Farmer Education and Farm Efficiency in Peru: The Role of Schooling, Extension Services, and Migration." EDT Discussion Paper 49. World Bank, Population and Human Resources Department, Washington, D.C. Processed.

Cotten, J. 1985. "Evaluation Research on the PL 480 School Feeding Program in Haiti." USA(1). Processed.

Cox-Edwards, Alejandra. 1989. "Understanding Differences in Wages Relative to Income per Capita: The Case of Teachers' Salaries." *Economics of Education Review* 8(2):197–203.

Crandall, David P., and Susan F. Loucks. 1983. *A Roadmap for School Improvement.* Andover, Mass.: The Network, Inc.

Creemers, Bert, Ton Peters, and Dave Reynolds, eds. 1989. *School Effectiveness and School Improvement*. Rockland, Mass.: Swets and Zeitlinger, Inc.

Csapo, Marg. 1987. "Special Educational Developments in Zambia." *International Journal of Educational Development* 7(2):107–12.

Cummings, William K. 1986. *Low-Cost Primary Education: Implementing an Innovation in Six Nations*. Ottawa, Ontario: International Development Research Center.

Cummings, William K., Maria Teresa Tatto, Dean Nielsen, G. B. Gunawardena, and N. G. Kularatne. 1990. BRIDGES Project Final Research Report on the Sri Lanka Study of the Effectiveness and Costs of Teacher Education for Elementary Teachers. Cambridge, Mass.: Harvard Institute for International Development.

Dall, Frank. 1989. "A Problem of Gender Access and Primary Education: Mali Case Study." *Harvard Institute of International Development Research Review* 2(4):7.

Demery, Lionel, and Tony Addison. 1987. *The Alleviation of Poverty under Structural Adjustment*. Washington, D.C.: World Bank.

Den Hartog Georgiades, William, and Howard Jones. 1989. "A Review of Research on Headmaster and School Principalship in Developing Countries." Education and Employment Division Background Paper PHREE/89/11. World Bank, Population and Human Resources Department, Washington, D.C. Processed.

Denison, Edward F. 1962. *The Sources of Economic Growth in the United States and the Alternatives before the U.S.* New York: Committee for Economic Development.

Deran, Elizabeth. 1965. "Earmarking and Expenditures: A Survey and a New Test." *National Tax Journal* 18(4):354–61.

Didonet, V. 1980. "Attendimento integrado de educaçao saude nutriçao, envolvimento comunitario do pre-escolar, atraves de metodologia de baixo custo e ampla cobertura: sintese da exposiçao." Paper presented at the World Assembly of Preschool Education, Quebec, 28 August.

Dillinger, William R. 1988. "Urban Property Taxation in Developing Countries." Policy, Planning and Research Working Paper 41. World Bank, Population and Human Resources Department, Washington, D.C. Processed.

Dorsey, Betty J. 1989. "Educational Development and Reform in Zimbabwe." *Comparative Education Review* 33(1):40–58.

Dossey, John A., Ina V. Mullis, Mary M. Lindquist, and Donald L. Chambers. 1988. *The Mathematics Report Card: Are We Measuring Up?* Princeton, N.J.: National Assessment of Educational Progress, Educational Testing Service.

Dove, Linda A. 1982. "The Deployment and Training of Teachers for Remote Rural Areas." *International Review of Education* 28(1):3–27.

_________. 1986. *Teachers and Teacher Education in Developing Countries*. London: Croom Helm.

Dunn, J., and G. Medeiros-Neto, eds. 1974. *Endemic Goiter and Cretinism: Continuing Threat to World Health*. Washington, D.C.: Pan American Health Organization.

Dutcher, Nadine. 1982. *The Use of First and Second Languages in Primary Education: Selected Case Studies*. World Bank Staff Working Paper 504. Washington, D.C.

Educational Testing Service. 1991. *Learning Around the World: Questions and Answers about IAEP*. Princeton, N.J.: Educational Testing Service.

Eisemon, Thomas O. 1988. *Benefiting from Basic Education, School Quality, and Functional Literacy in Kenya*. Comparative and International Education Series 2. Oxford: Pergamon Press.

_________. 1989. "Becoming a 'Modern' Farmer: The Impact of Primary Schooling on Agricultural Thinking and Practices in Kenya and Burundi." In D. Warren, L. Slikkerveer, and S. Titilola, eds., *Indigenous Knowledge Systems: Implications for Agriculture and International Development*, 41–67. Ames, Iowa: Iowa State University.

Eisemon, Thomas O., Robert Prouty, and John Schwille. 1989. "What Language Should Be Used for Teaching: Language Policy and School Reform in Burundi." *Journal of Multilingual and Multicultural Development* 10:473–97.

Eisemon, Thomas Owen, and John Schwille. 1991. "Primary Schooling in Burundi and Kenya: Preparation for Secondary Education or for Self-Employment?" *Elementary School Journal* 92(1):23–39.

Eisemon, Thomas O., John Schwille, and Robert Prouty. 1989. "Empirical Results and Conventional Wisdom: Strategies for Increasing Primary School Effectiveness in Burundi." BRIDGES Project draft report.

Elley, W. B., and Francis Mangubhai. 1983. "The Impact of Reading on Second Language Learning." *Reading Research Quarterly* 19(1):53–67.

Evans, David R. 1981. "The Educational Policy Dilemma for Rural Areas." *Comparative Education Review* 25(2):232–43.

Ezzaki, Abdelkader, Jennifer Spratt, and Daniel A. Wagner. 1987. "Childhood Literacy Acquisition among Children in Rural Morocco: Effects of Language Differences and Preschool Experience." In Daniel A. Wagner, ed., *The Future of Literacy in a Changing World*. London: Pergamon Press.

Fafunwa, A. Babatunde. 1987. "Education in the Mother Tongue: A Nigerian Experience." *Journal of African Studies*.

FAO (Food and Agriculture Organization). 1987. "Nutrition Country Profile: Tanzania." Rome.

_________. 1988. "Nutrition Country Profile: Ethiopia." Rome.

Farrell, Joseph P., and Stephen P. Heyneman, eds. 1989. *Textbooks in the Developing World: Economic and Educational Choices*. EDI Seminar Series. Washington, D.C.: World Bank.

Farrugia, Charles. 1986. "Career Choice and Sources of Occupational Satisfaction and Frustration among Teachers in Malta." *Comparative Education* 22(3):221–31.

Fine, Seymour H. 1981. *The Marketing of Ideas and Social Issues*. New York: Praeger.

Finn, Jeremy D., Loretta Dulberg, and Janet Reis. 1979. "Sex Differences in Educational Attainment: A Cross-National Perspective." *Harvard Educational Review* 49(4):477–503.

Florencio, Cecilia A. 1988. *Nutrition, Health, and Other Determinants of Academic Achievement and School-Related Behavior of Grades One to Six Pupils*. Quezon City: Center for Integrated and Development Studies, University of the Philippines.

_________. 1989. "Nutritional Welfare of Filipinos and the Philippine Food and Nutrition Program." *Study on Population, Human Resource Development and the Philippine Future*. Quezon City: Center for Integrated and Development Studies, University of the Philippines.

Fouilland, Jean-Louis. 1987. "Study of Regional and Local Financing of Primary Education in Mali." Draft.

Fraser, Barry J., Herbert J. Walberg, Wayne W. Welch, and John A. Hattie. 1987. "Syntheses of Educational Productivity Research." *International Journal of Educational Research* 11(2):147–252.

Friend, Jamesine. 1989. *Honduras Evaluation: First Grade Test.* Shelby, N.C.: Friend Dialogues, Inc.

Friend, Jamesine, Barbara Searle, and Patrick Suppes, eds. 1980. *The Radio Mathematics Project in Nicaragua, 1976–1977.* Stanford, Calif.: Institute for Mathematical Studies in the Social Studies, Stanford University.

Friesen, C. D. 1979. *The Results of Homework versus No-Homework Research Studies.* ERIC Reproduction Service Document ED 159174. Iowa City: University of Iowa.

Fryer, Michelle. 1989. *Radio Education Project: First Year Summative Evaluation.* Boston, Mass.: Education Development Center.

Fullan, Michael. 1982. *The Meaning of Educational Change.* New York: Columbia University, Teachers College Press.

Fullan, Michael, and Alan Pomfret. 1987. "Research on Curriculum and Instruction Implementation." *Review of Educational Research* 57(1):335–97.

Fuller, Bruce. 1987. "What Factors Raise Achievement in the Third-World." *Review of Educational Research* 57(3):255–92.

Fuller, Bruce, and Conrad W. Snyder, Jr. 1991. "Vocal Teachers, Silent Pupils? Life in Botswana Classrooms." *Comparative Education Review* 35(2):274–94.

Galda, Klaus, and José González. 1980. "Teacher's Promotion Decisions in Nicaraguan First Through Fourth Grades." In Jamesine Friend, Barbara Searle, and Pattrick Suppes, eds., *The Radio Mathematics Project in Nicaragua, 1976–1977.* Stanford, Calif.: Institute for Mathematical Studies in the Social Sciences, Stanford University.

Gallagher, Mark. 1989. "Fiscal Contraction and the Social Sectors in Developing Countries." Background paper for *World Development Report 1990.* World Bank, Africa Technical Department, Washington, D.C. Processed.

Galloway, Rae. 1989. "The Prevalence of Malnutrition and Parasites in School-Age Children: An Annotated Bibliography." Education and Employment Division Background Paper PHREE/89/24. World Bank, Population and Human Resources Department, Washington, D.C. Processed.

Gatawa, B. 1986. "ZINTEC." In C. B. Treffgarne, ed., *Education in Zimbabwe.* Occasional Paper 9. London: Department of International and Comparative Education, London University.

Ghani, Zainal. 1990. "Pre-Service Teacher Education in Developing Countries." In Val D. Rust and Per Dalin, eds., *Teachers and Teaching in the Developing World.* New York: Garland Press.

Gimeno, José B., and Ricardo M. Ibáñez. 1981. *The Education of Primary and Secondary School Teachers: An International Comparative Study.* Paris: Unesco (United Nations Educational, Scientific, and Cultural Organization).

Glass, Gene V., Leonard S. Cahen, Mary L. Smith, and Nikola N. Filby, eds. 1982. *School Class Size: Research and Policy.* Beverly Hills, Calif.: Sage Publications.

Glass, Gene V., Barry McGaw, and Mary Lee Smith. 1981. *Meta-Analysis in Social Research*. Beverly Hills, Calif.: Sage Publications.

Glass, Gene V., and Mary L. Smith. 1979. "Meta-Analysis of Research on Class Size and Achievement." *Education, Evaluation, and Policy Analysis* 1(1):2–16.

Glewwe, Paul. 1988. *The Distribution of Welfare in Côte d'Ivoire in 1985*. World Bank LSMS Working Paper 29. Washington, D.C.

González Suárez, Mirta. 1987. "Barriers to Female Achievement: Sex Stereotypes in Textbooks." Paper presented to the Comparative and International Education Society, Washington, D.C.

Gorman, Kathleen S., Susan D. Holloway, and Bruce Fuller. 1988. "Pre-School Quality in Mexico: Variation in Teachers, Organisation, and Child Activities." *Comparative Education* 24(1):91–101.

Guthrie, James W., Walker I. Garms, and Lawrence C. Pierce. 1988. *School Finance and Education Policy: Enhancing Educational Efficiency, Equality, and Choice*. Englewood Cliffs, N.J.: Prentice Hall.

Haddad, Wadi. 1979. *Educational and Economic Effects of Promotion and Repetition Practices*. World Bank Staff Working Paper 319. Washington, D.C.

_________. 1985. "Teacher Training: A Review of World Bank Experience." EDT Discussion Paper 21. World Bank, Population and Human Resources Department, Washington, D.C. Processed.

Haddad, Wadi, and Teri Demsky. Forthcoming. *Education Policy Analysis: The Burkina Case*. EDI Development Policy Case Series. Washington, D.C.: World Bank.

Haglund, Elaine. 1982. "The Problem of the Match: Cognitive Transition between Early Childhood and Primary School: Nigeria." *Journal of Developing Areas* 17(October):77–92.

Halpern, Robert, and Robert Myers. 1985. *Effects of Early Childhood Intervention on Primary School Progress and Performance in the Developing Countries*. Washington, D.C.: United States Agency for International Development.

Hanson, Mark E. 1986. *Education Reform and Administrative Development: The Cases of Colombia and Venezuela*. Stanford, Calif.: Hoover Institution Press.

Hanushek, Eric. 1986. "The Economics of Schooling: Production and Efficiency in Public Schools." *Journal of Economic Literature* 24:1141–77.

Harber, Clive. 1984. "Schooling for Bureaucracy in Nigeria." *International Journal of Educational Development* 4(2):145–54.

Harbison, Ralph W., and Eric A. Hanushek. Forthcoming. *Educational Performance of the Poor: Lessons from Rural Northeast Brazil*. New York: Oxford University Press.

Hartley, Michael J., and Eric V. Swanson. 1984. "Achievement and Wastage: An Analysis of the Retention of Basic Skills in Primary Education." Final report of The International Study of the Retention of Literacy and Numeracy: An Egyptian Case Study. RPO 671-55. World Bank, Washington, D.C. Processed.

Haskins, Ron. 1989. "Beyond Metaphor: The Efficacy of Early Childhood Education." *American Psychologist* 44(2):274–82.

Havelock, Ronald G., and A. M. Huberman. 1977. *Solving Educational Problems: The Theory and Reality of Innovations in Developing Countries*. Paris: Unesco (United Nations Educational, Scientific, and Cultural Organization).

Haverman, Robert H., and Barbara L. Wolfe. 1984. "Schooling and Economic Well-Being: The Role of Non-Market Effects." *Journal of Human Resources* 19(3):377–406.

Hawkridge, David, John Jaworski, and Henry McMahon. 1990. *Computers in Third World Schools: Examples, Experiences and Issues*. London: Macmillan Press.

Hetzel, Basil S., John T. Dunn, and John B. Stanbury. 1987. *The Prevention and Control of Iodine Deficiency Disorders*. New York: Elsevier Press.

Heyneman, Stephen P. 1980a. *The Evaluation of Human Capital in Malawi*. World Bank Staff Working Paper 420. Washington, D.C.

__________. 1980b. "Investment in Indian Education: Uneconomic?" *World Development* 8:145-63.

Heyneman, Stephen P., Joseph P. Farrell, and Manuel A. Sepúlveda-Stuardo. 1981. "Textbooks and Achievement in Developing Countries: What We Know." *Journal of Curriculum Studies* 13(3):227–46.

Heyneman, Stephen P., and Dean Jamison. 1980. "Student Learning in Uganda: Textbook Availability and Other Factors." *Comparative Education Review* 24(2):206–20.

Heyneman, Stephen P., Dean Jamison, and Xenia Montenegro. 1984. "Textbooks in the Philippines: Evaluation of the Pedagogical Impact of a Nationwide Investment." *Educational Evaluation and Policy Analysis* 6(2):139–50.

Heyneman, Stephen P., and William A. Loxley. 1983. "The Effect of Primary-School Quality on Academic Achievement across Twenty-Nine High and Low-Income Countries." *The American Journal of Sociology* 88(6):1162–94.

Hildebrand, G. 1986. *Human Capital Formation and Child Care Centers: An International Perspective*. East Lansing, Mich.: Michigan State University.

Hirschman, Albert O. 1967. *Development Projects Observed*. Washington, D.C.: The Brookings Institution.

Holsinger, Donald B., and John D. Kasarda. 1975. "Does Schooling Affect Birth Rates?" *School Review* 84(1):71–90.

Holsinger, Donald B., and Gary L. Theisen. 1977. "Education, Individual Modernity, and National Development: A Critical Appraisal." *Journal of Developing Areas* 11(3):315–33.

Horn, Robin, and Ana-María Arriagada. 1986. "The Educational Attainment of the World's Population: Three Decades of Progress." EDT Discussion Paper 37. World Bank, Population and Human Resources Department, Washington, D.C. Processed.

Huberman, A. Michael, and Matthew B. Miles. 1984. *Innovation Up Close: How School Improvement Works*. New York: Plenum Press.

Husén, Torsten, Lawrence J. Saha, and Richard Noonan. 1978. *Teacher Training and Student Achievement in Less Developed Countries*. World Bank Staff Working Paper 310. Washington, D.C.

Husén, Torsten, and T. Neville Postlethwaite, eds. 1991. *International Encyclopedia of Education*. Supplementary vol. 2. Oxford: Pergamon Press.

ICED (International Council for Educational Development). 1974. "Building New Educational Strategies to Serve Rural Children and Youth." Draft of a second report to UNICEF (United Nations Children's Fund).

IEA (International Association for the Evaluation of Educational Achievement). 1988. *Science Achievement in Seventeen Countries: A Preliminary Report*. Oxford: Pergamon Press.

_________. 1991. *Third International Mathematics and Science Study: An Introduction*. The Hague.

IEES (Improving the Efficiency of Educational Systems Project Sector Study). 1984a. *Indonesia*. Tallahassee, Fla.: Florida State University.

_________. 1984b. *Malaysia*. Tallahassee, Fla.: Florida State University.

_________. 1984c. *Somalia*. Tallahassee, Fla.: Florida State University.

_________. 1984d. *Yemen Arab Republic*. Tallahassee, Fla.: Florida State University.

_________. 1986. *Indonesia Education and Human Resources Sector Assessment. Tallahassee, Fla.: Florida State University*.

_________. 1987. *Haiti*. Tallahassee, Fla.: Florida State University.

_________. 1988. *Nepal*. Tallahassee, Fla.: Florida State University.

ILO (International Labour Office). 1988. *Yearbook of Labour Statistics*. Geneva.

Inkeles, Alex, and David H. Smith. 1974. *Becoming Modern: Individual Change in Six Developing Countries*. Cambridge, Mass.: Harvard University Press.

INSET (Inservice Training of Primary School Teachers) Project. 1982. Overview document. Bristol, Eng.: Bristol University, School of Education.

IREDU (Institut de Recherche sur l'Economie de l'Education). 1989. "External Aid for Primary Education in Developing Countries." University of Bourgogne, Dijon. Processed.

Jacobson, Stephen L. 1988. "The Effect of Pay Incentives on Teacher Absenteeism." *Journal of Human Resources* 24(2):280–87.

James, Estelle. 1991. "Public Policies toward Private Education: An International Comparison." *International Journal of Educational Research* 15(5): 359–76.

Jamison, Dean. 1982. "Reduced Class Size and Other Alternatives for Improving Schools: An Economist's View." In Gene V. Glass, Leonard S. Cahen, Mary L. Smith, and Nikola N. Filby, eds., *School Class Size: Research and Policy*. Beverly Hills, Calif.: Sage Publications.

_________. 1986. "Child Malnutrition and School Performance in China." *Journal of Developmental Economics* 20(March):299–309.

Jamison, Dean, and Marlaine Lockheed. 1987. "Participation in Schooling: Determinants and Learning Outcomes in Nepal." *Economic Development and Cultural Change* 35(2):279–306.

Jamison, Dean, and Peter Moock. 1984. "Farmer Education and Farm Efficiency in Nepal: The Role of Schooling, Extension Services, and Cognitive Skills." *World Development* 12:67–86.

Jamison, Dean, Barbara Searle, Klaus Galda, and Stephen Heyneman. 1981. "Improving Elementary Mathematics Education in Nicaragua: An Experimental Study of the Impact of Textbooks and Radio on Achievement." *Journal of Educational Psychology* 73(4):556–67.

Jennings-Wray, Zellyne D. 1984. "Implementing the 'Integrated Approach to Learning': Implications for Integration of the Curricula of Primary Schools in the Caribbean." *International Journal of Educational Development* 4:265–78.

Jimenez, Emmanuel. 1987. *Pricing Policy in the Social Sectors: Cost Recovery for Education and Health in Developing Countries*. Baltimore, Md.: Johns Hopkins University Press.

__________. 1990. "Social Sector Pricing Policy Revisited: A Survey of Some Recent Controversies in Developing Countries." *Proceedings of the World Bank Conference on Development Economics, 1989*.

Jimenez, Emmanuel, and Marlaine E. Lockheed. 1989. "Enhancing Girls' Learning through Single-Sex Education: Evidence and a Policy Conundrum." *Educational Evaluation and Policy Analysis* 11(2): 117–42.

Jimenez, Emmanuel, Marlaine E. Lockheed, Eduardo Luna, and Vicente Paqueo. 1991. "School Effects and Costs for Private and Public Schools in the Dominican Republic." *International Journal of Educational Research* 15(5):393–410.

Jimenez, Emmanuel, Marlaine E. Lockheed, and Vicente Paqueo. 1991. "The Relative Efficiency of Private and Public Schools in Developing Countries." *World Bank Research Observer* 6(2):205–18.

Jimenez, Emmanuel, Marlaine E. Lockheed, and Nongnuch Wattanawaha. 1988. "The Relative Efficiency of Private and Public Schools: The Case of Thailand." *World Bank Economic Review* 2(2):139–64.

Jimenez, Emmanuel, Vicente Paqueo, and María Lourdes de Vera. 1988. "Does Local Financing Make Primary Schools More Efficient?" Policy, Planning and Research Working Paper 69. World Bank, Population and Human Resources Department, Washington, D.C. Processed.

Jimenez, Emmanuel, and Jee-Peng Tan. 1987. "Decentralised and Private Education: The Case of Pakistan." *Comparative Education* 23(2):173–90.

Johnson, D. Gale, and Ronald D. Lee, eds. 1987. *Population Growth and Economic Development: Issues and Evidence*. Madison, Wis.: University of Wisconsin Press.

Johnston, Bruce F., and William C. Clark. 1982. *Redesigning Rural Development: A Strategic Perspective*. Baltimore, Md.: Johns Hopkins University Press.

Kagitcibasi, Cigdem. 1983. "Early Childhood Education and Preschool Intervention: Experiences in the World and in Turkey." IDRC Report MR 209-e. Ottawa: International Development Research Center.

Kakwani, Nanak, Elene Makonnen, and Jacques van der Gaag. 1990. "Structural Adjustment and Living Conditions in Developing Countries." Policy, Research and External Affairs Working Paper 467. World Bank, Population and Human Resources Department, Washington, D.C. Processed.

Kaluba, L. H. 1988. "Education and Community Self-Help in Zambia." In Mark Bray and Kevin Lillis, eds., *Community Financing of Education: Issues and Policy Implications in Less Developed Countries*. Comparative and International Education Series 5. Oxford: Pergamon Press.

Kanbargi, R. 1988. "Child Labour in India: The Carpet Industry of Varanasi." In Assefa Bequele and Jo Boyden, eds., *Combating Child Labour*. Geneva: International Labour Office.

Kellaghan, T., and Vincent Greaney. 1990. "Using Examinations To Improve Education: A Study in Fourteen African Countries." World Bank, Africa Technical Department, Washington, D.C. Processed.

Kelly, Gail P. 1987. "Setting State Policy on Women's Education in the Third World: Perspectives from Comparative Research." *Comparative Education* 23(1):95–102.

Kelly, Michael, Eileen Nkwanga, Henry Kaluba, Paul Achola, and K. Nilsson. 1986. "The Provision of Education for All: Towards the Implementation of Zambia's Educational Reforms under Demographic and Economic Constraints." University of Zambia, School of Education, Lusaka. Processed.

Kemmerer, Frances N. 1990. "An Integrated Approach to Primary Teacher Incentives." In David W. Chapman and Carol A. Carrier, eds., *Improving Educational Quality: A Global Perspective*. Westport, Conn.: Greenwood Press.

Kemmerer, Frances N., and Sivasailam Thiagarajan. 1989. *Teachers Incentive Systems: Final Report*. Tallahassee, Fla.: Florida State University.

Kerr, J. Graham. 1932. *A Scotsman's Heritage*. London: A. Maclehose.

Khan, Shahrukh R. Forthcoming. "South Asia." In Elizabeth M. King and M. Anne Hill, eds., *Women's Education in Developing Countries: Barriers, Benefits, and Policy*. Baltimore, Md.: Johns Hopkins University Press.

King, Elizabeth M. 1981. "Child Schooling and Time Allocation in Philippine Rural Households." Paper presented at meetings of the Population Association of America, Washington, D.C.

King, Elizabeth M., and M. Anne Hill, eds. Forthcoming. *Women's Education in Developing Countries: Barriers, Benefits, and Policy*. Baltimore, Md.: Johns Hopkins University Press.

King, Elizabeth M., and Lee A. Lillard. 1983. *Determinants of Schooling Attainment and Enrollment Rates in the Philippines*. N-1962-AID. Santa Monica, Calif.: Rand Corporation.

Knight, John B., and Richard H. Sabot. 1990. *Education, Productivity, and Inequality: The East African Natural Experiment*. New York: Oxford University Press.

Komenan, André. 1987. "World Education Indicators." EDT Discussion Paper 88. World Bank, Population and Human Resources Department, Washington, D.C. Processed.

Komenan, André, and Christiaan Grootaert. 1988. "Teachers–Non-Teachers Pay Differences in Côte d'Ivoire." Policy, Planning and Research Working Paper 12. World Bank, Washington, D.C. Processed.

Kulik, James A., and Chen-Lin Kulik. 1988. "Timing of Feedback and Verbal Learning." *Review of Educational Research* 58(1):79–97.

Lapointe, Archie E., Nancy A. Mead, and Gary Phillips. 1989. *A World of Differences: An International Assessment of Mathematics and Science*. Princeton, N.J.: Center for the Assessment of Educational Progress, Educational Testing Service.

Lau, Lawrence J., Dean T. Jamison, and Frederic F. Louat. 1991. "Education and Productivity in Developing Countries: An Aggregate Production Function Approach." Policy, Research and External Affairs Working Paper 612. World Bank, Office of the Vice President Development Economics, and the Population and Human Resources Department, Washington, D.C. Processed.

Lee, Valerie E., Jeanne Brooks-Gunn, and Elizabeth Schnur. 1989. "Does Head Start Work? A 1-Year Follow-Up Comparison of Disadvantaged Children Attending Head Start, No Preschool, and Other Preschool Programs." *Developmental Psychology* 24(2):210–22.

Lee, Valerie E., and Marlaine E. Lockheed. 1990. "The Effects of Single-Sex Schooling on Student Achievement and Attitudes in Nigeria." *Comparative Education Review* 34(2):209–32.

Leithwood, K. A. 1989. "A Review of Research on the School Principalship." Education and Employment Background Paper PHREE/89/07. World Bank, Population and Human Resources Department, Washington, D.C. Processed.

Levin, Henry M. 1985. "A Benefit-Cost Analysis of Nutritional Programs for Anemia Reduction." *World Bank Research Observer* 1(2):219–51.

_________. 1987. "What Have We Learned about Cost-Benefit and Cost-Effectiveness Analysis?" In David S. Cordray, Howard S. Bloom, and Richard I. Light, eds., *Education Practice in Review: New Series*. San Francisco: Jossey Bass.

Levin, Henry M., Gene V. Glass, and Gail Meister. 1984. *Cost Effectiveness of Four Educational Interventions*. Stanford, Calif.: Institute for Research on Education Finance and Governance.

Levin, Henry M., and Gail Meister. 1986. "Is CAI Cost-Effective?" *Phi Delta Kappan* 67(10):745–49.

Levin, Henry M., Ernesto Pollitt, Rae Galloway, and Judith McGuire. Forthcoming. "Micronutrient Deficiency Disorders." In Dean T. Jamison and W. T. Mosley, eds., *Disease Control Priorities in Developing Countries*. Washington, D.C.: World Bank.

Levinger, Beryl. 1983. "School Feeding Programs in Less Developed Countries: An Analysis of Actual and Potential Impact." Office of Evaluation, Bureau for Food and Voluntary Aid, Agency for International Development, Washington, D.C. Processed.

Lewin, Keith, and Dieter Berstecher. 1989. "The Costs of Recovery: Are User Fees the Answer?" *IDS Bulletin* 20(1):59–71.

Liberia, Ministry of Education. 1989. "Teacher Incentives Study." Monrovia, Liberia: Ministry of Education; and Tallahassee, Fla.: Florida State University.

Livingstone, Ian D. 1985. "Perceptions of the Intended and Implemented Mathematics Curriculum." New Zealand Council for Educational Research, Wellington. Processed.

Lockheed, Marlaine E., Josefina Fonacier, and Leonard J. Bianchi. 1989. "Effective Primary Level Science Teaching in the Philippines." Policy, Planning and Research Working Paper 208. World Bank, Population and Human Resources Department, Washington, D.C. Processed.

Lockheed, Marlaine E., Bruce Fuller, and Ronald Nyirongo. 1989. "Family Background and Student Achievement." *Sociology of Education* 62(4):239–56.

Lockheed, Marlaine E., and Eric Hanushek. 1988. "Improving Educational Efficiency in Developing Countries: What Do We Know?" *Compare* 18(1):21–38.

_________. 1991. "Concepts of Educational Efficiency and Effectiveness." In Torsten Husén and T. Neville Postlethwaite, eds., *International Encyclopedia of Education*. Supplementary vol. 2. Oxford: Pergamon Press.

Lockheed, Marlaine E., and Abigail M. Harris. 1982. "Classroom Interaction and Opportunities for Cross-Sex Peer Learning in Science." *Journal of Early Adolescence* 2(2):135–43.

Lockheed, Marlaine E., Dean T. Jamison, and Lawrence Lau. 1980. "Farmer Education and Farm Efficiency: A Survey." *Economic Development and Cultural Change* 29(1):37–76.

Lockheed, Marlaine E., and André Komenan. 1989. "Teaching Quality and School Effects on Student Achievement in Africa: The Case of Nigeria and Swaziland." *Teaching and Teacher Education* 5(2):93–113.

Lockheed, Marlaine E., and Nicholas T. Longford. 1989. *A Multilevel Model of School Effectiveness in a Developing Country*. World Bank Discussion Paper 69. Washington, D.C.

Lockheed, Marlaine E., John Middleton, and Greta Nettleton, eds. 1991. "Educational Technology: Sustainable and Effective Use." Education and Employment Background Paper PHREE/91/32. World Bank, Population and Human Resources Department, Washington, D.C. Processed.

Lockheed, Marlaine E., and Alastair G. Rodd. 1990. "World Bank Lending for Education Research, 1982–89." Policy, Research and External Affairs Working Paper 583. World Bank, Population and Human Resources Department, Washington, D.C. Processed.

Lockheed, Marlaine E., Stephen Vail, and Bruce Fuller. 1986. "How Textbooks Affect Achievement in Developing Countries: Evidence from Thailand." *Education, Evaluation, and Policy Analysis* 8(4):379–92.

Lourié, Sylvian. 1982. "Inequalities in Education in Rural Guatemala: Inequalities in Educational Development." Paper presented at an International Institute for Educational Planning seminar. Paris: Unesco (United Nations Educational, Scientific, and Cultural Organization).

Luna, Eduardo, and Sarah González. 1986. "The Underdevelopment of Mathematics Achievement: Comparison of Public and Private Schools in the Dominican Republic." Centro de Investigaciones UCMM, Santiago. Processed.

MacAdam, Colin. 1984. "Towards Democracy: The Literacy Crusade in Nicaragua." *International Review of Education* 30(3):359–68.

MacLennan, E., J. Fitz, and J. Sullivan. 1985. *Working Children*. London: Low Pay Unit. Cited in Assefa Bequele and Jo Boyden, eds., *Combating Child Labour*. Geneva: International Labour Office, 1988.

Malawi, Ministry of Education and Culture. 1986. "Education Statistics: Malawi, 1986." Lilongwe.

Mallon, N. 1989. "The Community Classroom." *World Education Reports* 28 (Spring): 15–18.

McCleary, William A. 1988. "Notes on the Principles and Practice of Earmarking." Processed.

McGinn, Noel F. 1988. "Foreword." In Mary B. Anderson, *Improving Access to School-*

ing in the Third World: An Overview. BRIDGES Research Report Series, Issue 1. Cambridge, Mass.: Harvard University.

McGinn, Noel F., Donald R. Snodgrass, Yung Boo Kim, Shin-Bok Kim, and Quee-Young Kim. 1980. *Education and Development in Korea*. East Asian monographs. Cambridge, Mass.: Council on East Asian Studies, Harvard University Press.

McGinn, Noel F., Donald P. Warwick, and Fernando Reimers. 1989. "Policy Choices to Improve School Effectiveness in Pakistan." Paper presented at the seventh World Congress of Comparative Education, Montreal, Canada.

McMahon, Walter W. 1984. "The Relation of Education and R&D to Productivity Growth." *Economics of Education Review* 3(4):299–313.

McSweeney, Brenda G., and Marion Freedman. 1980. "Lack of Time as an Obstacle to Women's Education: The Case of Upper Volta." *Comparative Education Review* 24(2):124–39.

Mertens, Sally, and Sam J. Yarger. 1987. "Teaching as a Profession: Leadership, Empowerment, and Involvement." *Journal of Teacher Education* 39(1):32–37.

Meyer, John W., Francisco O. Ramírez, and Y. Soysal. 1989. "World Expansion of Mass Education, 1870–1980." Paper presented at the annual meeting of the American Educational Research Association, San Francisco.

Michel, Andrée. 1986. *Down with Stereotypes! Eliminating Sexism from Children's Literature and School Textbooks*. Paris: Unesco (United Nations Educational, Scientific, and Cultural Organization).

Middleton, John, James Terry, and Deborah Bloch. 1989. "Building Educational Evaluation Capacity in Developing Countries." Policy, Planning and Research Working Paper 140. World Bank, Population and Human Resources Department, Washington, D.C. Processed.

Mingat, Alain, and Jee-Peng Tan. 1985. "Improving the Quantity-Quality Mix in Education: A Simulation of Policy Tradeoffs." EDT Discussion Paper 15. World Bank, Population and Human Resources Department, Washington, D.C. Processed.

__________. 1988. *Analytical Tools for Sector Work in Education*. Baltimore, Md.: Johns Hopkins University Press.

Moock, Peter R., and Joanne Leslie. 1986. "Childhood Malnutrition and Schooling Deficit in the Terai Region of Nepal." *Journal of Developmental Economics* 20(January–February):33–52.

Mulligan, James G. 1984. "A Classroom Production Function." *Economic Inquiry* 22(2):218–26.

Mundangepfupfu, Mwazwita R. 1988. "School Quality and Efficiency in Malawi." Report of the pilot of the primary school quality survey. Lilongwe, Liberia: Ministry of Education and Culture.

Murnane, Richard J., and David K. Cohen. 1986. "Merit Pay and the Evaluation of the Problem: Why Some Merit Plans Fail and Few Survive." *Harvard Educational Review* 56(1):1–17.

Murnane, Richard J., and Randall J. Olsen. 1990. "The Effects of Salaries and Opportunity Costs on Length of Stay in Teaching: Evidence from North Carolina." *Journal of Human Resources* 25(1):106–24.

Murphy, Michael J., and others. 1984. *Teacher Career Ladders in Britain: A Study of Their Structure and Impact*. Columbus, Ohio: University Council for Educational Administration.

Murray, Karen B. 1988. "Profile of the New Generation of Teachers in the Turkish Education System." *International Review of Education* 34(1):5–15.

Mwamwenda, Tuntufye S., and Bernadette B. Mwamwenda. 1987. "School Facilities and Student Academic Achievement." *Comparative Education* 23(2):225–36.

_________. 1989. "Teacher Characteristics and Pupils' Academic Achievement in Botswana Primary Education." *International Journal of Educational Development* 9(1):31–42.

Myers, Robert G. Forthcoming. *The Twelve Who Survive*. London: Routledge.

Myers, William. 1988. "Alternative Services for Street Children: The Brazilian Approach." In Assefa Bequele and Jo Boyden, eds., *Combating Child Labour*. Geneva: International Labour Office.

Narayan, A. 1988. "Child Labour Policies and Programmes: The Indian Experience." In Assefa Bequele and Jo Boyden, eds., *Combating Child Labour*. Geneva: International Labour Office.

National Board for Professional Teaching Standards. 1989. "Toward High and Rigorous Standards for the Teaching Profession." Detroit, Mich. Processed.

National Center for Education Statistics. 1988. *1988 Digest of Education Statistics*. Washington, D.C.: U.S. Department of Education.

Naumann, Jens. 1984. "The Volume and Structure of External Aid to Education." Paper presented at the International Institute for Educational Planning, Paris, November.

Nettleton, Greta S. 1991. "Uses and Costs of Educational Technology for Distance Education in Developing Countries: A Review of the Recent Literature." In Lockheed, Marlaine E., John Middleton, and Greta S. Nettleton, eds., "Educational Technology: Sustainable and Effective Use." Education and Employment Background Paper PHREE/91/32. World Bank, Population and Human Resources Department, Washington, D.C. Processed.

Neumann, Peter H. 1980. *Publishing for Schools: Textbooks and the Less Developed Countries*. World Bank Staff Working Paper 398. Washington, D.C.

Ngomba, Peter, and John Oxenham. 1989. "Adjusting Education to Economic Crisis." *IDS Bulletin* 20(January):1–10.

NIER (National Institute for Educational Research). 1986. *Assessment of Educational Achievement in Asia and the Pacific: Report of a Seminar*. Tokyo.

Nitsaisook, Malee, and Lorin W. Anderson. 1989. "An Experimental Investigation of the Effectiveness of In-Service Teacher Education in Thailand." *Teaching and Teacher Education* 5(4):287–302.

Nkinyangi, John A. 1982. "Access to Primary Education in Kenya: The Contradictions of Public Policy." *Comparative Education Review* 26(2):199–217.

Noah, H. J., and M. A. Eckstein. 1990. "Trade-Offs in Examination Policies: An International Perspective." In P. Broadfoot, R. Murphy, and H. Torrance, eds., *Changing Educational Assessment: International Perspectives and Trends*. London: Routledge.

Noss, Andrew. 1990. "Education and Adjustment: A Review of the Literature." Policy, Research and External Affairs Working Paper 701. World Bank, Population and Human Resources Department, Washington, D.C. Processed.

Oakes, Jeannie. 1986. *Educational Indicators: A Guide for Policymakers*. Published for the Center for Policy Research in Education. Santa Monica, Calif.: Rand Corporation.

_________. 1989. "What Educational Indicators? The Case for Assessing School Context." *Educational Evaluation and Policy Analysis* 11(2):181–99.

O'Connor, Sorca. 1988. "Women's Labor Force Participation and Preschool Enrollment: A Cross-National Perspective, 1965–1980." *Sociology of Education* 61(1):15–28.

Odebunmirey, Akin. 1983. *Why They Drop Out of School: A Nigerian Perspective*. ERIC Reproduction Service Document ED 236259. Iowa City: University of Iowa.

Ogundare, Samuel F. 1988. "Curriculum Development: A Description of the Development of the National Curriculum for Primary Social Studies in Nigeria." *Educational Studies* 14(1):43–50.

Oliveira, João B., and François Orivel. 1981. "Cost Effectiveness Analysis of an In-Service Teacher Training System: Logos II in Brazil." Population and Human Resources Division Discussion Paper 81-40. World Bank, Washington, D.C. Processed.

Onyango, P. 1988. "Child Labour Policies and Programmes in Kenya." In Assefa Bequele and Jo Boyden, eds., *Combating Child Labour*. Geneva: International Labour Office.

Orivel, François, and Jean Perrot. 1988. "Les performances de l'enseignement primaire en Afrique francophone." Institut de Recherche sur l'Economie de l'Education, Dijon. Processed.

Pakistan, Government. 1983. "Draft Comprehensive Report: Primary Education Report." Islamabad. Processed.

Palazolli, C. *Le Maroque Politique*. Paris: Le Seuil.

Paul, Samuel. 1982. *Managing Development Programs: The Lessons of Success*. Boulder, Colo.: Westview Press.

Paxman, B., Carmelle Denning, and Anthony Read. 1989. "Analysis of Research on Textbook Availability and Quality in Developing Countries." Education and Employment Background Paper PHREE/89/20. World Bank, Population and Human Resources Department, Washington, D.C. Processed.

Peaslee, Alexander L. 1965. "Elementary Education as a Pre-requisite for Economic Growth." *International Development Review* 7:19–24.

_________. 1969. "Education's Role in Development." *Economic Development and Cultural Change* 11(3):293–318.

Pellett, P. 1983. "Changing Concepts on World Malnutrition." *Ecology of Food and Nutrition* 13(2):115–25.

Peters, Thomas J., and Robert H. Waterman, Jr. 1988. *In Search of Excellence: Lessons from America's Best Run Companies*. New York: Harper and Row.

Pfau, Richard H. 1980. "The Comparative Study of Classroom Behaviors." *Comparative Education Review* 24(3):400–14.

Pfeffermann, Guy P. 1987. *Public Expenditure in Latin America*. World Bank Discussion Paper 5. Washington, D.C.

Pinstrup-Anderson, Per, M. Jamarillo, and Frances Stewart. 1987. "The Impact of Government Expenditure." In Giovanni Cornia, Richard Jolly, and Frances Stewart, eds., *Adjustment with a Human Face*. Oxford: Clarendon Press.

Pollitt, Ernesto. 1984a. *Child Development Reference Document*. Vol. 1: *Risk Factors in the Mental Development of Young Children in the Developing World*. New York: UNICEF (United Nations Children's Fund).

_________. 1984b. "Nutrition and Educational Performance." *Prospects*. Paris: Unesco (United Nations Educational, Scientific, and Cultural Organization).

_________. 1990. *Malnutrition and Infection in the Classroom*. Paris: Unesco (United Nations Educational, Scientific, and Cultural Organization).

Pollitt, Ernesto, and Lisa Gossin. 1989. "The Impact of Poor Nutrition and Disease on Educational Outcomes: Determinants Excluded from Educational Planning." University of California, Davis, Department of Applied Behavioral Sciences. Processed.

Pollitt, Ernesto, Jere D. Haas, and David A. Levitsky, eds. 1989. "International Conference on Iron Deficiency and Behavioral Development." *American Journal of Clinical Nutrition* 50(3):565–705.

Pollitt, Ernesto, Nita Lewis, Cutberto Garza, and Robert J. Shulman. 1983. "Fasting and Cognitive Function." *Journal of Psychiatric Research* 17(2):169–74.

Popkin, Barry M., and Marisol Lim-Ybáñez. 1982. "Nutrition and School Achievement." *Social Science and Medicine* 16(1):53–61.

Porter, Andrew C. 1991. "Creating a System of School Process Indicators." *Educational Evaluation and Policy Analysis* 13(1):13–29.

Potar, N. 1984. "Project RIT (Reduced Instructional Time)." Project RIT, Muang Mai School, Lopburi, Thailand. Processed.

Powell, C., S. Grantham-McGregor, and M. Elston. 1983. "An Evaluation of Giving the Jamaican Government School Meal to a Class of Children." *Human Nutrition: Clinical Nutrition* 37(5):381–88.

Psacharopoulos, George. 1975. "Earnings and Education in OECD Countries." Paris: Organisation for Economic Co-operation and Development.

_________. 1985. "Returns to Education: A Further International Update and Implications." *Journal of Human Resources* 20(4):584–604.

_________. 1987a. "Are Teachers Overpaid: Some Evidence from Brazil." *Teaching and Teacher Education* 3(4):315–28.

_________, ed. 1987b. *Economics of Education: Research and Studies*. New York: Pergamon Press.

Psacharopoulos, George, and Ana-María Arriagada. 1987. "School Participation, Grade Attainment, and Literacy in Brazil: A 1980 Census Analysis." EDT Discussion Paper 86. World Bank, Population and Human Resources Department, Washington, D.C. Processed.

_________. 1989. "The Determinants of Early Age Human Capital Formation: Evidence from Brazil." *Economic Development and Cultural Change* 37(4):683–708.

Psacharopoulos, George, and Maureen Woodhall. 1985. *Education for Development: An Analysis of Investment Choices*. New York: Oxford University Press.

Purkey, Stewart L., and Marshall S. Smith. 1983. "Effective Schools: A Review." *The Elementary School Journal* 83(4):427–52.

Qasem, K. A., ed. 1983. *Attitude of Parents Towards Schooling of Children.* Dhaka: Bangladesh Foundation for Research in Education and Planning and Development.

Querido, A., F. Delange, J. Dunn, R. Fierro-Benítez, H. Ibbertson, D. Koutras, and H. Perinetti. 1974. "Definitions of Endemic Goiter and Cretinism, Classification of Goiter Size and Severity of Endemias, and Survey Techniques." In J. Dunn and G. Medeiros-Neto, eds., *Endemic Goiter and Cretinism: Continuing Threat to World Health*. Washington, D.C.: Pan American Health Organization.

Raudenbush, Stephen W., and J. Douglas Willms, eds. 1991. *Schools, Classrooms and Pupils: International Studies of Schooling from a Multilevel Perspective*. New York: Academic Press.

Resnick, Lauren B. 1987. *Education and Learning to Think*. Washington, D.C.: National Academy Press.

Reynolds, David, ed. 1985. *Studying School Effectiveness*. London: Falmer Press.

Rickards, John P. 1982. "Homework." In H. Mitzel, ed., *Encyclopedia of Educational Research*, 5th ed. London: Free Press.

Roberto, Eduardo L. 1987. "Social Marketing and Its Applications: A State of the Art Review." Paper prepared for the World Bank, Washington, D.C. Processed.

Robinson, Brandon. 1977. "El Salvador Education Sector Analysis: Executive Summary and Status Report." Washington, D.C.: Agency for International Development.

Robinson, M., and others. 1984. "Third Annual Report on the Study of USAID Contributions to the Egyptian Basic Education Program II." Washington, D.C.: Creative Associates. Cited in Mary B. Anderson, *Improving Access to Schooling in the Third World: An Overview*. BRIDGES Research Report Series, Issue 1. Cambridge, Mass.: Harvard University, 1988.

Robitaille, David F., and Robert A. Garden, eds. 1989. *The IEA Study of Mathematics II: Contexts and Outcomes of School Mathematics*. New York: Pergamon Press.

Rojas, C. A., and Z. Castillo. 1988. "Evaluación del programa Escuela Nueva en Colombia." Instituto SER de Investigación, Bogotá. Processed.

Romain, Ralph. 1985. "Lending in Primary Education: Bank Performance Review, 1962–83." EDT Discussion Paper 20. World Bank, Population and Human Resources Department, Washington, D.C. Processed.

Rondinelli, Dennis A., John Middleton, and Adriaan Verspoor. 1990. *Planning Education Reforms in Developing Countries: A Contingency Approach*. Durham, N.C.: Duke University Press.

Rondinelli, Dennis A., and J. Nellis. 1986. "Assessing Decentralization Policies in Developing Countries: The Case for Cautious Optimism." *Development Policy Review* 4(March):3–23.

Rosenberg, Nathan. 1982. *Inside the Black Box: Technology and Economics*. Cambridge, Mass.: Cambridge University Press.

Rosenzweig, Mark R., and Robert E. Evenson. 1977. "Fertility, Schooling, and Economic Contribution of Children in Rural India: An Econometric Analysis." *Econometrica* 45(5):1065–79.

Rust, Val D., and Per Dalin, eds. 1990. *Teachers and Teaching in the Developing World*. New York: Garland Press.

Rutter, Michael. 1983. "School Effects on Pupil Progress: Research Findings and Policy Implication." *Child Development* 54(1):1–29.

Rutter, Michael, Barbara Maughan, Peter Mortimore, and Janet Ouston, with Alan Smith. 1979. *Fifteen Thousand Hours: Secondary Schools and Their Effects on Children*. Cambridge, Mass.: Harvard University Press.

Ryan, John W. 1972. "Educational Resources and Scholastic Outcomes: A Study of Rural Primary Schooling in Iran." Ph.D. diss., Stanford University, Stanford, Calif.

Ryoo, Jai-Kyung. 1988. "Changes in Rates of Return over Time: A Case of Korea." Ph.D. diss., Stanford University, Stanford, Calif.

Salmi, Jamil. 1987. "Language and Schooling in Morocco." *International Journal of Educational Development* 7(1):21–31.

Schiefelbein, Ernesto. 1987. "Education Costs and Financing Policies in Latin America." EDT Discussion Paper 60. World Bank, Population and Human Resources Department, Washington, D.C. Processed.

Schiefelbein, Ernesto, and John Simmons. 1981. *Determinants of School Achievement: A Review of Research for Developing Countries*. Ottawa: International Development Research Center.

Schroeder, Larry. 1988. "Intergovernmental Grants in Developing Countries." Policy, Planning and Research Working Paper 38. World Bank, Population and Human Resources Department, Washington, D.C. Processed.

Schultz, T. Paul. 1961. "Investment in Human Capital." *American Economic Review* 51(1):1–17.

_________. 1985. "School Expenditure and Enrollments, 1960–1980: The Effect of Income, Prices, and Population Growth." Economic Growth Center Discussion Paper 487. Yale University, New Haven, Conn. Processed.

Schwille, John, Thomas O. Eisemon, and Robert Prouty. 1989. "Between Policy and Students: The Reach of Implementation in Burundian Primary Schools." Paper presented at the annual meeting of the Comparative and International Education Society, Boston, April.

Scribner, Sylvia, and Michael Cole. 1981. *The Psychology of Literacy*. Cambridge, Mass.: Harvard University Press.

Searle, Barbara. 1985. "General Operational Review of Textbooks." EDT Discussion Paper 1. World Bank, Population and Human Resources Department, Washington, D.C. Processed.

Sembiring, R., and I. Livingstone. 1981. *National Assessment of the Quality of Indonesian Education*. Djakarta, Indonesia: Ministry of Education and Culture.

Semeo Regional Center for Education Innovation and Technology. 1987. "Writing to Read Program in Pinyahan Elementary School Evaluation." Quezon City, Philippines. Processed.

Shaeffer, Sheldon F. 1979. "Schooling in a Developing Society: A Case Study of Indonesian Primary Education." Dissertation Abstracts International 40/10-A. Ph.D. diss., Stanford University, Stanford, Calif.

Sharan, Shlomo, and Chana Shachar. 1988. *Language and Learning in the Cooperative Classroom*. New York: Springer-Verlag.

Shepard, Lorrie A., and Mary Lee Smith. 1989. *Flunking Grades: Research and Policies on Retention*. New York: Falmer Press.

Shulman, Lee S. 1987. "Knowledge and Teaching: Foundations of the New Reform." *Harvard Educational Review* 57(1):1–22.

Sibanda, Doreen. 1982. *The Zimbabwe Integrated Teacher Training Course*. Bristol, Eng.: INSET (Inservice Training of Primary School Teachers) Project, School of Education, Bristol University.

Sierra Leone. 1979. "Education and Development: Sector Study." SL/IDA Education Project Unit, Freetown.

Sigman, Marian, Charlotte Neumann, Ake A. Jansen, and Nimrod Bwibi. 1989. "Cognitive Abilities of Kenyan Children in Relation to Nutrition, Family Characteristics, and Education." *Child Development* 60(6):1463–74.

Sinclair, M. E., with Kevin Lillis. 1980. *School and Community in the Third World*. London: Croom Helm.

Slavin, Robert E., Marshall B. Leavey, and Nancy A. Madden. 1984. "Combining Cooperative Learning and Individualized Instruction: Effects on Student Mathematics Achievement, Attitudes, and Behaviors." *Elementary School Journal* 84(4):409–22.

Smith, Herbert L., and Paul P. Cheung. 1986. "Trends in the Effects of Family Background on Educational Attainment in the Philippines." *American Journal of Sociology* 91(6):1387–1408.

Smith, Peter C., and Paul P. Cheung. 1981. "Social Origins and Sex-Differential Schooling in the Philippines." *Comparative Education Review* 25(1):28–44.

Somerset, H. C. A. 1987. "Examination Reform in Kenya." EDT Discussion Paper 64. World Bank, Population and Human Resources Department, Washington, D.C. Processed.

Sommer, Alfred, Joanne Katz, and Ignatius Tarwotjo. 1984. "Increased Risk of Respiratory Diseases and Diarrhea in Children with Pre-Existing Mild Vitamin A Deficiency." *American Journal of Clinical Nutrition* 40(5):1090–95.

Sommer, Alfred, and others. 1986. "Impact of Vitamin A Supplementation on Childhood Mortality: A Randomized Controlled Community Trial." *Lancet* 1:1169–73.

Stallings, Jane, and D. Stipek. 1986. "Research on Early Childhood and Elementary School Teaching Programs." In Merlin C. Wittrock, ed., *Handbook of Research on Teaching*. New York: Macmillan Co.

Suryadi, Ace, Michael Green, and Douglas M. Windham. 1981. "Teacher Quality as a Determinant of Differential Mathematics Performance among Poor Rural Children in Indonesia Junior Secondary Schools." School of Education, State University of New York, Albany. Processed.

Swartland, J., and D. Taylor. 1988. "Community Financing of Schools in Botswana." In Mark Bray and Kevin Lillis, eds., *Community Financing of Education: Issues and Policy Implications in Less Developed Countries*. Oxford: Pergamon Press.

Tan, Jee-Peng, and Vicente Paqueo. 1989. "The Economic Returns to Education in the Philippines." *International Journal of Educational Development* 9(3):243–50.

Tarwotjo, Ignatius, Alfred Sommer, Keith P. West, Jr., Edi Djunaedi, Lisa Mele, and Barbara Hawkins. 1987. "Influence of Participation on Mortality in a Randomized Trial of Vitamin A Prophylaxis." *American Journal of Clinical Nutrition* 45(6):1466–71.

Tatto, María T., K. H. Dharmadasa, K. G. Kularatna, and D. Nielsen. 1990. "Effectiveness and Costs of Three Approaches to Train Elementary School Teachers in Sri Lanka." Research report for the BRIDGES Project, submitted to the Harvard Institute for International Development and the United States Agency for International Development. Processed.

Taylor, D. C. 1983. "The Cost-Effectiveness of Teacher Upgrading by Distance Teaching in Southern Africa." *International Journal of Educational Development* 3(1):19–31.

Teja, Ranjit S. 1988. "The Case of Earmarked Taxes: Theory and Example." Working Paper WP/88/18. International Monetary Fund, Washington, D.C.

Thailand, Ministry of Education. 1987. *National Evaluation Results for Sixth Grade Students*. Bangkok: Office of the National Primary Education Commission.

Thailand, Office of the Prime Minister. 1981. *An Evaluative Study of Primary School Efficiency in Thailand*. Bangkok.

Thompson, A. R. 1990. "Deployment Issues: Making the Best of the Staff We Already Have." In Val D. Rust and Per Dalin, eds., *Teachers and Teaching in the Developing World*. New York: Garland Press.

Tibi, Claude. 1990. "What Policies for Teachers?" International Institute for Educational Planning Working Paper. Paris: Unesco (United Nations Educational, Scientific, and Cultural Organization).

Tilak, Jandhyala B. G. 1989a. *Education and Its Relation to Economic Growth, Poverty, and Income Distribution: Past Evidence and Further Analysis*. World Bank Discussion Paper 46. Washington, D.C.

_________. 1989b. "Financing and Cost Recovery in Social Sectors in Malawi." World Bank, Southern Africa Department, Washington, D.C. Processed.

Tinker, Irene, and Michelle B. Bramsen, eds. 1975. *Women and World Development*. Washington, D.C.: American Association for the Advancement of Science.

Tragler, A. 1981. "A Study of Primary School Health in Bombay." *Indian Pediatrics* 18(8):551–56.

Tsang, Mun, and S. Kidchanapanish. Forthcoming. "Private Resources and the Quality of Primary Education in Thailand." *International Journal of Educational Research*.

Tsang, Mun, and Chris Wheeler. 1991. "Local Initiatives and Their Implications for a Multi-Level Approach to School Improvement in Thailand." In Henry M. Levin and Marlaine E. Lockheed, eds., "Effective Schools in Developing Countries." Education and Employment Background Paper PHREE/91/38. World Bank, Population and Human Resources Department, Washington, D.C. Processed.

Tyack, David, and Elizabeth Hansot. 1988. "Silence and Policy Talk: Historical Puzzles about Gender and Education." *Educational Researcher* 17(3):33–41.

Unesco (United Nations Educational, Scientific, and Cultural Organization). 1961. "Outline of a Plan of Action for African Educational Development." Paper prepared for the Regional Conference of African States on the Development of Education in Africa, Addis Ababa, July.

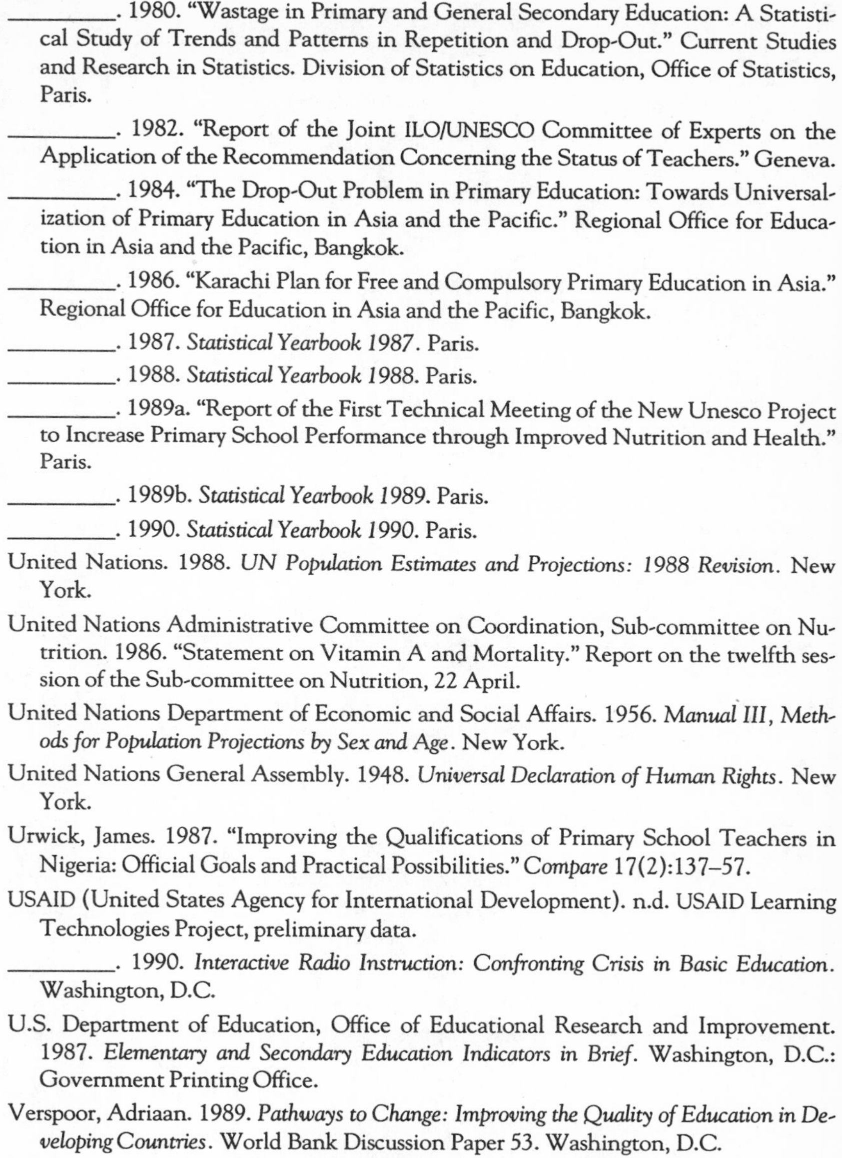

———. 1980. "Wastage in Primary and General Secondary Education: A Statistical Study of Trends and Patterns in Repetition and Drop-Out." Current Studies and Research in Statistics. Division of Statistics on Education, Office of Statistics, Paris.

———. 1982. "Report of the Joint ILO/UNESCO Committee of Experts on the Application of the Recommendation Concerning the Status of Teachers." Geneva.

———. 1984. "The Drop-Out Problem in Primary Education: Towards Universalization of Primary Education in Asia and the Pacific." Regional Office for Education in Asia and the Pacific, Bangkok.

———. 1986. "Karachi Plan for Free and Compulsory Primary Education in Asia." Regional Office for Education in Asia and the Pacific, Bangkok.

———. 1987. *Statistical Yearbook 1987*. Paris.

———. 1988. *Statistical Yearbook 1988*. Paris.

———. 1989a. "Report of the First Technical Meeting of the New Unesco Project to Increase Primary School Performance through Improved Nutrition and Health." Paris.

———. 1989b. *Statistical Yearbook 1989*. Paris.

———. 1990. *Statistical Yearbook 1990*. Paris.

United Nations. 1988. *UN Population Estimates and Projections: 1988 Revision*. New York.

United Nations Administrative Committee on Coordination, Sub-committee on Nutrition. 1986. "Statement on Vitamin A and Mortality." Report on the twelfth session of the Sub-committee on Nutrition, 22 April.

United Nations Department of Economic and Social Affairs. 1956. *Manual III, Methods for Population Projections by Sex and Age*. New York.

United Nations General Assembly. 1948. *Universal Declaration of Human Rights*. New York.

Urwick, James. 1987. "Improving the Qualifications of Primary School Teachers in Nigeria: Official Goals and Practical Possibilities." *Compare* 17(2):137–57.

USAID (United States Agency for International Development). n.d. USAID Learning Technologies Project, preliminary data.

———. 1990. *Interactive Radio Instruction: Confronting Crisis in Basic Education*. Washington, D.C.

U.S. Department of Education, Office of Educational Research and Improvement. 1987. *Elementary and Secondary Education Indicators in Brief*. Washington, D.C.: Government Printing Office.

Verspoor, Adriaan. 1989. *Pathways to Change: Improving the Quality of Education in Developing Countries*. World Bank Discussion Paper 53. Washington, D.C.

Wagner, Daniel A., ed. 1987. *The Future of Literacy in a Changing World*. London: Pergamon Press.

———. 1989. "In Support of Primary Schooling in Developing Countries: A New Look at Traditional Indigenous Schools." Education and Employment Background Paper PHREE/89/21. World Bank, Population and Human Resources Department, Washington, D.C. Processed.

Walberg, Herbert J., Roseanne A. Paschal, and Thomas Weinstein. 1985. "Homework's Powerful Effects on Learning." *Educational Leadership* 42(7):76–79.

_________. 1986. "Elementary School Mathematics Productivity in Twelve Countries." *British Educational Research Journal* 12(3):237–48.

Warren, Donald. 1990. *American Teachers: Histories of a Profession at Work*. New York: Macmillan Co.

Warwick, Donald P., Fernando Reimers, and Noel McGinn. 1989. "The Implementation of Reforms in the Primary Schools of Pakistan." Harvard Institute for International Development, Cambridge, Mass. Processed.

Waugh, Russell F., and Keith F. Punch. 1987. "Teacher Receptivity to Systemwide Change in the Implementation Stage." *Review of Educational Research* 57(3):237–54.

Weick, Karl E. 1976. "Educational Organizations as Loosely Coupled Systems." *Administrative Science Quarterly* 21:1–9.

Weitzman, Michael. 1987. "Excessive School Absences." *Advances in Developmental and Behavioral Pediatrics* 8:151–78.

Westview Special Studies in Science, Technology, and Public Policy. 1987. *Microcomputer Applications in Education and Training for Developing Countries*. Boulder, Colo.: Westview Press.

Wheeler, Christopher W., Stephen Raudenbush, Pragob Kunarak, and Aida Pasigna. 1989. *Policy Initiatives to Improve Primary School Quality in Thailand: An Essay on Implementation, Constraints, and Opportunities for Educational Improvement*. BRIDGES Research Report Series, Issue 5. Cambridge, Mass.: Harvard University.

Windham, Douglas M. 1988. *Indicators of Educational Effectiveness and Efficiency*. Tallahassee, Fla.: IEES Educational Efficiency Clearinghouse, Florida State University.

Winkler, Donald R. 1987. "Screening Models and Education." In George Psacharopoulos, ed., *Economics of Education: Research and Studies*. New York: Pergamon Press.

_________. 1989. "Decentralization in Education: An Economic Perspective." Policy, Planning and Research Working Paper 143. World Bank, Population and Human Resources Department, Washington, D.C. Processed.

Wittrock, Merlin C. 1986. "Students' Thought Processes." In Merlin C. Wittrock, ed., *Handbook of Research on Teaching*. New York: Macmillan Co.

Woodhall, Maureen. 1987. "Human Capital Concepts." In George Psacharopoulos, ed., *Economics of Education: Research and Studies*. New York: Pergamon Press.

World Bank. 1986. *Financing Education in Developing Countries: An Exploration of Policy Options*. Washington, D.C.

_________. 1987a. *Social Indicators of Development 1987*. Washington, D.C.

_________. 1987b. *World Development Report 1987*. New York: Oxford University Press.

_________. 1988a. *Adjustment Lending: An Evaluation of Ten Years of Experience*. Country Economics Department, Policy and Research Report 1. Washington, D.C.

_________. 1988b. *Education in Sub-Saharan Africa: Policies for Adjustment, Revitalization, and Expansion*. A World Bank Policy Study. Washington, D.C.

_________. 1988c. *World Development Report 1988*. New York: Oxford University Press.

_________. 1989. "Structural Adjustments and Living Conditions in Developing Countries." Population and Human Resources Department, Welfare and Human Resources Development Division, Washington, D.C. Processed.

_________. 1990a. *Primary Education*. A World Bank Policy Paper. Washington, D.C.

_________. 1990b. "Structural Adjustment and Poverty: A Conceptual, Empirical, and Policy Framework." Dimensions of Structural Adjustment Unit, Africa Region Report 8393-AFR. Washington, D.C.

World Conference on Education for All. 1990. *Meeting Basic Learning Needs: A Vision for the 1990s*. New York: Inter-Agency Commission for World Conference on Education for All.

Yan-you, Wang, and Yang Shu-hua. 1985a. "Improvement in Hearing among Otherwise Normal School Children in Iodine-Deficient Areas of Guizhou, China, Following Use of Iodized Salt." *Lancet* 8454:518–20.

_________. 1985b. "Occult Impaired Hearing among 'Normal' School Children in Endemic Goiter and Cretinism Areas due to Iodine Deficiency in Guizhou." *Chinese Medical Journal* 98(2):89–94.

Zigler, Edward F. 1987. "Formal Schooling for Four-Year-Olds? No." *American Psychologist* 42(3):254–60.

Zymelman, Manuel, and Joe DeStefano. 1989. *Primary School Teachers' Salaries in Sub-Saharan Africa*. World Bank Discussion Paper 45. Washington, D.C.

Index

Page numbers in boldface indicate boxes. Numbers preceded by the letter "n" refer to chapter notes.

Absenteeism: of students, 74, 76; of teachers, 59, 62, 101, 102–03, 104, 109, 112, 131

Access, equitable: defined, 145; factors contributing to, 146–54; promoting 154–67

Achievement, academic: of boys versus girls, 150; effect of family background on, 73–74; effect of group work on, 64–65; effect of multiple-shift system on, 158; effect of nutritional status on, 74–76; predictors of, 58

Achievement testing, 125–26, 138–41; by IEA, 12–16, 20 n9, 150; in Thailand, **140**

Addison, Tony, 35

Adjustment policies, government, 34–36, 38 n10

Administration, educational. *See* Management, educational system; Organizational structure; Supervision, school; Training, administrative

Administrative skill deficiencies, 123–24

Admission policy, 159

Africa: child labor in, 152; school enrollment in, 23, 148

Africa, Sub-Saharan: effect of government adjustment on, 35; improved educational inputs in, 41

Agarwal, D., 77, 79

Agricultural sector, 3–4, 5–6

Ahmed, Manzoor, 31, 38 n5

Aid: donors' neglect of primary education, 207, 214–15; sources of, 210–12, 215; statistics on, 207–10, 217, 218 n1–2; for subsectoral development programs, 216–17; use of, 212–14; ways to increase, 215–16

Akubue, Augustine, 67–68

Al-Hariri, Rafedi, 166

Ali, Anthony, 67–68

Allowances, hardship, 108

Altbach, Philip G., 48

Amadeo, E., 35

Anastasi, A., 138

Anderson, J., 150

Anderson, Lorin W., 63, 69

Anderson, Mary B., 63, 146, 150

Anzalone, Stephen J., 51, 53, 56, 70, 72

APEID (Asia and Pacific Programme of Educational Innovation for Development), 91, 101, 107, 109–10, 158

Armer, Michael, 5

Armitage, Jane, 50, 52, 53
Arriagada, Ana-María, 17, 65, 74, 147, 152
Arts curriculum, 45
Ashworth, A., 77
Associations, professional, 114–15
Attrition, teacher, 101–02, 103–04, 106
Au, Kathryn H., 64
Audiovisual tools, 51
Auerhan, J., 37
Austerity, 31–34
Austin, J. D., 89 n8
Authority lines, 121–23
Autonomy, school, 122–23
Avalos, Beatrice, 62, 67

Baker, Victoria J., 44, 51, 101, 109
Bangladesh: improving schooling for girls and rural children in, 164, **165**; improving school supervision in, **131**
Bangladesh Rural Advancement Committee, **165**
Bautista, Arturo, 75
Beeby, Clarence E., 107, 174, 219
Behrman, Jere R., 41, 182, 183, 206 n6
Benavot, Aaron, 3, 8, 45, 88 n3
Bennett, Nicholas, 112
Bequele, Assefa, 152
Berg, Alan D., 85
Berk, R. A., 138
Berman, Paul, 219, 223
Berstecher, Dieter, 173, 192
Bianchi, Leonard J., 14, 52, 65, 67, 147
Bianchi, P., 17
Bilingual education, 46, 167, **168**
Birdsall, Nancy, 41, 107, 171, 173, 183
Bleichrodt, N., 75
Blind. *See* Sight-impaired
Bloch, Deborah, 125, 127
Boarding schools, 163–64
Boissiere, M., 19 n6, 41, 183
Bolvin, John O., 143
Book flood experiment, **50**
Books. *See* Book flood experiment; Shared book method; Textbooks
Botswana: Ministry of Education and Culture, 107; Ministry of Finance and Development Planning, 52; taxes for education in, **191**
Boyden, Jo, 152
Bramsen, Michelle B., 149
Bray, Mark, 133, 157–58, 159, 173, 191, 194, 196
Brazil: child labor in, 152; educational investment in, 175, **176**; enrollment and dropout rates in, 147; taxes for education in, **190**; textbooks in, 49–50, 52; urban and rural enrollment in, 148
Breakfast, school, 75, 84
Brems, Susan, 85
Brenner, M., 152
Brooks-Gunn, Jeanne, 83
Brown, Byron W., 58
Buchanan, James, **192–93**
Burden, Paul R., 114
Bureaucracy, 109, 113, 118
Burkina Faso: labor-saving technology in, 164; rural education centers in, **32**
Butterworth, Barbara, 70

Camargo, J. M., 35
Cameron, John, 92, 95
Campbell, Patricia, 145
Career ladder plans, 113, **114**
Carnoy, M., 17
Carraher, David W., 9
Carraher, Terrezinha, 9
Castells, M., 17
Castillo, Z., 161
Centralization, of educational systems, 118–20
CERID/WEI (Research Center for Educational Innovation and Development), 146
Chalk and chalkboards, 51
Chall, Jeanne S., 8
Chamie, Mary, 148, 154, 158
Chang, Patricia M., 38 n2
Chapman, David W., 123–24
Chen, Lincoln C., 77
Chernichovsky, Dov, 74
Chesterfield, Ray, 168
Cheung, Paul P., 73, 152
Child labor, 41, 44, 152–53, 164, 182; in India and Kenya, **162–63**

China: encouraging girls' school attendance in, 164; gender equality program in, 166; taxes for education in, **191**
Chivore, B. R. S., 92, 102
Chung, Fay, 172, 195
Clark, Elmer J., 147
Clark, William C., 223
Class size, 89 n6, 109, 148; as a factor in learning, 40; inadvisability of lowering, 60–61, 86–87, 179
Clusters. *See* Housing clusters; School clusters
Cochrane, Susan H., 4
Cognitive skills: development of, xv, 7–10; importance of, 5–6; nutrition and, 74; relation to promotion and completion, 183. *See also* Higher-order thinking; Literacy; Numeracy
Cohen, David K., 112, 223
Cohen, Elizabeth G., 64
Colbert de Arboleda, Clara Victoria, 161
Colclough, Christopher, 37
Cole, Michael, 5
Cole, Nancy, 64
Coleman, Albert B., 147
Colletta, Nat J., 164
Colombia: decentralized administration in, 118; improving rural education in, **160–61**; multigrade teaching in, 159
Commonwealth Secretariat, 92, 113
Community involvement: in financing education, 191–200; in schools, 108, 115, 134, **160–61**, 166, 188
Community Junior Secondary Schools, Botswana, 188
Completion, school, 11–12, 37, 41, 181–85. *See also* Dropout; Repetition
Compulsory education, 22, 38 n2
Computers: availability of, 53; cost of providing, 56–57, 86; as learning aids, 50–51
Construction, school, 154–55, 168, 178–79
Coombs, Philip H., 31, 38 n5
Cope, Janet, 46
Cornia, Giovanni A., 35, 37, 191
Cost: of educating primary students, 150, 152, 161–62, 181–85; effect of double- or multiple-shift system on, 156–58, 180; of health interventions, 86; of improving primary education, 174–78; monitoring, 141; of interactive radio instruction, 71; of nutritional interventions, 84–85; of providing computers, 56–57; reducing direct educational, 161–64; reducing indirect educational, 164; reducing teacher salary, 180; of school lunches, 85; of teacher training, 57, 70, 95–97, 102, 179–80; of testing, 139
Cost-effectiveness: as a consideration in selecting educational inputs, 40, 86–88, 172, 178–80; of computers, 51, 88; of distance education, 70–71; of group work, 64; of health and nutrition interventions, 84–86; improving primary education's, 172, 194; of interactive radio instruction, 40, 71; lack of research on, 40, 88, 204 n3; of peer tutoring, 65, **66;** of reducing class size, 61, 86–87, 179; of school lunches, 85; in use of resources for primary education, 173, 174, 198, 229
Côte d'Ivoire: educational television in, 31; school access in, 146
Cotlear, Daniel, 6
Cotten, J., 75
Cox-Edwards, Alejandra, 107
Crandall, David P., 219, 223
Creemers, Bert, 19 n2
Csapo, Marg, 169
Cummings, William, 71, 72, 159
Curriculum: aid for reform of, 212; content of, 8, 45–46; in developing countries, 39, 86; intended versus implemented, 46–47; objectives of, 7; relevance of, 21, 30–31, 148; role of textbooks in, 48; staff role in developing, 114, 115, 122, 123

Dalin, Per, 94
Dall, Frank, 152
Demand for education, 145–46, 160–66

Demery, Lionel, 35
Demsky, Teri, 32
Den Hartog Georgiades, William, 44, 125
Denison, Edward F., 2
Denning, Carmelle, 45, 52, 53
Deolalikar, Anil B., 182, 206 n6
Deran, Elizabeth, 189
DeStefano, Joe, 67, 92, 95, 106, 108, 110–11
de Vera, María L., 194
Didonet, V., 81
Dillinger, William R., 198
Disabled. *See* Handicapped
Disadvantaged Schools Program, Zimbabwe, **194–95**
Discrimination: against minority groups, 153–54, 166; gender, 148–50, **151,** 159, 166; in language policy, 153–54, 166–67; redressing, 169
Distance education, **69,** 69–71
Dominican Republic, interactive radio in, 71, **73,** 159–60
Dorsey, Betty J., 94
Dossey, John A., 9, 10
Dove, Linda A., 66, 108, 113
Drenth, P., 75
Dropout, 181; causes of, 11, 153, 180; costs of, 182–83; in developing countries, **82;** effect of, 146; reducing, 41, 183–85; in rural areas, 146–47, 159
Dulberg, Loretta, 150
Dunn, John T., 77
Dutcher, Nadine, 167

Earmarking funds, 189–90, **190–91, 192–93,** 206 n12–13
Earnings: effect of education on, 3, 5, 204 n1; as indirect measure of productivity, 3
Economic policies. *See* Fiscal policy reform; Privatization effects; Stabilization policies; Structural adjustment reforms; Subsidies
Economic and social development: contribution of education to, 1, 2–3, 18, 22–23, 171, 174, 227, 231; educated labor force as a condition for, 2–3
EDSAC 1, Ghana, 115
Education, primary: aid for, 207–16; availability of, 21, 22–29, 154–60; benefits of, 1–6, 19 n2, 22, 174, 204 n1; cost of improving, 174–78; compulsory, 22, 38 n2; demand for, 145–46, 160–66; effects of austerity on, 21–22, 31–37, 171; financing of, 171–206; purposes of, xv, 1, 7–10, 181; reform in low-income countries, 224–27; reform in middle-income countries, 227–231; relevance of, 21, 29–31; in rural areas, **32,** 146–48, **160–61,** 162–164, **165**; shortcomings of in developing countries, 10–16; spending for, 31–37, 88, 171. *See also* Organizational structure; Management, educational system
Education projects: financing for, 30; of World Bank, 53, 55
Education, teacher. *See* Teachers, education of; Training, teacher
Educational research institutes, Korea, **143**
Educational Testing Service, 139
Egypt, school attendance in, 146
Efficiency, education system: centralization and, 118, 119–20; cost-sharing as a means to improve, 188; difficulties in improving, 39–40; external, 30, 41, 88 n2; importance of, 30, 37; internal, 30, 180, 187, 203, 205 n4; measuring, 139, 141
Efficiency in financing education, 187, 188, 190–94, 197–200
Eisemon, Thomas O., 6, 16, 60, 121, 124, 167
Electronic learning aids, 51
Elley, W. B., 50
El Salvador: educational television in, 31; rural and urban students in, 147
Elston, M., 75, 84
Enrollment: in Asia, 148; of boys versus girls, 148, 229; effect of gender discrimination on, 148–50; effect of latrines on, 155; effect of poverty on, 150–53; fees for, 174; of girls, 27, 148, 164, **165**, 229; growth in, 21, 24, 145; in newly industrialized

countries, 3; proximity to school as a determinant of, 146; in rural areas, 146–48, **165**; rates of total, 23–24; ratios of gross, 25–28, 224; ratios of net, 28, 37, 224; in upper-middle-income countries, 229; in Zimbabwe, **93**
Entrepreneurial skill deficiencies, 124
Equity: definition of educational, 145; in financing education, 188, 200–02, 203; social, 30. *See also* Access, equitable
Escuela Nueva, Colombia, 159, **160–61**
Ethnic groups, 153, **168**
Evans, David R., 150
Evenson, Robert E., 205–06 n5
Ezzaki, Abdelkader, 82

Fafunwa, A. Babatunde, 168
Family. *See* Home environment; Parents
FAO (Food and Agriculture Organization), 79
Farmers' use of education, 3–4, 5–6, **6**
Farrell, Joseph P., 49
Farrugia, Charles, 115
Feedback effect, 65
Feeding programs, school, 84, 85, 212
Female Education Scholarship Program, Bangladesh, **165**
Females: benefit of multiple-shift system for, 158; discrimination in school against, 149–50, **151**; effect of education on, 2, 4; obstacles to education for, 148. *See also* Girls
Fertility, 4, 204 n1
Financing, educational: through budget reallocation, 185–87; difficulties of, 171–72; equity in, 173, 200–02; government role in, 173–74, 202; importance of, 171; in Korea, 203, **204–05;** local resources for, 190–200, 200–01; national resources for, 189–90, 193, 201; objectives of, 172–73; shortcomings of, 172. *See also* Subsidies; Taxes
Fine, Seymour H., 206 n16
Finn, Jeremy D., 150
Fiscal policy reform, 189. *See also* Taxes
Fitz, J., 153
Florencio, Cecelia A., 74, 80
Fonacier, Josefina, 14, 52, 65, 67, 147
Fouilland, Jean-Louis, 196
Fox, M. Louise, 107
Fraser, Barry J., 45
Freedman, Marion, 164
Friend, Jamesine, 73
Friesen, C. D., 89 n8
Fryer, Michelle, 73
Fullan, Michael, 124, 136, 219
Fuller, Bruce, 52, 63, 67, 68, 74, 81, 98, 102, 109, 115, 125

Galda, Klaus, 41
Gallagher, Mark, 35
Garden, Robert A., 14
Garms, Walker J., 201
Gatawa, B., 93
Gender: differences in achievement, 150; differences in enrollment, 148, 229; discrimination, 148–50, **151**
Ghani, Zainal, 95
Gimeno, José B., 92, 95
Girls: academic achievement of, 150; discrimination against, 149–50, **151**; enrollment and school attendance of, 1, 27, 148–49, 155–56, 164, 174; schooling in Bangladesh for, **165**; schooling in Saudi Arabia for, 166
Glass, Gene V., 51, 61, 66, 88
Glewwe, Paul, 146, 152
González, José, 41
González, Sarah, 52
González-Suárez, Mirta, 151
Gorman, Kathleen S., 81
Government role: in developing teaching force, 91–92, 101, 115–16; in educating the handicapped, 167, **169**; in encouraging educational equity, 202; in financing education, 173–74, 188; in providing primary school education, 156, 173–74; in recommending educational policy, 40
Grantham-McGregor, S., 75, 84
Greaney, Vincent, 126
Green, Michael, 58
Grootaert, Christiaan, 108

Group work, 64–65
Guatemala: bilingual education in, **168**; enrollment and dropout rates in, 147; minority access to schooling in, 153
Guinea, taxes for education in, **191**
Guinea Bissau, taxes for education in, **199**
Guthrie, James W., 201

Haas, Jere D., 75
Haddad, Wadi, 32, 62, 98, 100, 183, 206 n6
Haglund, Elaine, 77
Halpern, Robert, 81, 83
Handicapped: educational opportunities for, 154, 167; education in Zambia for, **169**. *See also* Hearing-impaired; Sight-impaired
Hanson, Mark E., 118, 127
Hansot, Elizabeth, 159
Hanushek, Eric, 39–40, 41, 51, 73, 88, 125, 127, 176, 204 n3
Harambee schools, Kenya, 188
Harber, Clive, 94, 101, 109
Harbison, Ralph W., 41, 51, 176
Harris, Abigail M., 67
Hartley, Michael J., 17
Haskins, Ron, 81, 83
Havelock, Ronald G., 219
Haverman, Robert H., 4
Hawkridge, David, 50, 53
Head Start, United States, 81, **82–83**
Health, child: effect of mother's education on, 4; interventions to improve, 80–86; relationship between school performance and, 72, 180
Hearing-impaired, 80, 86, 87, **169**
Helminths, 80
Herencia, C., 150
Hetzel, Basil S., 77
Heyneman, Stephen P., 14, 15, 49, 52, 58, 63, 125
Higher-order thinking, 8, 10, 46, 126, 219
Hildebrand, G., 84
Hirschman, Albert O., 223
Holloway, Susan D., 81
Holsinger, Donald B., 4, 5
Home environment, 73–74, 76–77, 123; child labor within, 152. *See also* Parents
Homework, 64, 89 n8
Horn, Robin, 17
Housing clusters, 108
Huberman, A. M., 219, 223
Human capital: importance of primary education for, 17–18; underdevelopment of, 16–17
Hunger, 75, 77
Husén, Torsten, 62

Ibáñez, Ricardo M., 92, 95
Ibrahim, R., 138
ICED (International Council for Educational Development), 146
IEA (International Association for the Evaluation of Educational Achievement), 12, 14, 16, 139, 150
IEES (Improving the Efficiency of Educational Systems Project Sector Study), 91, 94, 95, 98, 101, 103, 107–10
IEL Project, Liberia, 72
IMPACT Project, Philippines, 72
Income groupings of countries, 19 n1
India: educating child workers in, 152, **162**; lowering dropout rate in, 155
Indicators, education, 127, 141
Indonesia: multigrade teaching in, 158–59; textbook project in, **129**
Information systems, 116 n7; importance of, 125–27; strengthening of, 138–42. *See also* Monitoring systems
Initiation aux Travaux Manuels, Tunisia, 30
Inkeles, Alex, 5
INSET Project, 113
INSPIRE Project, Malaysia, 72
Institutional capability, 117–18
Instruction. *See* Teaching
Instructional materials. *See* Learning materials
Interactive radio instruction, 31, 40, 71, **73**, 128–30
Iodine deficiency, 75, 80
IREDU (Institut de Recherche sur l'Economie de l'Education), 218 n2

IRI. *See* Interactive radio instruction
Iron deficiency, 75, 77, 79
Islamic schools. *See* Koranic schools

Jacobson, Stephen L., 112
Jamarillo, M., 35
James, Estelle, 188
Jamison, Dean T., 3, 5, 49, 52, 61, 63, 74, 77, 152, 162
Jaworski, John, 50, 53
Jennings-Wray, Zellyne D., 67
Jimenez, Emmanuel, 14, 170 n3, 171, 188, 192–93, 194, 206 n10
Johnston, Bruce F., 223
Jolly, Richard, 35, 37, 191
Jones, Howard, 44, 125

Kagitcibasi, Cigdem, 81
Kamens, David, 8, 45, 88 n3
Kanbargi, R., 152
Karmacharya, D. M., 70
Kasarda, John D., 4
Kasemsestha, W., 138
Katz, Joanne, 76, 85
Kawanobe, Satoshi, 138
Kellaghan, T., 126
Kelly, Gail P., 148, 152
Kelly, Michael, 98
Kemmerer, Frances N., 62, 94, 102, 103, 104, 106
Kenya: achievement testing in, 139; certification examination in, 138; child labor in, **162–63**; educational spending cuts in, 37; private schools in, 188
Kenya Radio Language Arts Project, **73**
Kerr, J. Graham, 19 n7
Kidchanapanish, S., 191, 196
King, Elizabeth M., 152, 205 n5
Klein, Susan, 145
Knight, John B., 19 n6, 41, 183
Komenan, André, 17, 58, 65, 108
Koranic schools, 31, 81–82, 149, 156, 170 n2
Korea, Republic of: educational research institutes in, **143**; education tax in, **190**; financing of universal primary education in, 203, **204–05**
Korean Educational Development Institute, **143**
Korean Institute for Research in Behavioral Sciences, **143**
Kulik, Chen-Lin, 65
Kulik, James A., 65

Labor force: in developing countries, 16–17; effect of educated, 2–3. *See also* Child labor
Language: of instruction, 31, 146, 153–54, 166–67, **168**; teaching of, 8, 45–46; types of official, 88 n3
Lapointe, Archie E., 9, 10
Lau, Lawrence J., 3
Learning: contribution of school environment to, 43; as a determinant of student promotion, 41; factors contributing to, 41; by group work, 67; importance of, xv; measurement of, 138–41; teaching practices not conducive to, 67. *See also* Teaching
Learning capacity of children, 41, 43, 72–80
Learning materials: availability of, 52–53; importance of developing, 128; programmed, 31, 72; provision of, 53–57, 130, 212; return on investment in, 175, **176**; role of, 47–51
Learning time: availability of, 58–60; defined, 89 n5; in developing versus industrialized countries, 39; importance of, 41, 57–58; strategies to increase, 60–62
Leavey, Marshall B., 64
Lee, Valerie E., 83, 170 n3
Leithwood, K. A., 136, 137
Leslie, Joanne, 77
Levin, Henry M., 40, 51, 65, 66, 85, 88
Levinger, Beryl, 85
Levitsky, David A., 75
Lewin, Keith, 37, 192, 173
Liberia, Ministry of Education, 92, 101, 102, 103, 104
Life expectancy, 4
Lillard, Lee A., 205 n5
Lillis, Kevin, 173, 191, 194, 196
Lim-Ybáñez, Marisol, 75

Literacy: basic, 8–9, 16, 159, 167; dropout and, 146, 181; higher-order, 10; in newly industrialized economies, 3; life expectancy and, 4; programs in India and Kenya, **162–63**; schooling required to achieve, 20 n8; value of, 5. *See also* Language; Reading
Livingstone, Ian D., 14, 125
Lockheed, Marlaine E., 3, 14, 39–40, 52, 58, 65, 67, 73, 74, 88, 115, 125, 127, 147, 152, 162, 170 n3, 188, 204 n3
Logos II project, Brazil, **69**, 71
Louat, Frederic F., 3
Loucks, Susan F., 219, 223
Lourié, Sylvian, 147, 153
Loxley, William A., 49, 58, 125
Luna, Eduardo, 52
Lunch, school, 85
Lussier, Karen, 38 n2

MacAdam, Colin, 92
MacLennan, E., 153
Madden, Nancy, 64
Malawi, Ministry of Education and Culture, 148
Malaysia: educational television in, 31; management training in, **135**
Mali: child labor in, 152; financing education in, 200
Mallon, N., 165
Malnutrition, 39; prevalence of, 77–79; school performance and, 74–76; school programs to eliminate, 84–85
Management, classroom, 43, 63–65, **131**. *See also* Teaching
Management, educational system, 117–20, 127–134, 222–24: competence of staff in, 120–22, 123–24, 134–36; role of information systems in, 125–27, 138–42. *See also* Organizational structure; Training, administrative
Management, school, 117; role of community in, 134; role of staff in, 44, 122–23, 124–25, 133–34, 136–37. *See also* Principal; Supervision, school; Training, administrative
Mangubhai, Francis, 50
Martin, R., 70
Mathematics: achievement, 9; curriculum, 45; textbooks, 46, **48–49**
McCleary, William A., 189, 192–93, 198
McGinn, Noel F., 63, 64, 131, 146, 156, 195, 204–5
McLaughlin, Milbrey W., 219, 223
McLaughlin, Stephen D., 51
McMahon, Henry, 50, 53
McMahon, Walter W., 2, 3
McSweeney, Brenda G., 164
Mead, Nancy A., 9, 10
Meesook, Oey A., 74
Meister, Gail, 51, 65, 66, 88
Merit pay, 111–12
Meyer, John W., 25
Michel, Andrée, 150
Micronutrient deprivation, 75–76, 77–79, 84–85
Middleton, John, 125, 127, 128, 232
Miguel, M. M., 138
Miles, Matthew B., 223
Mingat, Alain, 174
Modernity, 5, 50
Monitoring systems, 125, 127; criteria for strong, 141; standardized achievement testing and, 126, 139, 141
Montenegro, Xenia, 49, 52
Moock, Peter, 5, 77
Morocco: languages in, 170 n1; unequal access to schooling in, 153–54
Motivation, teacher, 90, 91, 92, 101–10
Mulligan, James G., 61
Multigrade classes, 158–59, **160–61**
Multiple-shift system, 156–58
Mundangepfupfu, Mwazwita R., 52, 53
Murnane, Richard J., 106, 112, 223
Murphy, Michael J., 114
Murray, Karen B., 92, 95
Mwamwenda, Bernadette B., 102, 115
Mwamwenda, Tuntufye S., 102, 115
Myers, Robert G., 81, 83, 163

Narayan, A., 163
National Board for Professional Teaching Standards, 90

National Institute of Educational Evaluation, Korea, 143
National Institute of Educational Management, Malaysia, **135**
Naumann, Jens, 207, 218 n1
Nellis, J., 232
Nepal: school attendance in, 146, 155–56; taxes for education in, **191**
Nettleton, Greta S., 70
Neumann, Peter H., 47
Ngomba, Peter, 35
NIER (National Institute for Educational Research), 138
Nigeria, bilingual education in, 168
Nitsaisook, Malee, 67, 69
Nkinyangi, John A., 163–64
Nonformal Primary Education Program, Bangladesh, **165**
Nonsalary benefits, teacher, 107–08
Noonan, Richard, 62
Noss, Andrew, 37
Numeracy: basic, 9–10, 159; dropout and, 181; higher-order, 10; value of, 5
Nutrition: effect of education on, 4, 204 n1; interventions to improve, **82,** 84–85; relationship between school performance and, 72, 74–76, 180
Nyerere, Julius, 30
Nyirongo, Ronald, 74

Oakes, Jeannie, 127
O'Connor, Sorca, 81
Odebunmirey, Akin, 152
Office of the National Primary Education Commission, Thailand, **140**
Ogundare, Samuel F., 47
Oil price shock, 32
Oliveira, João B., 69, 71
Olsen, Randall J., 106
Onyango, P., 163
Opportunity cost: of nonattendance at school, 166; of schooling, 19 n4, 145, 150, 158, 182; to teachers, 70, 106, 113
Orderly school environment, 43
Organizational structure: centralized versus decentralized, 118–20; improving, 127–28; levels in, 120–23; nature of, 117–18; school clusters as a form of, 132, **133**. *See also* Management, educational system; Supervision, school; Training, administrative
Orivel, François, 15, 69, 71
Oxenham, John, 35

Pakistan: improving school supervision in, **131**; shelterless schools in, 155; taxes for education in, **190**
Pakistan Government, 46
Palazolli, C., 170 n1
PAMONG Project, Indonesia, 72
Paqueo, Vicente, 14, 188, 194, 182
Parasitic infection, 74, 76, 80, 86
Parents, 173; attitude toward bilingual education, **168;** effect of education on, 2, 4; involvement in school, **82,** 117, 166, 194; relations between teachers and, 123; role in child teachability, learning, and school completion, 73–74, 76–77, 84, 153, 180, 183; tendency to send children to school, 2, 146, 150–52, 166, 173–74; willingness to pay for education, 150–53, 182, **194–95,** 195–96, **204–05**
Paschal, Roseanne A., 89
Paul, Samuel, 223
Paxman, B., 52, 53
Peaslee, Alexander L., 2–3
Pedagogy. *See* Teaching
Peer tutoring, 31, 65, **66**
Pellett, P., 77
Pencils and paper, 51, 53
Perrot, Jean, 15
Peters, Thomas J., 44
Peters, Ton, 19 n2
Pfau, Richard H., 67
Pfeffermann, Guy P., 35
Philippines: child labor in, 152; spending for primary education in, 187; taxes for education in, **191**; textbook project in, **129**; urban and rural students in, 147

Phillips, Gary, 9, 10
Pierce, Lawrence C., 201
Pinstrup-Anderson, Per, 35
Pollitt, Ernesto, 74, 75, 76, 77, 80, 84
Poll tax, 198, **199**
Pomfret, Alan, 124
Popkin, Barry M., 75
Population, school-age: defined, 28; enrollment of, 24–29; growth of, 21, 24, 90–91, 145; out of school, 38 n1, 145; as a proportion of total population, 33
Population growth: as a constraint on development, 4; enrollment challenges and, 145, 180, 225; teacher supply and, 90–91, 110
Porter, Andrew C., 141
Potar, N., 72
Poverty: child labor and, 152; education and, 1–2, **82–83**, 174; school enrollment and, 150–53
Powell, C., 75, 84
Practice teaching, 100
Preschools, 81–84. *See also* Head Start, United States
Primary school cycle, 7
PRIMER Project, Jamaica, 72
Principal, school: authority of, 109, 132–34; characteristics of effective, 44; role as school manager, 136–37; role as teacher supervisor, 124; selection of, 44, 125, 136; training of, 125, 136–37; weak managerial skills of, 124–25
Privatization effects, 187
PROAPE program, Brazil, 81
Problem-solving skills, 10, 15; of farmers, 6; importance of, 2, 5, 8; in mathematics, 9–10, 75; in spelling, 75; teaching of, 7, 46–47, 53, 61, 64, 227, 229
Professional development, continuing, 137
Programmed materials, 31, 72
Promotion: of students 41, 183; of teachers, 109, 113–14, **114**
Property tax, **199**
Prosser, Roy, 31, 38n5
Prouty, Robert, 60, 124, 167
Psacharopoulos, George, 3, 19 n2, 74, 107, 147, 152
Public choice theory, **192–93**
Punch, Keith F., 124
Purkey, Stewart L., 43

Qasem, K. A., 155
Querido, A., 75

Radio. *See* Interactive radio instruction
Ramírez, Francisco O., 25
Raudenbush, Stephen W., 19 n2
Read, Anthony, 52, 53
Reading, 8–9, 10, 14–15, **48–49, 50,** 51
Recruitment: of female and rural students, 162–64, **165;** of teachers, 67, 92, **93,** 94, 108, 155, **165**
Reimers, Fernando, 63, 64, 131, 156
Reis, Janet, 150
Religion, 153, 166. *See also* Koranic schools
Repetition: causes of, 180; costs of, 182–83; effect of, 175, 206 n6; incidence of, 12; reducing, 41, **82,** 183–85
Research capacity, 127, 141–42, **143**
Resnick, Lauren B., 7–8, 10
Reynolds, David, 19 n2, 43
Ribeiro, Lucie, 45
Rickards, John P., 89 n8
RIT Project, Thailand, 72
Robinson, Brandon, 147
Robinson, M., 146, 152
Robitaille, David F., 14
Rodd, Alastair G., 127
Rojas, C. A., 161
Romain, Ralph, 56
Rondinelli, Dennis A., 128, 232
Rosenberg, Nathan, 17
Rosenzweig, Mark R., 205–06 n5
Rural population: educational opportunities for, 4, 30; enrollment and dropout among, 146–47; malnutrition among, 77; need for preschool services for, 87; programs in Bangladesh for, **165**; reducing educa-

tional costs for, 162–64; relevance of education to, 21, 30–31
Rural schools: in Bangladesh, **165;** in Burkina Faso, **37;** in Colombia, **160–61;** staffing of, 108, 113, 132; urban versus, 147–48
Rust, Val D., 94
Rutter, Michael, 43
Ryan, Doris W., 63
Ryan, John W., 63
Ryoo, Jai-Kyung, 3

Sabot, Richard H., 19 n6, 41, 183
Saha, Lawrence J., 62
Saks, Daniel H., 58
Salaries, teacher: augmenting, 107; decline of, 37, 104–06; importance of adequate, 62, 102, 110; inadequacy of, 92, 102–03, 106–08; increases in, 109–110, 112; irregular payment of, 103–04; in Korea, **205;** linkages between civil service pay scale and, 110–11, **114;** of rural versus urban teachers, 108; spending for, 36–37, 180. *See also* Merit pay; Nonsalary benefits, teacher
Salmi, Jamil, 154
Scheduling, 164, **165.** *See also* Multiple shifts
Schiefelbein, Ernesto, 62–63, 74, 188
Schistosomiasis, 80
Schliemann, Analucia D., 9
Schnur, Elizabeth, 83
School clusters, 132, **133**
School improvement funds, 130
Schools, primary: alternatives to traditional, 31, 159–60, **160–61, 162–63;** boarding, 163–64; characteristics of effective, 43–44; construction and renovation of, 154–55, 168, 178–79; in developing versus industrialized countries, 39; encouraging initiatives of local, 130; feeding programs in, 84, 85, 212; Koranic, 31, 81–82, 149, 156; objectives of, 40–41; private, 156, 188; single-sex, 159; supervision of, 121, 123–24, 130–34, **131**, 136, **161**; variation in quality of, 14. *See also* Preschools
Schroeder, Larry, 198
Schultz, T. Paul, 2, 19 n2, 105, 171
Schwille, John, 59, 60, 121, 124, 167
Science, 16, 45
Scribner, Sylvia, 5
Scrimshaw, Nevin S., 77
Searle, Barbara, 129
Seelye, H. Ned, 168
Sembiring, R., 125
Semeo Regional Center, 51
Sepúlveda-Stuardo, Manuel A., 49
Shachar, Chana, 64
Shapiro, B. J., 63
Sharan, Shlomo, 64
Shared book method, **50**
Shelterless schools, 39, 155
Shepard, Lorrie A., 183, 206 n6
Shu-hua, Yang, 80
Shulman, Lee S., 98, 100
Shyam, M., 70
Sibanda, Doreen, 93
Sight-impaired, 79, 80, 86, 87, **169**
Sigman, Marian, 74, 77
Simmons, John, 62–63, 74
Sinclair, M. E., 32
Slavin, Robert E., 64
Smith, David H., 5
Smith, Herbert L., 73, 152
Smith, Marshall S., 43
Smith, Mary Lee, 61, 183, 206 n6
Smith, Peter C., 73
Snyder, Conrad W., Jr., 52, 67, 68
Social studies curriculum, 45
Somerset, H. C. A., 138, 139
Sommer, Alfred, 76, 85
Soysal, Y., 25
Special education, 167, **169**
Spending: of donor aid, 212–14, 217; by government per primary student, 33–34, 175–77; effect of adjustment on educational, 35–37. *See also* Cost; Financing, educational
Spratt, Jennifer, 82
Stabilization policies, 34
Stages of educational development, 219–220, 224
Stallings, Jane, 64

Stanbury, John B., 77
Stewart, Frances, 35, 37, 191
Stipek, D., 64
Structural adjustment reforms, 34–36, 38 n10
Subsectoral development programs, 216–17
Subsidies, 172, 174, 187, 188
Sullivan, J., 153
Supervision: school, 121, 123–24, 130–34, **131**, 136, **161;** teacher, 109, 122, 124
Supplements, in-kind, 108
Supply, educational, 154–60
Suryadi, Ace, 58
Sutton, Margaret, 164
Swanson, Eric V., 17
Swartland, J., 188

Tan, Jee-Peng, 174, 182, 192
Tarwotjo, Ignatius, 76, 85
Taxes: earmarking, 189–90, **190–91, 192–93;** education sector's reliance on, 172, 173, 189; local education, 196–99; national education, 189–90; poll, 198; property, **199**
Taylor, D. C., 69, 188
Teachability, child. *See* Learning capacity of children
Teacher guides, 50, 53
Teachers: absenteeism of, 59, 62, 101–04, 109, 112; attrition of 101–02; career advancement of, 109, 113–14, 116; characteristics of effective, 63; compensation for, 37, 62, 102–08, 110–12, 116; deployment of, 108, 112–13, 132, 155–56; education of, 53, 62, 65–67, 90, 92–96; encouraging professionalism of, 113–15; motivation of, 90, 91, 92, 101–10; recruitment of, 67, 92, **93,** 113, 155–56, **165;** status of, 91–92; supervision of, 109, 124; supply of, 90–91, **93,** 110, 155, 225; training of, 68–71, 95–100, **160–61,** 179–80; unions for, 113–14, 117; working conditions for, 109, 113, 115. *See also* Salaries, teacher
Teaching: deficiencies in developing countries, 67–68; as a determinant of student learning, 90, 115; multigrade classes, 158–59; practices for effective, 63–65; skills, 97–98, 100; through interactive radio, 31, 66, 71, **73;** through programmed materials, 31, 72. *See also* Training, teacher; Tutoring
Teaching materials. *See* Learning materials
Technology, 1, 2, 5–6, 16, 17, 19–20 n7, 229, 231
Teja, Ranjit S., 193, 206 n12
Television, 31, 88–89 n4
Terry, James, 125, 127
Tests: deficiencies in, 126; diagnostic, 50; IEA, 12–16, 20 n9, 150; importance of, 65, 125, 141; improving, 138–41; in Korea, **143;** lack of, 67–68; textbooks and performance on, 49–50; in Thailand, **140;** types of, 126
Textbooks: as aids to learning, 46–50, 115; availability of, 49, 52–53, 54–55; distribution of, 54–55, 57, 128; gender discrimination in, 149–50, **151**; in Indonesia, **129;** mathematics, **48–49**; in the Philippines, **129;** production of, 57, 123, 128, **129;** quality of, 46–47, 52–53, 57; reading, **48–49** *See also* Book flood experiment
Thailand: achievement testing in, **140**; in-service training for principals in, 137; Ministry of Education, 140; Office of the Prime Minister, 147; school clusters in, 132, **133**; urban and rural students in, 147
Theisen, Gary L., 5
Thiagarajan, Sivasailam, 94, 102, 103, 104, 106
Thompson, A. R., 113
Tibi, Claude, 106
Tilak, Jandhyala B., 19 n3
Time. *See* Learning time
Tinker, Irene, 149
Tragler, A., 77

Training, administrative, 134; aid for, 217; of central administrators, 123, 135–36; of intermediate-level staff, 123–24, 136; in Malaysia, **135;** of principals, 125, 136–37
Training, teacher, 90–91, 94; aid for, 130, 212; content of, 95, 98; cost of, 57, 70, 95–97, 102, 179–80; to develop pedagogical skills, 97–100, 116; through distance education, **69,** 69–71; in-service, 68–71, 113, **131,** 179–80; length of, 95, 96–97; as nonsalary benefit, 108; requirements for, 67, 93–94, 100, 111; in rural areas, 70; in special education, 167; strategies to improve, 99–100, 115–16; in Zimbabwe, **93**. *See also* Teachers, education of
Tsang, Mun, 132, 191, 196
Turkey: target for class size in, 179; taxes for education in, **190–91**
Tutoring, 65, **66**, 130
Tyack, David, 159

Undugu Society, Kenya, **163**
Unesco (United Nations Educational, Scientific, and Cultural Organization), 84, 86, 94, 95, 108, 153, 156, 164, 183, 206 n6
Unions, teacher, 113–14, 117
United Nations Development Programme, 30
United Nations General Assembly, 22
United States: aid for education by, 211; Head Start program in, 81, **82–83**; taxes for education in, **190**
University of Ife, Nigeria, **168**
University of Zambia, **169**
UPE/IMPACT Project, Bangladesh, 72
Urban schooling, 146–48
Urwick, James, 94
USAID (United States Agency for International Development), 56, 57, 151

Vail, Stephen, 115
Ventrasca, Marc, 38n2
Verspoor, Adriaan, 68, 127, 128, 130, 131, 219, 232
Vitamin A deficiency, 76, 79
Vocational education, 45, **163**

Wagner, Daniel A., 82, 84, 156
Walberg, Herbert J., 89 n8
Warren, Donald, 92
Warwick, Donald P., 63, 64, 131, 156
Waterman, Robert H., Jr., 44
Waugh, Russell F., 124
Weick, Karl E., 222
Weinstein, Thomas, 89 n8
Weitzman, Michael, 76
Wheeler, Christopher W., 132, 137, 140
Willms, J. Douglas, 19 n2
Windham, Douglas M., 58, 127
Winkler, Donald R., 19 n2, 119, 192, 201, 202
Wittrock, Merlin C., 64
Wolfe, Barbara L., 4
Women. *See* Females
Woodhall, Maureen, 19 n2
World Bank, 3, 19 n1, 34, 35, 36, 37, 38 n6, 7, 155, 171, 172, 174, 186, 188, 190–91, 199; aid for education by, 210–11; education projects supported by, 130; textbook projects sponsored by, **129**
World Food Programme, 212
Worms, parasitic, 80, 84

Yan-you, Wang, 80
Youtz, Robert, 5
Yugoslavia, self-governing interested communities of education in, 202

Zambia: educating handicapped children in, **169**; ratio of teachers' salaries to total spending in, 37
Zambia Science Teachers Education Project, 115
Zigler, Edward F., 81
Zimbabwe: financing education in, **194–95**; teacher education course in, **93**
Zymelman, Manuel, 67, 92, 95, 106, 108, 110–11